Business Analysis Using Regression

A Casebook

Springer
New York
Berlin
Heidelberg
Hong Kong
London
Milan
Paris
Tokyo

Business Analysis Using Regression

A Casebook

Dean P. Foster
Robert A. Stine
Richard P. Waterman
University of Pennsylvania
Wharton School

Springer

Dean P. Foster
Robert A. Stine
Richard P. Waterman
Department of Statistics
Wharton School
University of Pennsylvania
Philadelphia, PA 19104
USA

Library of Congress Cataloging-in-Publication Data

Foster, Dean P.
Business analysis using regression: a casebook / Dean P. Foster, Robert A. Stine, Richard P. Waterman.
p. cm.
Includes index.
ISBN 0-387-98356-2 (softcover: alk. paper)
1. Regression analysis. 2. Social sciences—Statistical methods. 3. Commercial statistics. I. Stine, Robert A. II. Waterman, Richard P. III. Title.
HA31.3.F67 1998
300',1'.519536–dc21 97-39339

ISBN 0-387-98356-2 Printed on acid-free paper.

Printed in the United States of America. (MVY)

9 8 7 SPIN 11013983

Springer-Verlag is a part of *Springer Science+Business Media*

springeronline.com

Preface

Statistics is seldom the most eagerly anticipated course of a business student. It typically has the reputation of being a boring, complicated, and confusing mix of mathematical formulas and computers. Our goal in writing this casebook and the companion volume (*Basic Business Statistics)* was to change that impression by showing how statistics gives insights and answers interesting business questions. Rather than dwell on underlying formulas, we show how to use statistics to answer questions. Each case study begins with a business question and concludes with an answer. Formulas appear only as needed to address the questions, and we focus on the insights into the problem provided by the mathematics. The mathematics serves a purpose.

The material is organized into 12 "classes" of related case studies that develop a single, key idea of statistics. The analysis of data using statistics is seldom very straightforward, and each analysis has many nuances. Part of the appeal of statistics is this richness, this blending of substantive theories and mathematics. For a newcomer, however, this blend is too rich and they are easily overwhelmed and unable to sort out the important ideas from nuances. Although later cases in these notes suggest this complexity, we do not begin that way. Each class has one main idea, something big like standard error. Wc begin a class by discussing an application chosen to motivate this key concept, and introduce the necessary terminology. All of the cases in that class explore and develop that idea, with perhaps a suggestion of the idea waiting in the next class. Time in thc classroom is a limited commodity, and we use it to apply statistics rather than talk about statistics. We refer the students to a supplemental text for reference to formulas, calculations, and further motivation. In class, we do the data analysis of these cases in class using a computer projection system. This allows us to explore tangents and gives students a chance to see the flow of data analysis. These casebooks remove some of the note-taking burden so that students can follow along without trying to draw plots or scribble down tables. That said, we have seen that students still seem to fill every margin with notes. The course seems to work. It has been very highly rated by MBA students, and some have even asked about majoring in statistics!

It would be easy to claim that you can use any statistics software with these casebooks, but that's not entirely true. Before we developed these notes, we needed to choose the software that we would use to do the analyses and generate the figures that appear in this book. The preferences of our students demanded that we needed something that ran on PC's and Mac's; our own needs required that the software be able to handle large data sets, offer modern interactive graphics (plot linking, brushing), and include tools for the beginner up through logistic regression. JMP (whose student version is named JMP-IN) was the best fit to our criteria. Each of its analysis procedures includes a graph as part of the output, and even the student version allows an unlimited data set size (up to the capabilities of the computer). It's also very fast and was designed from the start with a graphical user interface. To top it off, the developers of JMP, led by John Sall at SAS, have been very responsive to our suggestions and have added crucial features that we needed. We're aware, though, that many will prefer to use another package, particularly Minitab. Minitab finished a close second to JMP in our comparisons of statistical software and is well-suited to accompany these notes. An appendix describes the commands from the student version of Minitab that provide the needed functionality.

The data used in the cases are a mixture of real and simulated data. We often use simulated or artificial data when we want to focus on a particular technique or issue. Real data is seldom so simple and typically has numerous nuances that distract from the point being made. The resulting analysis can become so convoluted that everyone tends to get lost. The included assignments were used recently when we taught this course (we have to change the assignments annually). You may want to require these or modify them to suit your emphasis. All of the data sets used in these examples are available over the Internet from StatLib (http://lib.stat.cmu.edu) and from our departmental web site (http://www-stat.wharton.upenn.edu). The data are available in both JMP format and ascii (text) format.

Acknowledgments

We have benefited from the help of many colleagues in preparing this material. Two have been very involved and we want to thank them here. Paul Shaman kept us honest and focused, shared his remarkable editorial skills, and as the chairman of our department provided valuable resources. Dave Hildebrand offered numerous suggestions, and is the source of the data for many of our examples. These include the cleaning crews and benefits data used in Class 1 as well as both data sets used to illustrate the two-way analysis of variance in Class 10 and the computer sales example of Class 12. We thank him for his generosity and encouragement along the way. Via Dave, Kathy Szabat brought the real estate data for Philadelphia in Chapter 2 to our attention.

A Brief Overview of the Cases

These casebooks have been used in the MBA program at Wharton as the lecture notes to accompany an 18-hour course consisting of 12 1.5 hour lectures. As in the previous casebook, there is not time to cover every example in class so we have placed the most important first in the examples for each class. Generally, though, there is more material in these notes than can be adequately covered in even the two-hour classes. At a minimum, you will want to cover the first example for each class since this example generally introduces the methods with more discussion; later examples for each class offer repetition and explore tangential (important, but still tangential) ideas. The material is inherently cumulative. Generally, our plan for the moving through this material is to introduce ideas like leverage, outliers and assumptions in simple regression where the plots are nice, and then get to multiple regression as quickly as possible. In the cases, we have edited the output shown here o that only the relevant material appears. We expand the output slowly over several days on a "need to know" basis rather than introduce all of the voluminous regression output at one time.

The remainder of this preface discusses the material in the classes, with particular emphasis on the points that we try to make in each. We typically begin each class by reviewing the overview material that introduces the lecture, covering the various concepts and terminology. This "blackboard time" lays out the key ideas for the class so that students have a road map for the day. Without this introduction, we have found that some will lose sight of the concepts while focusing instead on the use of the software. Once we have the ideas down, we turn to the examples for the day, emphasizing the use of statistics to answer important business questions. Each example begins with a question, offers some relevant data, and applies statistics to answer the question. We also assign a traditional textbook for the course. Although we present the basic ideas for each lecture here, students frequently benefit from further reading material. We discuss the text readings as part of the introduction the class.

A particularly important exercise that we have used when teaching from this material at Wharton is an individualized project. All too often, students flounder with regression projects because they choose a problem that is either too easy or too hard or just not suited to regression. We have approached this by giving each of them a unique data set that deals with a cost accounting model for a hypothetical firm. We construct 1000 data sets (which we assign by SS number) from a collection of several "true" models. The forms of these models are not revealed to the students, but are useful in grading the projects. All 1000 data sets pertain to the same problem, but each has novel features. Students can freely share ideas, but each is responsible for the nuances of an individual analysis.

Class 1

We depart from convention by starting our development of regression with the challenge of choosing the form of the model. Typically, the issue of transformations is deferred for the end of a course and often "falls off the end" in the crunch at the end of the term. We think is too important, and we cover it first. Later statistical considerations (p-values, confidence intervals, and such) are meaningless without specifying the model appropriately, so we begin here. The liquor sales and cellular telephone examples are revised in Class 3 and Classes 2 and 12, respectively.

Scatterplot smoothing has become quite common in theoretical statistics, but it has been slow moving into software used at this level and is often not covered. We use this powerful tool to help recognize the presence of nonlinearity, and JMP has a good implementation, though it lacks an interactive control like the kernel density smoothing tool available for smoothing histograms. Scatterplot smoothing is seldom found in texts at this level, so we have supplemented the output with additional discussion that we would generally defer to the course book.

We make extensive use of JMP's ability to plot nonlinear relationships on the original axes in building regression fits. We find that overlaid views of the models in the original coordinate system makes it easier to appreciate the effects of transformations. The cellular telephone example, though, exposes a weakness of JMP's implementation, though we are able to exploit it to make an important point. The weakness is that the transformation dialog used when fitting simple regressions provides only a few of the most common transformations, and you may need others (as in this example). This limitation forces us to transform the data explicitly and plot it in the transformed coordinates. One can then see that when the transformation works, plots of the transformed data in the new coordinate system are linear.

Class 2

Before we can compute inferential statistics, we need a model. This class introduces the linear regression model and its assumptions. For a bit of calibration, we begin with a simulated data set that shows what good data look like. Each of the subsequent cases illustrates a single, specific problem in regression. We have found it useful in developing the material for this class to focus on examples with a single problem. It is easy to be overwhelmed when the data have a multitude of complexities. Although the real world is indeed complex, this complexity makes it hard to recognize each regression assumption. This class has quite a few cases, so plan on skipping some or perhaps using them as study exercises. Two of the cases, the housing construction and liquor sales examples, are expanded in Class 3.

The intent of so many outlier examples in this class is to show that the impact of removing an outlier on a regression is not routine. The R^2 may either increase or decrease, and the predictions can become either more or less accurate, depending upon the placement of the outlier. The question of how to handle an outlier is important in the last three cases. We have tried to avoid the knee-jerk reaction of excluding outliers right away and focused the discussion around prediction. Keeping an outlier might be the right thing to do if, as in the cottage example, that's the only data that you have. Doing so, though, represents a gamble without some extra information.

Perhaps the right approach is something more like what is done in the Philadelphia housing example: set the outlier aside for the purpose of fitting and use it to validate the model chosen. This complex example also illustrates another aspect of outlier analysis: Is the outlier real or is it due to some missed nonlinear structure? The additional topics mentioned in this case, aggregation effects and causation, can become subjects for much larger class discussion than the tidbits offered here. One can also discuss the issue of weighting in this example. The communities are very different in population, so that the average price from one community may be based on many more home sales than in another.

Class 3

This class introduces the inferential statistics that one obtains from a regression model, building on two previously used data sets, the housing construction and liquor sales examples from Class 2. In the latter case, we not only show how confidence intervals for the optimal amount of shelf space computed in Class 1 transform nicely (a very appealing property of intervals not shared by standard errors) but also consider the issue of sensitivity. A crucial choice made in Class 1 for modeling this data is the selection of the log transformation to obtain the best fit. Though this choice seems "innocent" in that class, the effect is very dramatic when extrapolating.

The emphasis in these cases is more upon prediction intervals at the expense of confidence intervals. Confidence intervals for parameters are generally more well understood and familiar from prior classes prediction intervals. Few newcomers will have seen prediction intervals for new observations. Since prediction is a popular use of regression, we give it additional attention.

Class 4

We begin our treatment of multiple regression in this class and emphasize the issue of interpretation, particularly the differences between partial and marginal regression slopes. Since most will know the names of the cars in this data, point labeling in the plots is particularly effective to understand outliers. This data set is quite rich, with many more variables, compared to the smaller data sets used with simple regression, so it might be useful to cover some aspects of this case as preliminary homework. For example, it is useful to have studied the needed transformation in this problem before taking on the development of a multiple regression model.

The overall and partial F test are introduced in this class, though details and further discussion are deferred to Class 5. This test seems to be a hard one for many students to learn (and it is not a part of the standard output), so we have included it at an early point to allow for some incubation.

Class 5

Collinearity complicates the interpretation of regression coefficients, and our first case in this class deals with the most extreme case of collinearity that we could construct from real data without losing any sense of reality. The impact of correlation among the predictors is dramatic here. The second example shows less collinearity, and offers another chance to discuss the differences between marginal and partial regression slopes. This second example also introduces elasticities as the interpretation of the slopes in a log-log or multiplicative regression model. We typically assign aspects of this second case as an assignment, focusing class time on leftovers from the previous class and the first example.

A good exercise at this point is to try to build a multiple regression "by hand". While this expression once meant the tedious calculation of regression coefficients by inverting a matrix, we mean it in the sense of building a multiple regression from a series of simple regressions. In particular, you can build a partial regression by first regressing out all but one of the covariates from response and the remaining predictor, and then fit the simple regression between the residuals. In the two-predictor case with response Y and predictors X_1 and X_2 one fits the model

$$(\text{residuals from } Y \text{ on } X_2) = \beta_0 + \beta_1 (\text{residuals from } X_1 \text{ on } X_2)$$

Many will be surprised to see that the fitted slope in this equation is the multiple regression slope for X_1 from a model with Y regressed on both X_1 and X_2. This exercise makes it clear what is meant by the loose phrase often heard in regression that a slope measures the effect of one variable "holding the other variables fixed". Particularly relevant for this class, partial regression also makes it clear what goes wrong when one predictor is almost perfectly explained by another (high collinearity) — there's no variation left over to model the response. This is a tedious exercise, but JMP makes it easy to save the residuals and fit the partial regression. The final reward is to note that JMP does this task for you when it constructs the leverage plots.

Classes 6 and 7

This pair of classes are perhaps the hardest for both instructor and students. The first class attempts to lay the foundations in the two-group setting which are expanded in Class 7 to problems with more than two groups. JMP makes it very easy to add categorical factors to a regression model since the software automatically builds the underlying categorical variables (and easily forms interactions as well). However, this automation makes the results a bit more mystical since one does not manipulate the underlying variables. As often noted, a useful exercise is to do it once manually: construct dummy variables for the categorical factors "by hand" and add them to the regression. This will help clarify matters as well as help one to appreciate what's being done automatically.

Categorical variables are never easy to introduce, and JMP's coding convention using -1/+1 (rather than the more common 0/1 coding) makes it more difficult as well. This sort of coding is useful since comparisons are to an overall mean rather than some arbitrary category, but it does make for a rough start. The labeling in the output is also particularly hard to overcome; you just want to read something in the output labeled as "a-b" as the difference between these categories.

In Class 7, the wage discrimination example revisits an example from our other collection of cases (*Basic Business Statistics*). Using regression, the results shown here identify a significant difference, whereas a comparison based on subsets using a t-test is not significant due to the loss of data. The second example in Class 7 is designed to help with the accompanying project, which we typically have students doing at this part of our course (see the appendix for the assignments and the project description).

Class 8

The examples in this class attempt to convey a sense of what happens in realistic applications of regression. Most users try to do the careful, graphical, step-at-a-time approach, at least for a while. Then the pressures of time and too much data lead them to use stepwise regression or a similar technique. The first case illustrates the difficulties and wrong turns of building a regression model, and the second example shows the pitfalls of careless stepwise modeling. The fitted stepwise model looks good (big R^2, significant F), but turns out to have very poor predictive ability. We had long avoided teaching stepwise regression in our course, preferring to encourage our students to think more carefully about which factors belong in their models rather than rely on computational algorithms. We have learned over time, though, that students will discover stepwise regression and use it — it's simply too convenient in many problems. As a result, we figured that it was better to teach them how to use it carefully rather than pretend it didn't exist.

Class 9

This class builds on the issue of overfitting introduced in Class 8. The focus in this class turns to the use of post hoc comparisons in the one-way analysis of variance. The ideas are similar to those found when dealing with overfitting in regression: letting the data generate your hypotheses leads to problems. Fortunately, in a one-way analysis of variance, you have some useful tools that control the effects of selection bias, and JMP includes the standard anova methods for handling multiple comparisons. We focus on these rather than the overall anova table and its sums of squares.

The tables generated by JMP for the various multiple comparison methods can be very confusing. The simplest explanation we have found that works is to point out that the tables show one endpoint of a confidence interval for a difference in two means. If this endpoint is positive (or negative, depending on the context), the mean difference is significant since the associated interval would not include zero.

The third example is supplemental to this class. It shows how to take the ideas from the analysis of variance and use them to build a handy diagnostic for nonlinearity in regression. The case illustrates how one can take advantage of repeated observations to measure the degree of departures from linearity in regression.

Class 10

These examples illustrate two-way analysis of variance, first without and then with interaction. The example with no interaction is relatively straightforward. Compared to the subtleties of multiple regression, students can appreciate the benefits of a balanced experimental design. An interesting aspect which is only noted briefly here is that a balanced design insures that the coefficients are orthogonal. That is, there is no collinearity in this problem. If you remove one of the factors, the coefficients of those that remain are unchanged. The second example with interactions is more fun and challenging. Here, you get to see why it is important to check for interactions in statistical models. When the model is fit without interactions, the terms are not significant and it appears that none of the modeled factors affect the response. Adding the interactions shows what's going on very clearly and that the factors are truly very useful. In keeping with Class 9, we continue to address multiple comparisons issues since one will inevitably want to find the best combination of the factors.

Classes 11 and 12

These final two classes extend the ideas of regression modeling into special situations: modeling a categorical response (Class 11) and modeling time series data (Class 12). Neither class depends on the other, so you can pick and choose what to cover. We have found that marketing students find logistic regression particularly useful, though it can be hard to cover in just one class. Our treatment of time series avoids some of the more esoteric methods, like ARIMA models, and emphasizes what can be accomplished with simple lagged variables and regression.

We should note that the Challenger case used in Class 11 to illustrate logistic regression is particularly dramatic for students. Many seem to have been watching this launch and are keenly interested in the analysis of this data. We have also used this case to make connections to management courses that discuss decision making and judgmental errors in times of intense pressure.

LECTURE TEMPLATE

Quick recap of previous class

Overview and/or key application

Definitions

These won't always make sense until you have seen some data, but at least you have them written down.

Concepts

A brief overview of the new ideas we will see in each class.

Heuristics

Colloquial language, rules of thumb, etc.

Potential Confusers

Nip these in the bud.

Contents

Class 1. Fitting Equations to Data

This class introduces regression analysis as a means for building models that describe how variation in one set of measurements affects variation in another. The simplest of these models are linear relationships. In models having a linear structure, the slope of the fitted line is crucial; it quantifies how variation in one variable (the predictor or covariate) translates into variation in the other (the response). The slope helps us interpret how variation in the predictor affects variation in the response.

More complex models capture decreasing returns to scale and other nonlinear features of data. Regression analysis provides a systematic method for building equations that summarize these relationships, and all of these equations can be used for prediction.

Topics

Scatterplots and scatterplot smoothing
Least squares fitting criterion
Linear equations, transformations, and polynomials
Point prediction from a regression equation
Estimates in regression: intercept, slope

Examples

1. Efficiency of cleaning crews (linear equation).
2. Liquor sales and display space (nonlinear transformation).
3. Managing benefits costs (polynomial model).
4. Predicting cellular phone use (transformation via a formula).

Overview

We live in the age of data. From supermarket scanners to the world's stock exchanges, from medical imaging to credit scoring, data in massive amounts are accumulating around us. Taken by themselves, data are usually worthless. We need to turn the data into something useful, understandable, and most importantly communicable. Fortunately, our ability to model and interpret data has also grown, and a vital part of this course is to learn such skills.

In general terms, our main objective is to describe the associations among two or more sets of variables. In the simplest case we have just two variables, for example, *Time* and the value of the *S&P500.* If we were to record values once a minute, then a year's worth of data would have 120,000 data points, figuring 250 eight-hour days of trading. That's a spreadsheet with 2 columns and 120,000 rows. Just take a minute to think about how such a data set could have been comprehended 50 years ago. It would have been almost impossible. Imagine plotting it by hand. If you could plot a point a second, it would take over 33 hours to get the job done!

One of the cardinal rules of data analysis is to *always, always plot your data.* There is no excuse not to, and the human eye is probably the most well-developed machine in existence to recognize trends, curvatures, and patterns – representing the types of association that we are trying to understand and exploit. It seems quite obvious that a good way to represent the data from the example above would plot *Time* along the horizontal axis and the value of the *S&P500* along the vertical axis. You can then allow your eye to pick out pertinent features in the data; a feat that would have taken hours just a few years ago can now be accomplished with a click of the mouse.

Fortunately we have the computer to aid us, and recent advances in dynamic and linked statistical graphics enable us to get a very good idea of the behavior of such data sets. We have the tools available to extract the valuable information from the mass of data. When successful, we can exploit that information in the decision making process. You should never underestimate the value of the graphical tools we have available, and we will make extensive use of them in the examples that follow.

Definitions

Predictor and response. In regression, the predictor is the factor (X) that we suspect has an influence on the value of the response (Y). Typically in applications, we will want to predict the response using a given value of the predictor, such as predicting sales from advertising.

Intercept. In a regression equation, the intercept is the value of the response Y expected when the predictor X equals zero; that is, the intercept is the prediction of Y when X is zero. When interpreting the intercept, beware that it may be an extrapolation from the range of the observed data.

Scatterplot. A graphical display which plots as (x,y) pairs the values of two variables. For a spreadsheet with two columns, each row has two values in it. These two values are the horizontal (x) and vertical (y) coordinates of the points in the scatterplot. Convention dictates that the predictor X goes on the horizontal axis, with the response Y on the vertical axis.

Smoothing spline. A smooth curve that mimics a flexible drafting tool, emphasizing the systematic pattern in a scatterplot. Splines are very useful for focusing one's attention on trends in scatterplots and time series plots. Though a smoothing spline itself does not give an interpretable model, it does serve as a guide in helping us build a model that is.

Slope. The slope of a line that depicts the association between two variables, with X on the horizontal axis and Y on the vertical, is defined as the "change in Y for every one unit change in X." Big slopes say the line is very steep; small slopes say the line is almost flat. Positive slopes indicate that the line slopes from bottom left to top right. Negative slopes indicate that the line slopes from top left to bottom right. In regression, the slope is the expected change in the response when the predictor increases by one unit.

Concepts

Scatterplot smoothing. Smoothing a scatterplot is an attempt to summarize the way that the vertical coordinates depend upon the horizontal coordinates. The idea is that instead of looking at all the points which possess random variation, you could instead look at the "scatterplot smooth," a curve that depicts the association between the two variables. Smoothing attempts to separate the systematic pattern from the random variation.

There are very many ways of smoothing a scatterplot. For example, drawing the best fitting straight line is one way of smoothing. Drawing the best fitting quadratic curve or other polynomial is another. The use of *smoothing splines* is yet another way of depicting this average relationship. Splines are much more adaptable to the data than a line can ever be. Using splines allows you to detect small wiggles or bends in the association, something a line can never do. The catch, however, is that the choice of how much to smooth the data in this way is subjective (at least in the manner implemented in the statistical software package JMP that we will use here).

Least squares. Whatever smoothing technique we choose, be it splines or lines, we need a criterion to choose the "best" fitting curve. "Best" is not a very well defined word; we have to decide best for what? One criterion for deciding which is best has many desirable properties, and historically this criterion has been the most important. This definition of "best" is called the *least squares criterion*. For choosing the line which best fits data, the least squares criterion says that we need to find the line that minimizes the sum of the squared *vertical* distances from the points to the line. If you have not seen the least squares criterion before, it is not remotely obvious how you are going to find this special line. But that's the beauty of least squares: it turns out to be an easy problem.

The least squares criterion takes the ambiguity out of the line fitting process. If we were all to draw the best line in a scatterplot by eye, we would get similar, but definitely different answers. Least squares, if we all decide to use it, removes this subjectivity and ambiguity from the line fitting process. Finally, it's important to realize that even though least squares is exceptionally popular as a definition of "best," there is no fundamental reason why we should always use it. It became popular because it was easy to work with, not because it was necessarily the right thing to do. We will see many examples in which some of its weaknesses come through. For example, the sample mean $\overline{X}$ is the least squares estimate of the center of a population. Just as outliers may heavily influence the sample mean, outlying values may also affect the estimates determined by least squares, such as the intercept and slope of a regression line. The least squares line follows the unusual observations, wherever they choose to go. We'll see a lot more of outliers in Class 2.

Prediction. By separating the systematic structure in a scatterplot from the random variation, smoothing a scatterplot allows us to predict new observations. For example, if you want to predict the response Y at for some value of the predictor X, all you do is go up to the smooth and read off the height of the curve. You can also do the reverse, given a vertical Y value you can go back to the smooth and read down for an X value (though it's not usually set up to work as well this way — the least squares line minimizes vertical, not horizontal deviations).

An important aspect of prediction is the difference between extrapolation and interpolation. Extrapolation occurs when you try to figure out what is going on *outside* the range of the data you have collected. This is the essence of financial forecasting: we don't care about what happened yesterday other than it might be able to suggest what is going to happen tomorrow, which is outside the range of our data. Interpolation by contrast involves figuring out what is happening *inside* the range of your data, though not necessarily at a point for which you have data. Process engineers are very interested in doing this to find the optimal combination of inputs which maximizes output. It turns out that interpolation is a lot less risky than extrapolation.

Heuristics

The slope is a rate and has units attached to it.

The slope of a fitted line has units since it measures the change in Y for each unit change in X. It translates changes in X into changes in Y. When you need to answer a question of the form "How much does something change when I change a factor under my control?", you need a slope.

Potential Confusers

Why don't we make the best fitting line go though the origin when we know that the line has to go through this point? (For example, the relationship between weight and height.)

Two reasons. First, what does it gain us? If indeed the origin "should" be zero, then the intercept of the fitted regression ought to be close to zero. If it's not, then we have learned that something in the data does not match our expectations, and we need to find out why. Second, all of the important diagnostics that we use to evaluate a regression model presume that the line is not forced though the origin.

Not all relationships are straight lines.

A line is a nice summary, but may not be appropriate. The linear model says that a change of one unit in X has on average the same impact on the response Y, no matter how large X has become. For example, a linear equation for the dependence of sales on advertising means that increasing advertising by $100,000 has a fixed impact on sales, regardless of the level of advertising. That doesn't make too much sense in some situations. Does increasing advertising from $0 to $100,000 have the same impact on sales as going from $4,000,000 to $4,100,000? It's possible, but you might expect the effect to diminish as the level of advertising increases (decreasing returns to scale). For another example, the relationship between number of units produced and average cost might be a "U-shaped" curve, or roughly a quadratic polynomial.

Which variable is X and which is Y?

The *X* variable is the factor that we think that we can perhaps manipulate to affect the response *Y*, such as changing the level of promotion to affect sales. In some cases, *X* represents what we know and *Y* is what we want to predict, such as in a model of sales (*Y*) as a function of time (*X*). Often, you might think of *X* as the "cause" of *Y*. Though sometimes appealing, remember that regression — like correlation — is only a reflection of association.

Efficiency of Cleaning Crews

Cleaning.jmp

A building maintenance company plans to submit a bid on a contract to clean 40 corporate offices scattered throughout an office complex. The costs incurred by the company are proportional to the number of cleaning crews needed for this task. How many crews will be enough?

Recently collected observations record the number of rooms that were cleaned by varying numbers of crews. For a sample of 53 days, records were kept of the number of crews used and the number of rooms that were cleaned by those crews. The data appear in the scatterplot below.

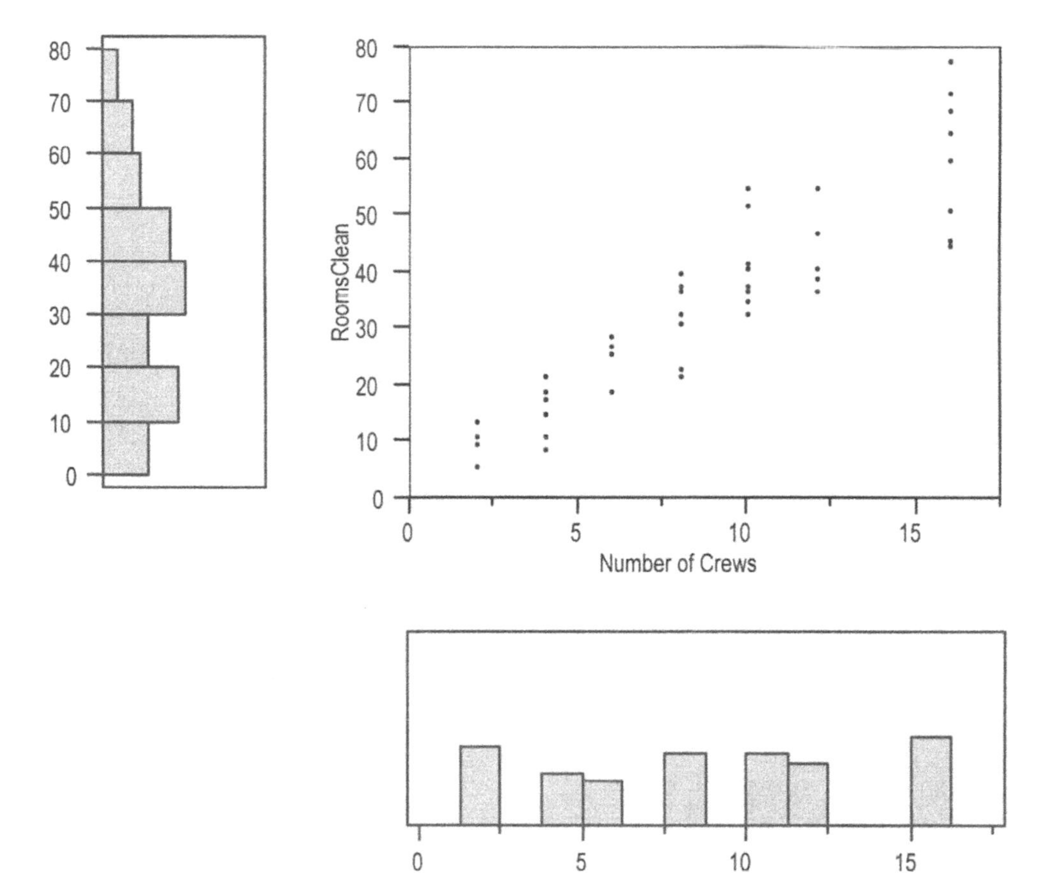

A regression line summarizes as an equation the linear pattern that is clear in the scatterplot. We generated this scatterplot using the *Fit Y by X* command from the *Analyze* menu. The fitted line is added to the scatterplot by clicking with the mouse first on the *red triangular* button that appears near the label at the top of the window showing the scatterplot on the computer screen. Then select the *Fit line* option.

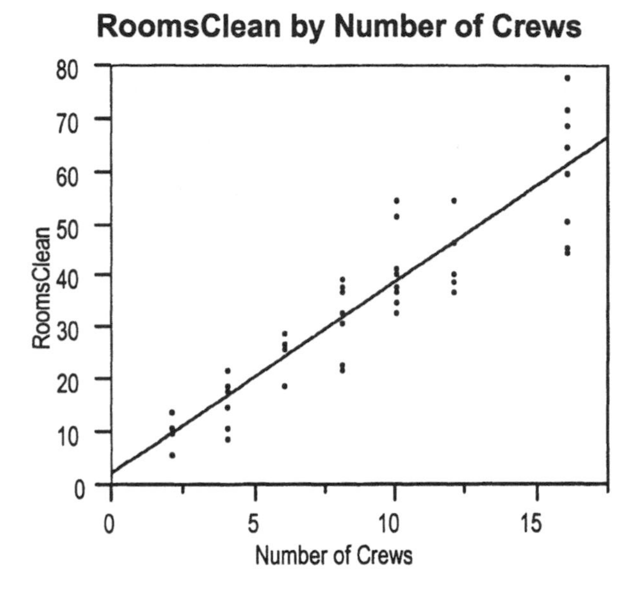

A summary of the equation that defines the line shown in the scatterplot appears in the summary output shown on the next page. The line is chosen as that which minimizes the *sum of squared vertical deviations*, and hence it is known as the least squares line. That is, we need to pick an intercept b_0 and a slope b_1 in order to minimize the deviations, expressed as

$$\min_{b_0, b_1} \sum_{i=1}^{n} (Y_i - b_0 - b_1 X_i)^2 ,$$

where we have generically labeled the ith value of the response as Y_i and the ith value of the predictor X_i. The computer software finds resulting choices for the intercept and slope which are shown on the next page.

With the estimates rounded to one decimal place, JMP gives this summary:

Parameter Estimates

Term	Estimate	...more discussed later
Intercept	1.8	
Number of Crews	3.7	(this term is the slope)

That is,

$$\text{RoomsClean} \approx 1.8 + 3.7 * \text{Number of Crews}\ .$$

What are the interpretations of the slope and intercept in this example?

When interpreting the intercept and slope in a fitted regression, remember that these estimates have units attached to them. The intercept can be tricky. It is measured in the units of the response, here the number of rooms cleaned. The intercept gives the value of the response (rooms cleaned) expected when the predictor is zero, here suggesting that 1.8 rooms are cleaned when no crews go out! We'll see later that this deviation from a sensible answer is consistent with random variation.

The slope, which translates variation on the *X* scale to the *Y* scale, is measured on the scale of (Units of *Y*)/(Units of *X*), or in this example

$$3.7 \frac{\text{Rooms Cleaned}}{\text{Cleaning Crew}}\ .$$

Written out in this way, the interpretation is immediate as the number of rooms cleaned per cleaning crew.

Since the intercept and slope have units, you must be cautious in making too much of their size. The size of the intercept and slope, unlike a correlation, depends on how we choose to measure the variables. Suppose, for example, that we measured the number of square feet cleaned rather than the number of rooms, and that each "room" was worth 400 sq.ft. Then look at what happens to the regression fit.

The plot is clearly the same. We have just changed the labeling on the *Y* axis.

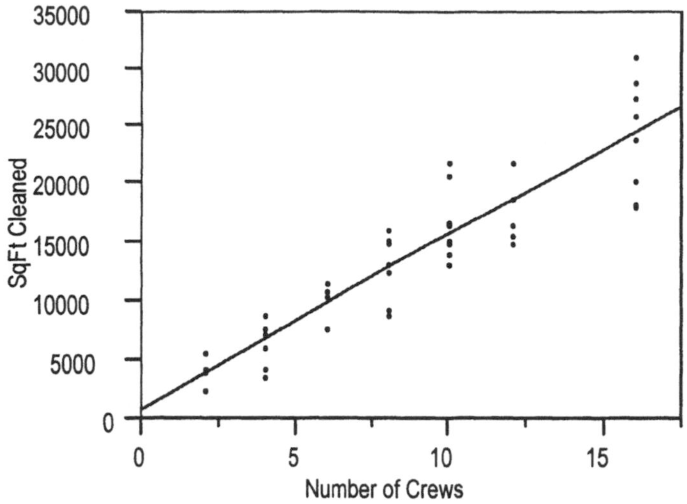

The coefficients of the fitted regression, however, appear to be much more impressive (i.e., much bigger numbers)

Term	Estimate	...more
Intercept	713.9	
Number of Crews	1480.4	

The intercept and slope are both 400 times larger. Don't let large values for the slope and intercept impress you. You can make these values as large or as small as you like by changing the units of measurement.

> A building maintenance company is planning to submit a bid on a contract to clean 40 corporate offices scattered throughout an office complex. The costs incurred by the maintenance company are proportional to the number of cleaning crews needed for this task.
>
> How many crews will be enough?

From the equation, we can solve for the number of crews needed to clean 40 rooms, on average. We have

$$40 = 1.8 + 3.7\,x \qquad \Rightarrow \qquad x \approx 10.3\,.$$

The equation confirms what the plot makes obvious: roughly 10 crews are needed. The plot also makes it clear, however, that sometimes 10 crews might not be enough (or might be more than enough). On average, 10 crews get the job done, but that does not imply certainty. We need a standard error for the slope, and we will build one after discussing the underlying assumptions in Class 2. The plot with the regression line superimposed also suggests that the variability of number of rooms cleaned increases with larger numbers of crews. The lack of constant variation about the regression fit is important, and we will return to this example in Class 2.

Having never seen regression or thought about scatterplots, you'd probably approach this problem differently. If asked how many rooms are cleaned on average by the crews, you might be tempted to take the average of the ratio

$$\frac{\text{Rooms Cleaned}}{\text{Number of Crews}}\,.$$

The average ratio of rooms to number of crews is about 4 (3.97 if you're compulsive about these things). This is pretty close to the regression slope, which is 3.7. Which is better? Right now, we don't have the machinery needed to justify a choice. However, when the modeling assumptions expected by a regression model hold, then the regression slope is better. We'll cover those assumptions in Class 2. Both the regression slope and the average ratio are weighted averages of the number of rooms cleaned, and one can show that the best weights are those used by regression. (Here "best" means that the standard error of the regression estimate is smaller than that for the average ratio.) As we'll see in Class 2, the usual assumptions don't hold in this case.

Liquor Sales and Display Space

Display.jmp

A large chain of liquor stores (such as the one supervised by the Pennsylvania Liquor Control Board) would like to know how much display space in its stores to devote to a new wine. Management believes that most products generate about $50 in sales per linear shelf-foot per month.

How much shelf space should be devoted to the new wine?

The chain has collected sales and display space data from 47 of its stores that do comparable levels of business. The data appear in the following scatterplot. From the graph, we can see that where one or two feet are displayed, the wine generates about $100 per foot and sells better than most.

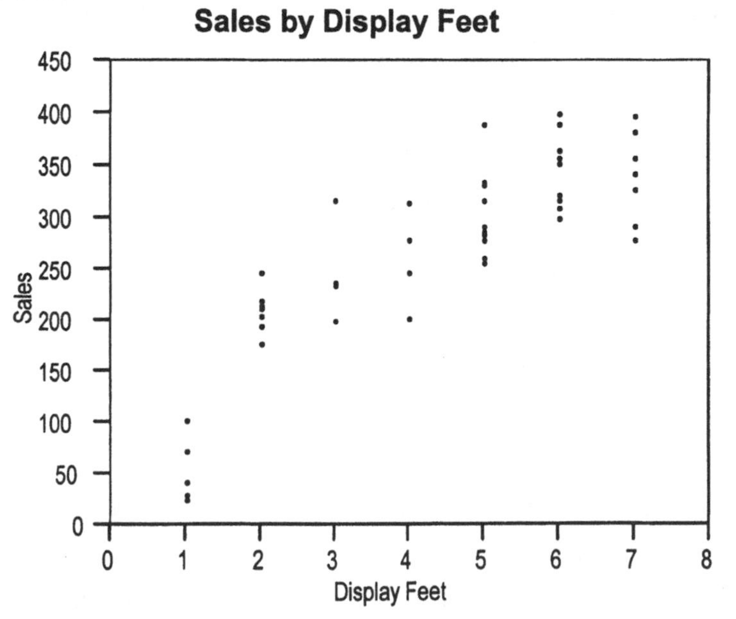

Before going on, pause for a moment and sketch in the curve that you think best summarizes the relationship between *Display Feet* and *Sales*.

Scatterplot smoothing computes local averages of the values to produce a curve that follows nonlinear trends. The smooth in the next plot uses a smoothing spline [1]with "smoothing parameter," labeled "lambda" in JMP, set to the value 10.

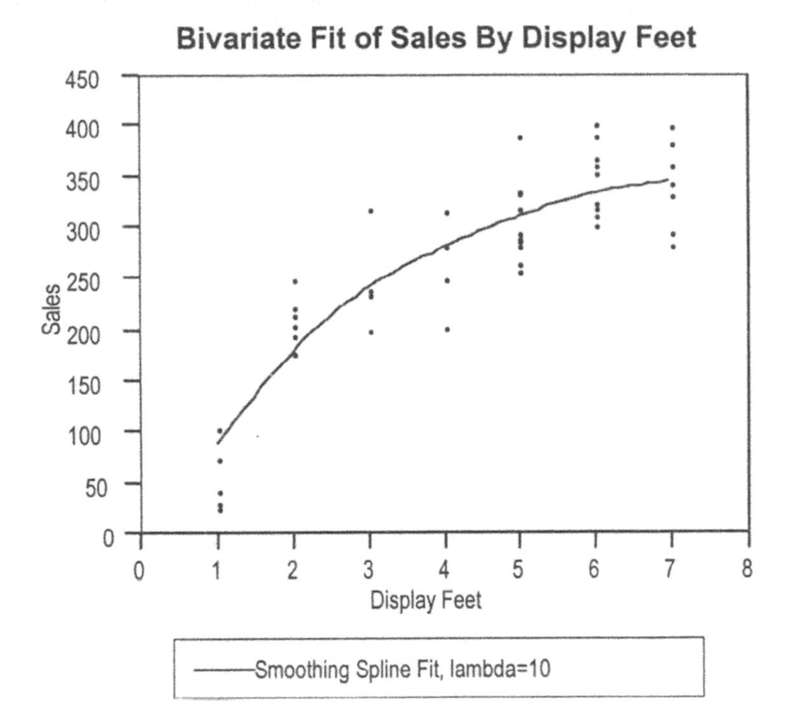

The degree of smoothness of the smoothing spline is adjustable through the choice of the smoothing parameter lambda. Smaller values of lambda give rougher, more flexible curves; larger values give smoother curves.

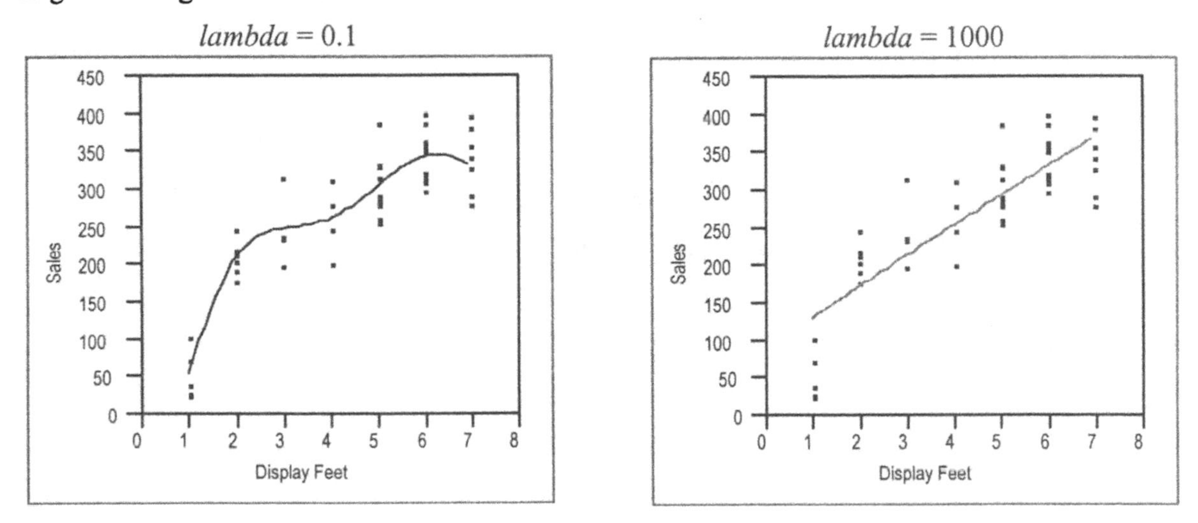

[1] The "Fit spline" command offered by the popup menu (red triangle) in a Fit Y by X view of the data adds a smoothing spline of chosen smoothness to the plot.

Comparison of the initial smooth curve to a linear fit makes the curvature in the data apparent. The spline will be used in this and other examples as a guide for selecting a transformation of the data to use in a model. The smoothing spline itself does not provide a model that is easily interpretable.

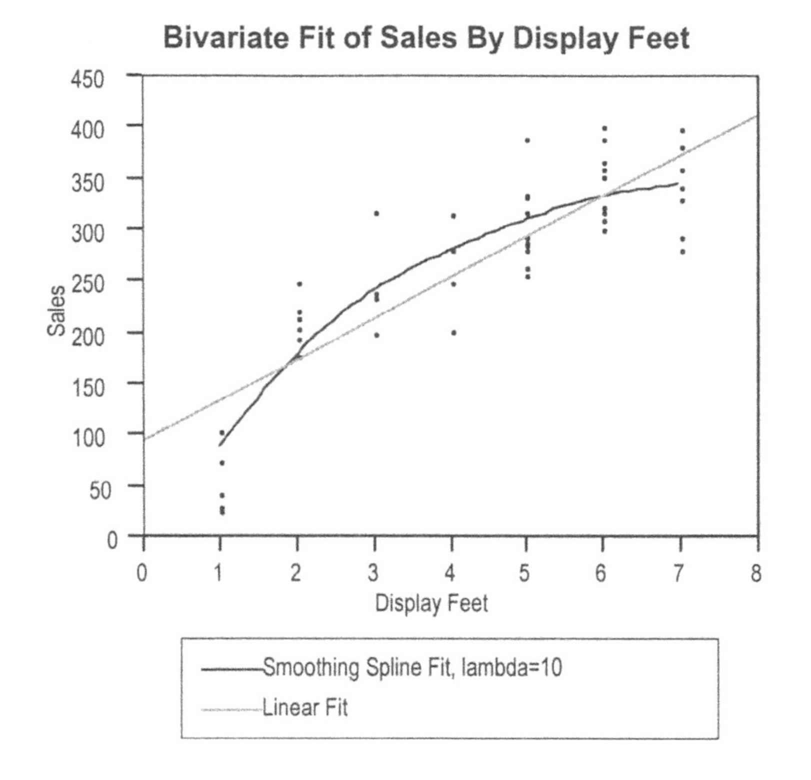

What sort of transformation should we use in a model to capture this nonlinear relationship? The data show decreasing returns (in terms of sales per month) for each added foot of display space. For example, an increase from one to two feet of display space appears to have a much larger impact on sales than an increase from five to six feet of display space.

Tukey's "bulging rule" is a simple device for selecting a nonlinear equation. The idea is to match the curvature in the data to the shape of one of the curves drawn in the four quadrants of the figure below. Then use the associated transformations, selecting one for either *X*, *Y*, or both. The process is approximate and does not work every time, but it does get you started in the right direction. (If none of these transformations help, you may need to try a different type of model, such as the polynomial transformations considered in the next case.)

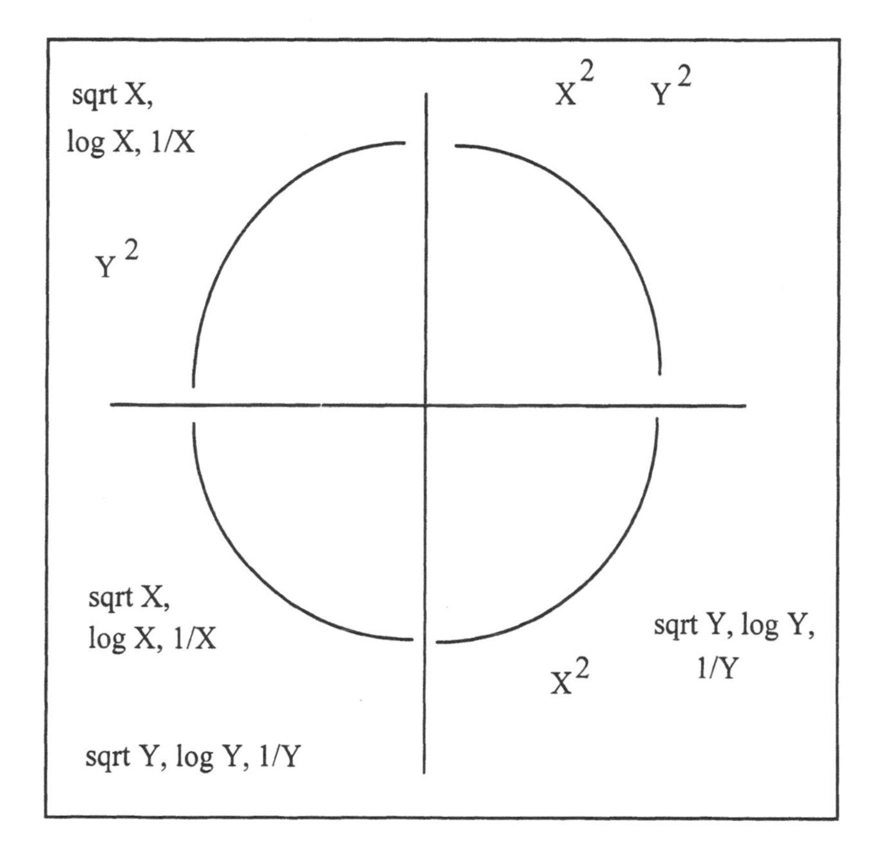

In this example, the initial curvature suggests that we are in the upper left quadrant and can consider lowering the power of *X*, such as via $X^{1/2}$, X^0 as log *X*, or X^{-1}, or increasing the power of *Y*.

Square roots, logs and reciprocals of the predictor *Display Feet* all capture the notion of *decreasing returns to scale,* albeit in different ways that have implications for extrapolation. That is, these transformations capture features seen in the plots: the slope relating changes in *Sales* to changes in *Display Feet* changes itself; the slope appears to get smaller as the number of display feet used increases.

The plots on this and the next two pages show regression models that use these transformations of the number of display feet. The square root transformation shown in the plot below seems to lack enough curvature. The fit using a model with the square root of display footage seems too linear in the following sense: the incremental gain in sales per added display foot drops off more rapidly in the data than the fit using a square root is able to capture.[2]

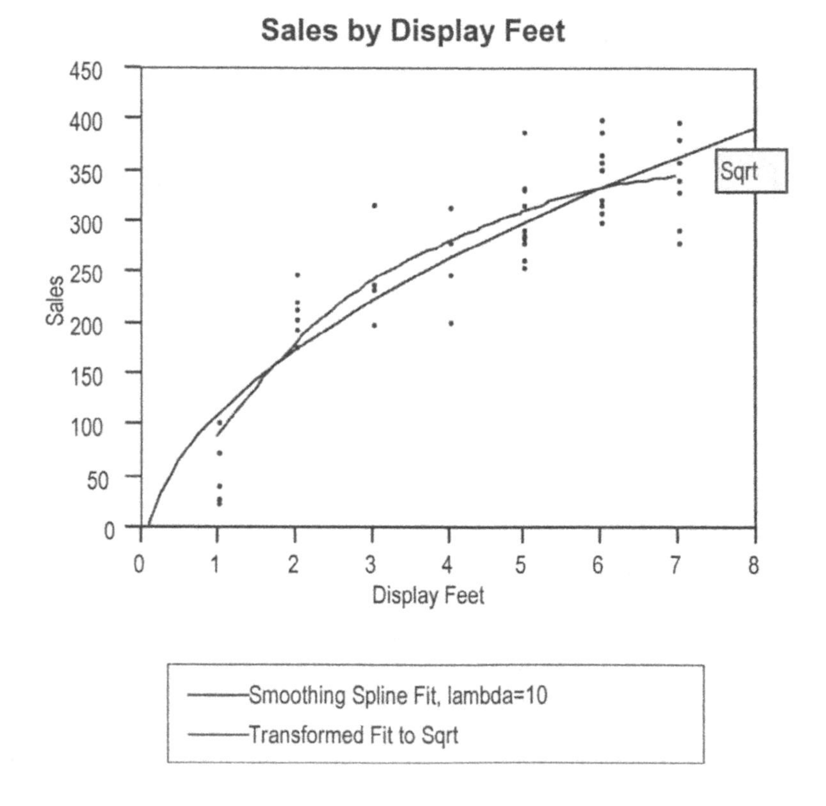

[2] In order to add transformations directly to the scatterplot in JMP, use the Fit Special... command offered by the popup command button (red triangle) near the top of the plot. Upon selecting this item from the menu, pick the transformations that you want to try. If you don't like the result, use the button labeling the fit to remove the offending curve.

Using the log of the number of display feet gives a fit that closely matches the pattern of curvature revealed by the smoothing spline. JMP automatically uses the so-called natural log, sometimes denoted by "ln". We'll also use the notation $\log_e$ when the distinction is important.

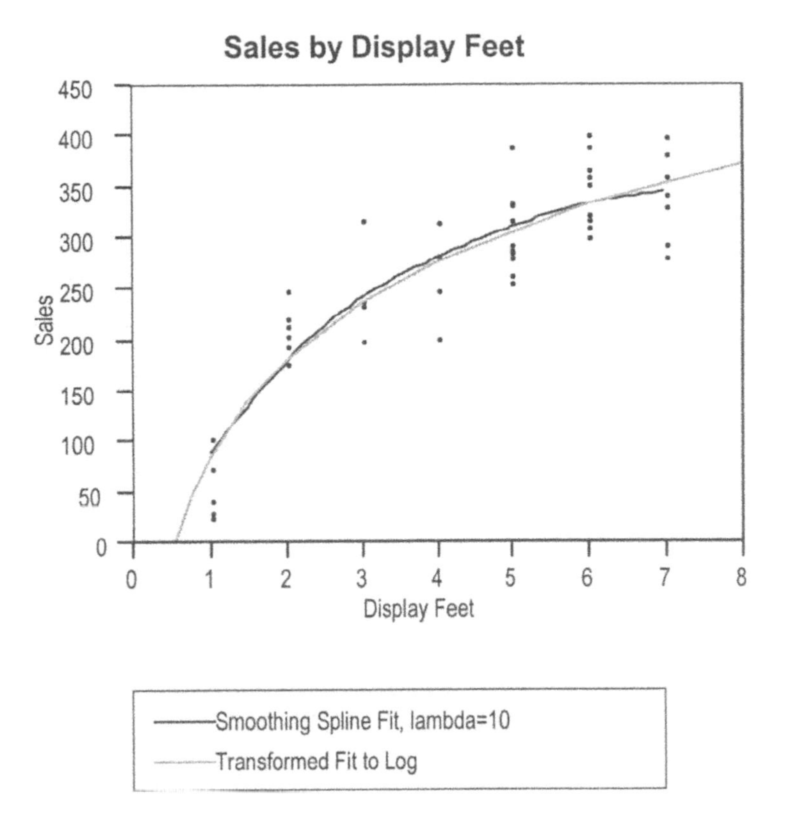

Using the reciprocal of display feet appears to go too far. Again, the fit shows the curvature observed in the data, but perhaps is too curved initially and too flat later at higher values of *Display Feet*, at least if we believe that the smoothing spline is doing the right thing.

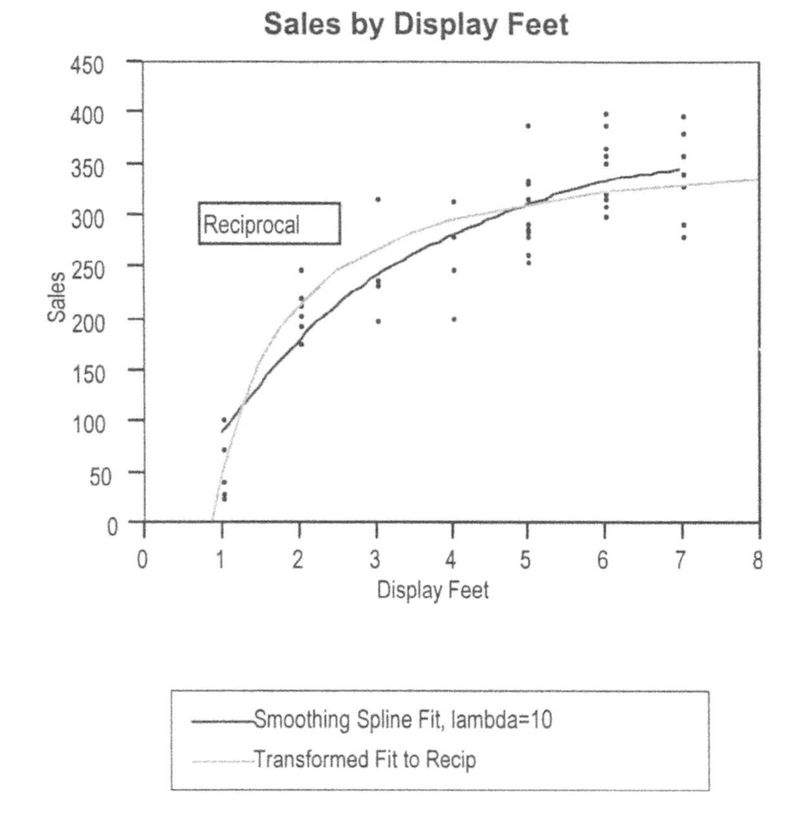

A large chain of liquor stores (such as the one supervised by the Pennsylvania Liquor Control Board) would like to know how much display space in its stores to devote to a new wine. Management believes that most products generate about $50 in sales per linear shelf-foot per month.

How much shelf space should be devoted to the new wine?

Visually, the log model seems to work best (at least in the sense that it agrees most with the spline fit). If we accept this model, we can use the fitted equation to determine the optimal amount of shelf space to use for the new wine. Here is the summary of the fitted equation:

Parameter Estimates

Term	**Estimate**	**....more**
Intercept	83.6	
Log(Display Feet)	138.6	

That is,

$$\text{Fitted sales} \approx 83.6 + 138.6\,(\log_e \text{Display Feet})\,.$$

The anticipated gain in sales over using the "typical" product by using x display feet devoted to this new wine is

$$\begin{aligned} \text{gain}(x) &= (\text{sales of this product for } x \text{ display feet}) - (\text{typical sales of } x \text{ feet}) \\ &\approx (83.6 + 138.6 * \log_e x) - (50\,x)\,. \end{aligned}$$

The figure at the top of the next page shows the size of the gain. We are trying to find the value for *Display Feet* that maximizes the gain in sales, the height of the shaded region between the logarithmic fit and the line $y = 50x$.

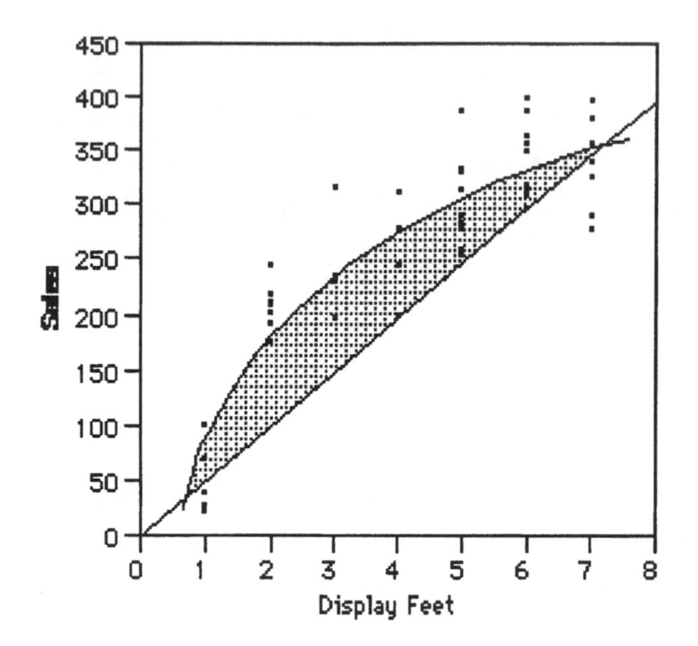

Maximizing the expression for the gain in sales requires that we differentiate it as a function of x, with x denoting the number of display feet used for the wine.

$$\frac{\partial\, gain(x)}{\partial x} = \frac{138.6}{x} - 50$$

Setting this derivative to zero gives the optimal number of display feet.

$$\frac{\partial\, gain(x)}{\partial x} = 0 \quad \Rightarrow \quad \frac{138.6}{x} - 50 = 0 \quad \Rightarrow x = 2.77 \text{ feet}$$

The slope in a regression model that uses the log of the predictor X has an important substantive interpretation. In particular, we can interpret the coefficient of $\log_e X$ divided by 100 as *the expected change in Y for each 1 percent change in X.* That is, every one percent change in the predictor X on average produces a change in the response that is 1% of the slope. To get a sense why this interpretation holds (and justify the use of "natural logs"), consider the difference between fitted values at two slightly different values of X:. At the value X_1, the formula for the fitted value is

$$\hat{Y}_1 = a + b \log_e X_1$$

Now consider the fitted value nearby at $X_2 = 1.01(X_1)$, one percent larger value on the predictor scale. In this case, the fitted value may be rewritten as

$$\begin{aligned}\hat{Y}_2 &= a + b\log_e X_2 = a + b\log_e(1.01\,X_1)\\ &= (a + b\log_e X_1) + b\log_e(1.01)\\ &= \hat{Y}_1 + b\log_e(1.01)\\ &\approx \hat{Y}_1 + b(0.01)\end{aligned}$$

Thus, increasing X by 1% (going from X_1 to 1.01 X_1) is expected to increase the size of the response by about 1% of the slope. In this example, for each 1% increase in display footage, we expect a $1.386 increase in sales. (This is true only for small percentage changes, not real big ones.) One reason that logarithmic transformations using the so-called "natural logs" are popular in regression modeling is this interpretation of the slope in terms of percentage changes rather than absolute changes.

The intercept also has a special interpretation when the predictor is transformed. The intercept in any regression model is the expected value of the response when the predictor is zero. In this case, though, the predictor is the log of the number of feet.

$$\text{Predictor} = 0 \quad \Rightarrow \quad \log_e \text{Feet} = 0 \;\text{ OR }\; \text{Feet} = 1\,.$$

The intercept is the expected level of sales when there is one foot of display space (which is at the left edge of the prior plots).

As always when we make a subjective choice, it makes sense to assess how sensitive our results are to that choice. Here, we need to find out how sensitive our conclusion regarding the optimal amount of display space is to the choice of a logarithmic transformation. The reciprocal transformation, though its fit is rather different, leads to a similar estimate of the optimal number of display feet. The parameters of the fitted model using the reciprocal of the number of display feet are

Parameter Estimates

Term	Estimate	...more
Intercept	376.7	
Recip(Display Feet)	-329.7	

These estimates for the slope and intercept imply a different model for how the amount of display space affects sales

$$Sales = 376.7 - \frac{329.7}{\text{Display Feet}}$$

The estimate of the gain in sales generated by using x feet of display space is thus

$$gain(x) = 376.7 - \frac{329.7}{x} - 50x$$

The optimal gain occurs at a slightly smaller value of x than with the logarithmic model since the reciprocal curve flattens out more rapidly than the logarithmic curve. Nonetheless, the results are quite similar:

$$\frac{\partial\, gain(x)}{\partial x} = 0 \;\Rightarrow\; \frac{329.7}{x^2} - 50 = 0 \;\Rightarrow x = 2.57 \text{ feet}$$

It is comforting when two equations that both seem to fit reasonably well give comparable results. This lack of sensitivity to a subjective choice lends more credibility to the resulting estimate for the optimal amount of display space. In the end, we are likely to recommend that the store use 2.5 to 3 feet of display space, a range that is easy to comprehend.

Managing Benefits Costs

Insure.jmp

> Management is trying to renegotiate a contract with labor. Younger employees are happy about a new family leave policy, but elderly employees in the labor union have been reluctant to settle. A consultant has suggested to management that offering to pay for life insurance costs will appeal to older employees. Based on correlation and regression analysis, the consultant found that the cost of life insurance benefits to an employee increases with the age of the employee, so paying these costs would make the older employees happy. The fitted model was given as
>
> LifeCost = 258.877 + 5.07321 Age
>
> Is the consultant right? Will this strategy appease the older workers?

A sample of 61 employees was gathered. For each employee, management determined the age in years and cost of the life insurance benefit. This example appears as a Case Study at the end of Chapter 11 in the book *Statistics for Managers* by Hildebrand and Ott (pages 476-478).

To verify the claim of the consultant, the correlation between age and cost was computed to be 0.27, and indeed it is positive (significantly so, with p-value 0.035).

Correlations

Variable	Age	LifeCost
Age	1.00	0.27
LifeCost	0.27	1.00

The JMP summary also agrees with the consultant's fitted line.

Term	Estimate	more...
Intercept	258.9	
Age	5.1	

A plot of the data, however, suggests that the correlation summary is misleading. The relationship is not only nonlinear, but has a peak in the cost of the life insurance benefit near 45 years of age.

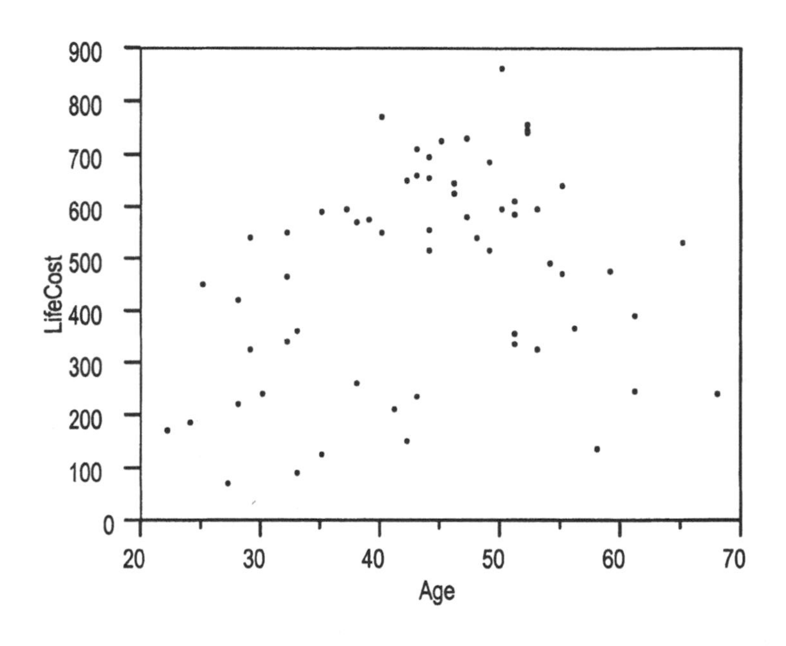

What sort of curve would you sketch to summarize the dependence of the life insurance costs on the age of the employee?

A fitted smoothing spline (in this figure, with smoothing parameter lambda set to 1000) appears to capture the nonlinear trend in costs. With the smoother added, the peak appears to be between 45 and 50 years of age. (You can use JMP's "crosshair" tool to find the peak in the graph.)

Recall that correlation is a measure of the degree of *linear* dependence between two variables. Thus, correlation is associated with a linear fit to the data. The figure shown next with the linear fit added makes it clear that such a linear summary is inappropriate in this case, whether measured by a correlation or a fitted line.

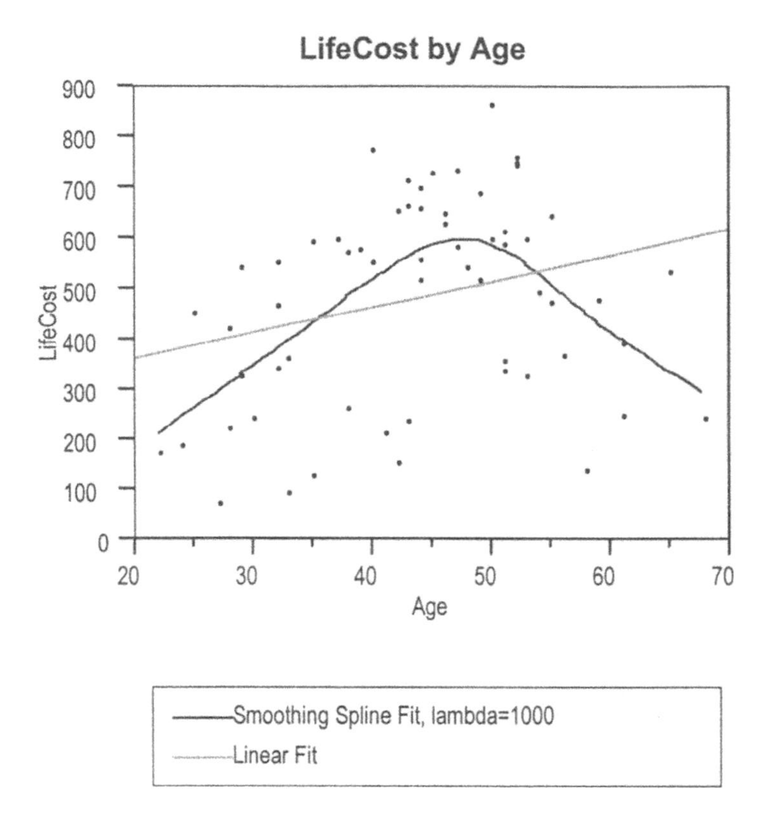

Once again, we need to use a nonlinear fit to capture the trend in the cost. The smooth curve first rises, then falls as age increases. This curvature does not match the pattern in any of the four quadrants used in Tukey's bulging rule, implying that none of these basic transformations will capture this type of nonlinearity. The power transformations associated with the bulging rule only approximate *monotone* nonlinear patterns, those that consistently either rise or fall. Since the cost of the life insurance benefit rises and falls, none of these simple transformations works.

For this type of curvature, a quadratic polynomial fit works quite well. We can fit a quadratic by selecting the *Fit Polynomial* item from the menu offered by red triangular button near the top of the scatterplot generated by the *Fit Y by X* command.

A plot of the fitted quadratic model with the same smooth spline fit shows reasonable agreement, at least for locating where the peak occurs. Both curves reach a peak at an age somewhere between 45 and 50, and the differences between the two curves are small relative to the size of the variation in the data around either. A "piecewise" linear fit (formed by joining two line segments near age 45 or 50) might also do well.

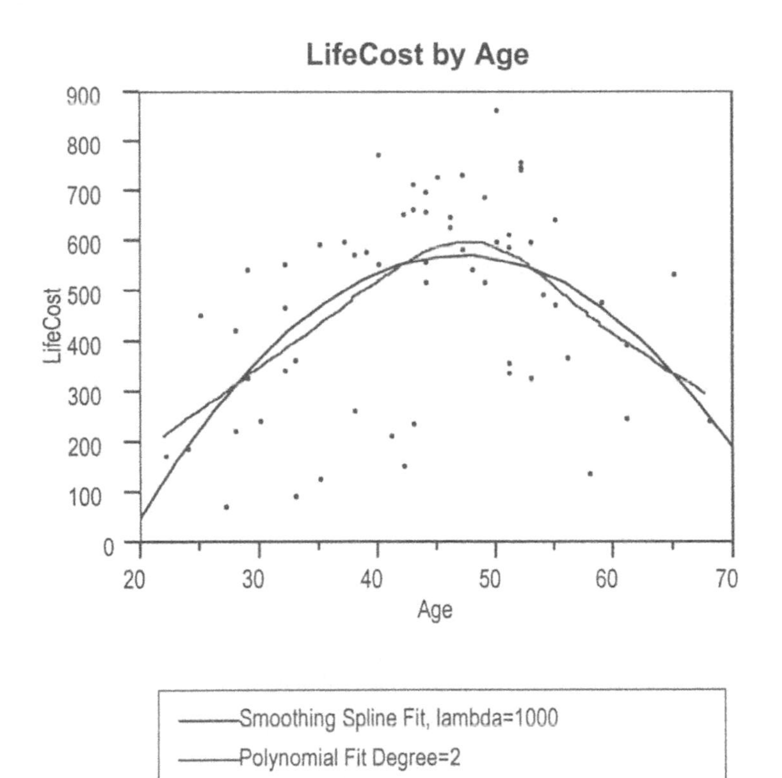

Management is trying to renegotiate a contract with labor. Younger employees are happy about a new family leave policy, but elderly employees in the labor union have been reluctant to settle. A consultant has suggested to management that offering to pay for life insurance costs will appeal to older employees. Based on a correlation analysis, the consultant found that the cost of life insurance benefits to an employee increases with the age of the employee, so paying these costs would make the older employees happy.

Is the consultant right? Would this strategy appease the older workers?

Although the correlation is positive, a linear summary is inappropriate for these data. The consultant calculated the correlation correctly, but it leads to the wrong conclusion. Since the cost of the life insurance benefit is actually quite small for the older workers, this "carrot" is unlikely to appeal to these employees and management must seek another strategy.

From the scatterplot, the peak cost occurs for employees who are between 45 and 50 years old. If we want a numerical value for the peak (rather than having to guess a range from the plot), we can use the equation for the quadratic model. The summary of this model is

Term	Estimate	...more
Intercept	-1015	
Age	4.61	
(Age-43.8)^2	-0.72	

When fitting polynomial models, JMP automatically "centers" the powers of the predictor by subtracting off the average value of the predictor, here 43.8 years. (The reason for this will become clear later, but observe for now that it keeps the squared values from getting quite so large.) The fitted equation for life insurance costs in terms of age is thus

$$LifeCost \approx -1015 + 4.61\,Age - 0.72\,(Age\text{-}43.8)^2 .$$

With this equation in hand, we can find the location of the maximum algebraically. Taking the derivative of the equation with respect to *Age* implies that the cost peaks at the value of *Age* which satisfies the equation

$$0 = 4.61 - 1.44\,(Age\text{-}43.8) .$$

The peak is thus quite near our graphical guess, at *Age* = 47.0 years.

Predicting Cellular Phone Use

Cellular.jmp

The use of cellular phones in the US grew dramatically, and the number of subscribers approached 35 million by the end of 1995.

Given this data, how many subscribers would you expect by the end of 1996? How many by the end of the 20th century?

These data show the number of subscribers to cellular phone service providers in the US. Each data point represents the total number of subscribers at the end of a six-month period. The data span eleven years, from the end of 1984 to the end of 1995. There are 23 observations, beginning with 91,600 at the end of 1984 (Period = 1) to nearly 35 million 11 years later at the end of 1995 (Period = 23).

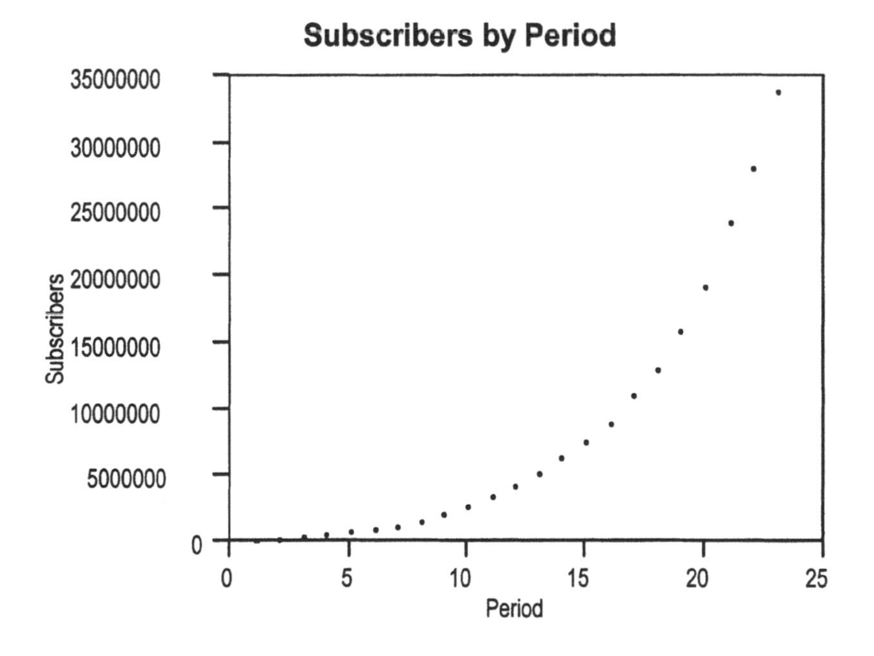

The curvature in the growth of the number of subscribers suggests that we need either to raise the power of X or lower the power of Y. We will take the latter approach in this example. Moving first to the square root of the response (reducing the power of Y from 1 to $^1/_2$), we see that this transformation is "not enough." It does not capture the rapid growth and lags behind the pattern of the data.[3]

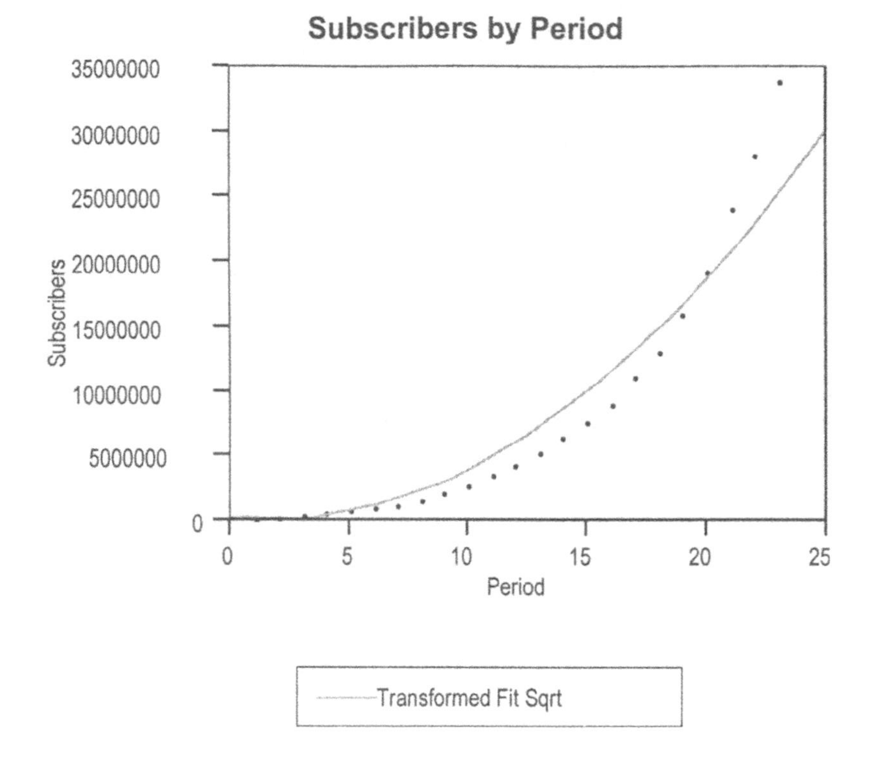

Sqrt(Subscribers) = -447.6 + 237.1 Period

[3] We obtain this fit by using the Fit Special item that appears in the pop-up menu offered when you click on the red triangle that appears near the top of the window that shows the scatterplot.

On the other hand, moving to the log transformation (in the ladder of transformations, the log transformation occupies the zero position) goes too far. Evidently, the transformation we need lies somewhere between these two, with an exponent somewhere between 0 (logs) and $^1/_2$ (square roots).

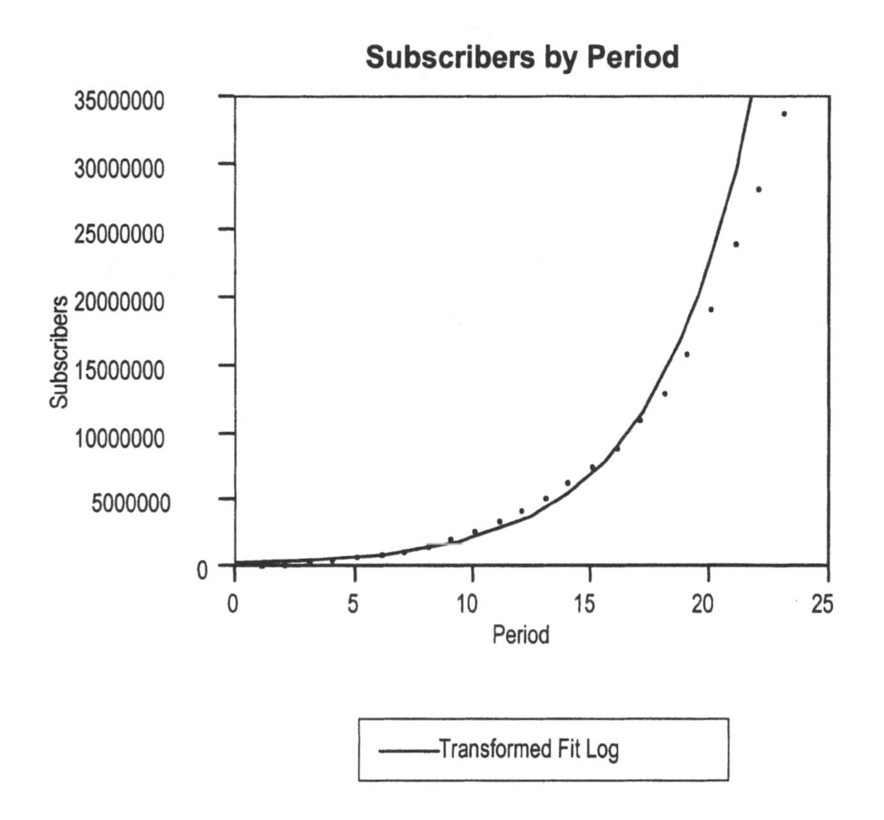

Log(Subscribers) = 12.1 + 0.24 Period

The transformations provided by the JMP dialog are the most common and generally do what is needed. In some cases, such as this one, however, we need to use other exponents. The offered transformations are either too weak or too strong.

To use other transformations, we have to use the formula editor and build the response variable ourselves. The process is less convenient, but teaches a valuable lesson: the goal of choosing a transformation is to obtain new data which when plotted are *linear on the transformed scale*. Rather than look at the data plotted on the original scale, we will need to look at plots on the transformed scale. Relationships that are nonlinear in the original data appear linear when we have chosen a good transformation.

A first guess at a new transformation does well for these data. Recall that logs (zero power) went too far and square roots (half-power) were not enough. Shown below is a plot of the quarter-power *Subscribers*$^{1/4}$ versus the period. It is quite linear — a line summarizes the relationship quite well.

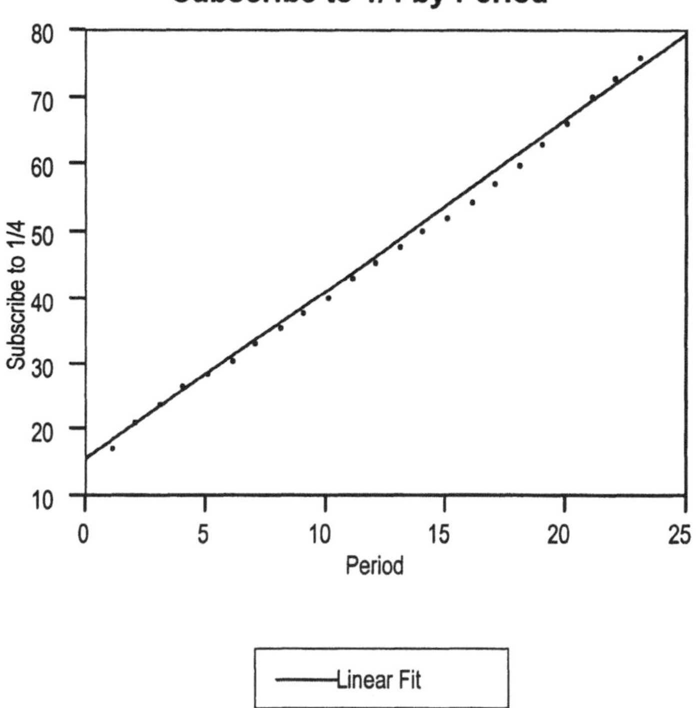

Subscribe to 1/4 = 15.4 + 2.55 Period

For the sake of comparison, here are plots of the square roots and logs, shown on transformed axes. Note the departures from linearity.

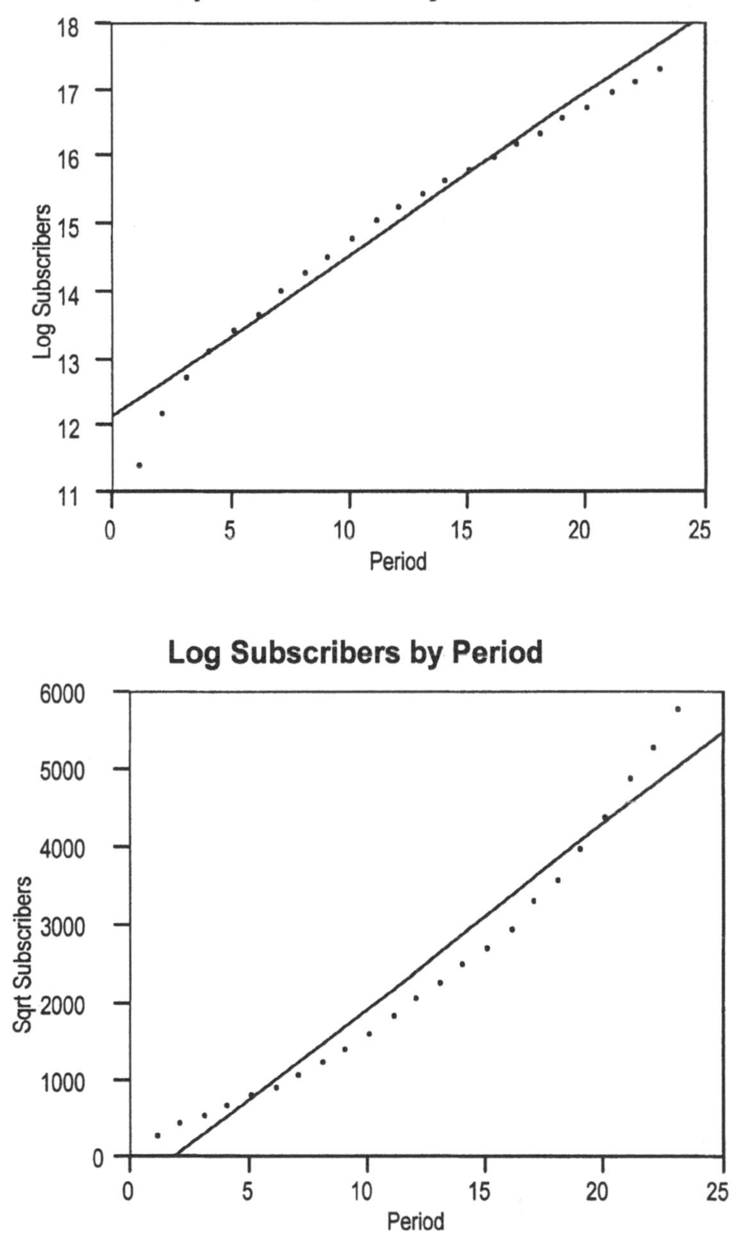

As you might have guessed from the original plot, polynomials do quite well for this data as well. For example, a fourth-order polynomial trend in period (a model using X, X^2, X^3, and X^4) gives a very good fit, as shown below. This fit is hardly parsimonious and it is difficult to interpret, but it certainly tracks the data well! Polynomials are able to track virtually any sort of trend, but seldom offer accurate extrapolations outside the range of observation.

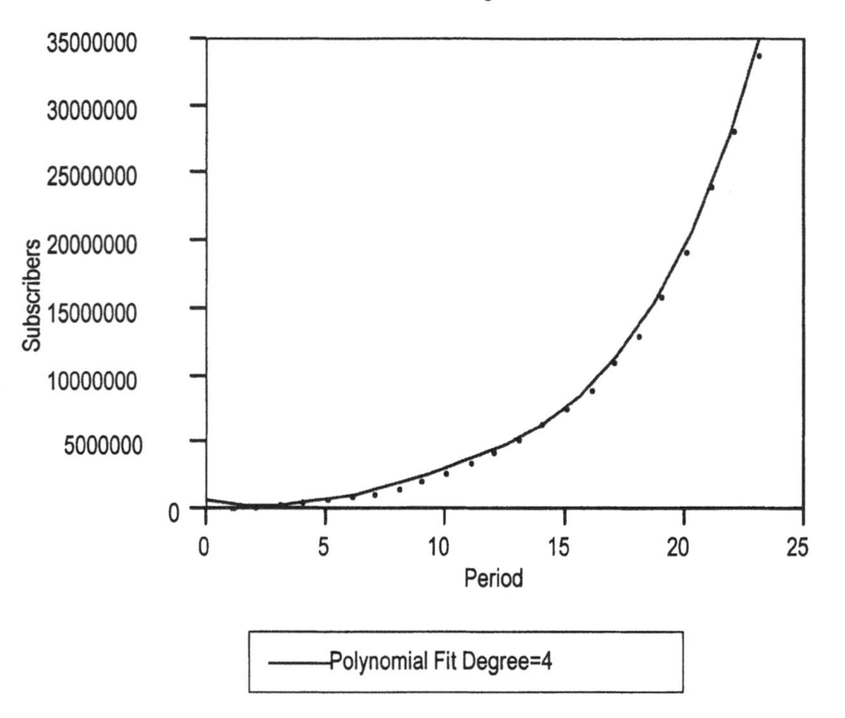

The polynomial fit produced by JMP is "centered" in that the mean of the predictor, here the mean of *Period* (12) is subtracted prior to raising the predictor to a power. This centering (the resulting predictor has average zero) keeps the power terms from getting quite so large.

Subscribers = -5.661e+6 + 821647.3 Period + 71171.668 (Period-12)^2 + 5844.2803 (Period-12)^3 + 289.36228 (Period-12)^4

The use of cellular phones in the US grew dramatically, and the number of subscribers approached 35 million by the end of 1995.

Given this data, how many subscribers would you expect by the end of 1996? How many by the end of the 20th century?

The quarter-power requires only a slope and intercept and seems to work well in yielding a linear relationship. The parameters of this model from the JMP output are

Term	Estimate	more...
Intercept	15.4	
Period	2.55	

The fitted equation is thus

$$Subscribers^{1/4} \approx 15.4 + 2.55\ Period.$$

In order to use this fit to predict the number of subscribers by the end of 1996 and the end of the 20th century, we need to compute the periods for these two. For the end of 1996, the period is 25. As a result, the prediction is

$$\text{Estimated \# by end of 1996} = (15.4 + 2.55 \times 25)^4 = 79.15^4 = \mathbf{39.2\ million}.$$

For the end of the 20th century (December, 1999) the period is 31 giving

$$\text{Estimated \# by end of 1999} = (15.4 + 2.55 \times 31)^4 = 94.45^4 = \mathbf{79.6\ million}.$$

What assumptions are implicit in these predictions? Does predicting the number at the end of next year seem reasonable? What about extrapolating five years out? For problems such as these, it is important to have a notion of the accuracy of the prediction. We return to this example in Class 2 and do that.

An Alternative Analysis

When modeling time series with strong growth such as this one, we might proceed as in our analysis of stock prices in Class 2 of *Basis Business Statistics*. Rather than model the raw data, the idea is to model the rate of growth. For the cellular telephone data, a simple transformation captures the pattern in the growth rates.

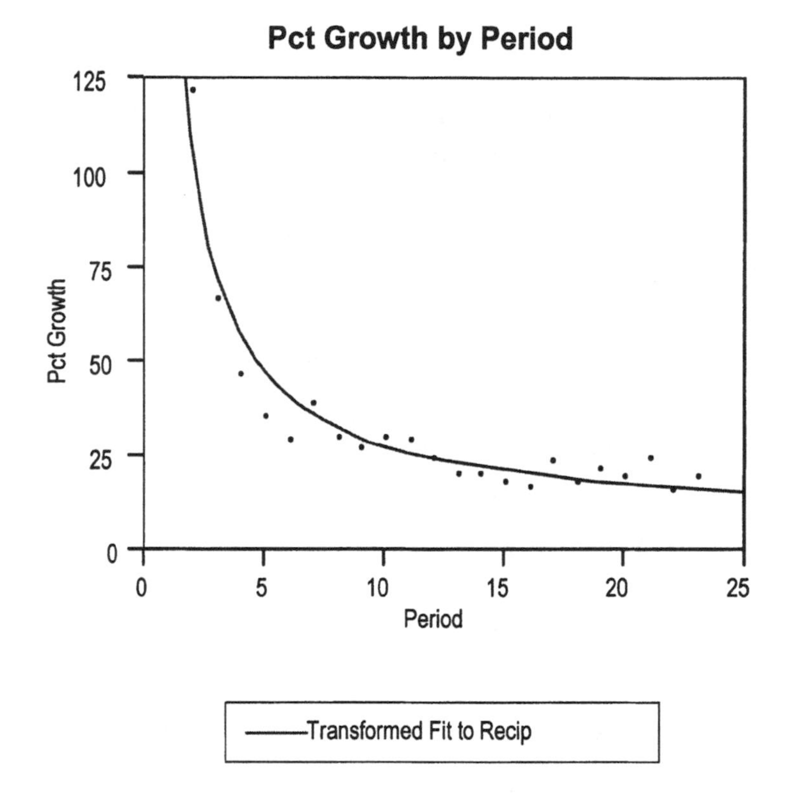

Pct Growth = 7.50205 + 199.656 Recip(Period)

The fitted model uses the reciprocal of period and seems to capture the data well, though at the right it seems to be consistently under predicting the rate of growth (important for extrapolation). From the output, the model fitted is

$$\text{Pct Growth} \approx 7.5 + \frac{200}{\text{Period}} \; .$$

This model predicts an ultimate growth rate of 7.5%.

To get a prediction for the number of subscribers at the end of 1996, we simply use this estimate of the rate of growth to compound the last value. For June 1996, we get

$$\text{Predicted for end of June 1996} = 33{,}785{,}661 \times (1 + (7.5 + \frac{200}{24}) / 100)$$
$$= 33{,}785{,}661 \times 1.1583 \approx 39.1 \text{ million.}$$

For the end of 1996, we continue to expect about 15% growth (the base rate 7.5% plus 200/25%) in the number of subscribers

$$\text{Predicted for end of December 1996} = 39{,}100{,}000 \times (1 + (7.5 + \frac{200}{25}) / 100)$$
$$= 39{,}100{,}000 \times 1.155 \approx \mathbf{45.2\ million}.$$

This prediction is quite a bit higher than the estimate from the quarter-power model. The predictions lack a measure of uncertainty, however, and to build intervals for the predictions, we need a more rigorous model.

To see how these predictions compare to the actual totals for the cellular industry, the data file includes more recent data that was not used in the previous calculations. We can use this data to see how well the predictions turned out. Here is a small table of the results. The reciprocal model does well in the short term, then overestimates the growth of the market.

Date	**Millions of Subscribers**	**1/4 Power Predictions**	**Reciprocal Predictions**
June 30, 1996	38.2	34.3	39.1
Dec 31, 1996	44.0	39.2	45.2
Dec 31, 1999	86.0	79.6	102

The plot on the next page shows what went "wrong" with the reciprocal model. The growth rates in the later data drop off faster than this model anticipated.

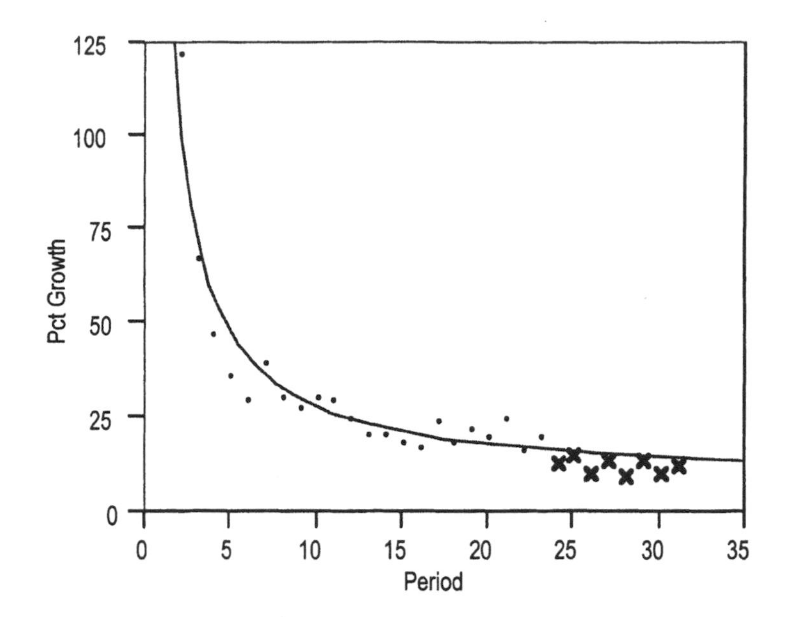

In the long run, the more conservative predictions provided by the linear model using just the quarter-power transformation do better, even though the predictions of this model are less accurate in the short term. What do you think would happen to the estimated slopes and intercepts if these models were refit to the expanded data set?

Class 2. Assumptions in Regression Modeling

Effective use of regression requires a thorough understanding of the assumptions behind regression models. Most of these assumptions resemble those that are needed for tests of means and confidence intervals (as discussed in the preceding volume of cases *Basic Business Statistics*). For example, regression analysis assumes that the observations are independent with equal variance.

We introduce the assumptions of regression analysis in the context of an idealized model. Particularly important deviations from the idealized model for regression analysis are the effects of dependence, skewness, and outlying values. In this class, we will consider special tools that identify such problem areas and quantify their effects. These methods rely on inspection of the observed deviations from the fitted equation to detect problems in the fitted model. These deviations are known as residuals. Scatterplots of the data and residuals graphically convey problems in the model. We return to the effects of dependence in Class 12.

Topics

Idealized regression model and assumptions
Estimates in regression: intercept, slope, and error variation
Fitted values and residuals
Outliers and their effects on a regression
Cyclic error patterns (dependence among errors)
Heteroscedasticity (lack of constant variance)
Normal quantile plots of residuals

Examples

1. The ideal regression model (simulated data)
2. Predicting cellular phone use, revisited (lack of independence)
3. Efficiency of cleaning crews, revisited (lack of constant variance)
4. Housing prices and crime rates (influential outlier)
5. Direct mail advertising and sales (an outlier that is not leveraged)
6. Housing construction (leveraged, but not influential)

Overview

The objective of fitting a regression line usually requires more than just a summary of the data. Sometimes we want to do inference based on the regression line and make a claim about a sampled population. Most often, such inference relies on the slope of the fitted regression line. Remember that a slope of zero indicates that the line is flat and that the two variables under study are not related. Regression output gives us an estimate of the slope along with an associated standard error, and a confidence interval for the slope provides a more informative statement than a mere "best guess" or point estimate. If we want a confidence interval for the slope, then we have to make some assumptions about the entire process that generates the data we observe. The underlying abstraction is that there is a "true" regression line and that the data is generated by adding random noise to it. This noise is the reason we see the points scattered about the line rather than lying exactly on it.

To buy into the concept of a confidence interval, you need to make some important additional assumptions about how the noise gets added to the line. *The typical assumptions we make are that the average size of the noise is zero, being normally distributed with the same variance at all data points and independent from one data point to the next.* Thus it becomes critical to check whether or not these assumptions are credible, and hence the remainder of the class is devoted to checking these fundamental assumptions. We will check the assumptions graphically, so you will see many plots, particularly residual plots, whose purpose is to evaluate the credibility of the assumptions. Such plots, like the normal quantile plot introduced in *Basic Business Statistics* are "diagnostics" because, just as in medicine, they serve the purpose of finding out things that have gone wrong, in our case the violation of the regression assumptions.

This class also introduces tools that locate outlying points. Though an outlier is easy to spot with only two variables in the analysis, it becomes a much harder job later when we move into more complex models. As mentioned previously, these tools can never tell you whether or not you want to keep the particular point. Are you prepared to have your whole analysis and related decisions driven by a single observation? It's a very difficult question, but typical of the complexity we will face during the coming weeks.

Key application

Identifying points that control the fitted regression line.

In Class 1, we explored the value of smoothing a scatterplot, be it through smoothing splines or by fitting a regression line. The value comes from the visual summary it provides of the data and its usefulness for prediction. In this class, we concentrate on the regression line summary of the scatterplot and gain more insight into what determines this line.

Assumptions are very important. In addition, a single point can influence the regression and you, as the analyst, must to be able to recognize this fact and then decide whether or not this is a good thing. Statistics offers many available tools to help identify single points that exert high influence on the regression line, but of course statistics says nothing about whether or not these influential points are useful to have. That's a matter for the subject area specialists. To make things concrete, consider the problem of determining the most cost-effective way for a city to sell its municipal bonds. The critical question is simply "What is the optimal issue size to minimize average cost?" This is a straightforward question, and one in which a statistical analysis of the data, average cost versus issue size, provides valuable insights. For this particular city, one issue size was almost three times larger than any other issue. Including this issue in the analysis leads one to conclude that the optimal issue size is twice as large as the issue size when this point is excluded from the analysis. Such a point is, by definition enormously influential — it matters both statistically and practically. It drives the decision you make.

Definitions

Autocorrelation. Sequential dependence between adjacent errors about the fitted relationship. In other words, the underlying errors are not independent, with one error typically depending on the previous one. Most often, autocorrelation appears when modeling time series, as discussed later in Class 12.

Heteroscedasticity. Lack of constant variance; the noise process is more variable for some values of X than others. Common when the data measure observations (e.g., companies) of different size.

Outlier. An atypical observation, either in the horizontal (X) or vertical (Y) direction. Observations that are unusual in the X direction are leveraged. Leveraged values which also possess substantial residual value in the vertical direction (*Y*) are influential.

Regression diagnostics. Methods for picking apart a regression analysis, in particular, finding influential data points and checking regression assumptions about the noise process. Most often these are graphical and rely upon the residuals.

Residual. The vertical deviation of a point from the fitted regression line.

RMSE (Root mean squared error). The square root of the average squared residual, an estimator of the standard deviation of the assumed errors around the underlying true regression line.

Heuristics

Where do the errors come from in a regression model? What are they?

It's rare that we have a problem for which we can think of one only variable to use as a predictor (*X*) of the response (*Y*). We may have data for only one predictor, but surely we can think of other relevant factors. A useful way to think of the errors in regression is that they are the sum of all of those other factors. Little events don't formally get taken into account with our simple models, but they show up in the data as small fluctuations about the trend, producing the noise or error about the line. Thinking back, we have noted (labeled the Central Limit Theorem) that sums tend to be normally distributed. Thus, it is often reasonable to expect the errors in a regression to be normally distributed.

What gives rise to autocorrelation?

Most often, autocorrelation arises from factors that are omitted from a regression model. Suppose we are trying to predict the daily S&P500 . During "bullish" times, our model might predict that the price will increase fairly consistently along a somewhat linear trend. It does not grow exactly along a straight line, though. Too many factors influence what's bought and sold, and our models are unlikely to include all of the relevant factors. For example, suppose Alan Greenspan mumbles something about interest rates on a Monday morning. Though most likely not a part of our predictive model, his thoughts are going to affect the market on Monday afternoon, deviating from our predicted trend. On Tuesday, the influence of Greenspan's remarks will not have disappeared, although they might not be quite so influential. The same goes for Wednesday and for some time into the future. These residuals, or deviations from our model, have all been influenced by the same event, Greenspan's musings, and so are dependent. In other words, they are correlated with each other, and since this correlation occurs over time, the residuals are *autocorrelated.*

Potential Confuser

What's the difference between leverage and influence?

Leverage comes from an observation lying on the fringe of the horizontal axis, at the extreme left or right side of the plot. It's in a position to make the fitted line tilt in either direction and has the *potential* to affect the line. Going further, a point is influential if the regression fit changes "a lot" (in the sense that the slope changes by a considerable amount) when that point is removed. A leveraged observation may be influential, but need not if the line would pass though that location without the observation.

Keep in mind, though, that regression involves more than just a slope. Removing a leveraged point that is not influential in the strict sense might still have a very large impact on the goodness of fit. Outlying values may affect the estimated error variation around the fitted line, so taking them out can lead to a very different impression of how well the fitted model is doing.

Summary of the Ideal Regression Model

Structure: $Y_i = \beta_0 + \beta_1 X_i + \varepsilon_i$

where: (1) observations are independent,

(2) errors have mean zero and constant variance:

$$E\,\varepsilon_i = 0, \qquad \text{Var}\,\varepsilon_i = \sigma^2, \quad \text{and}$$

(3) errors are normally distributed.

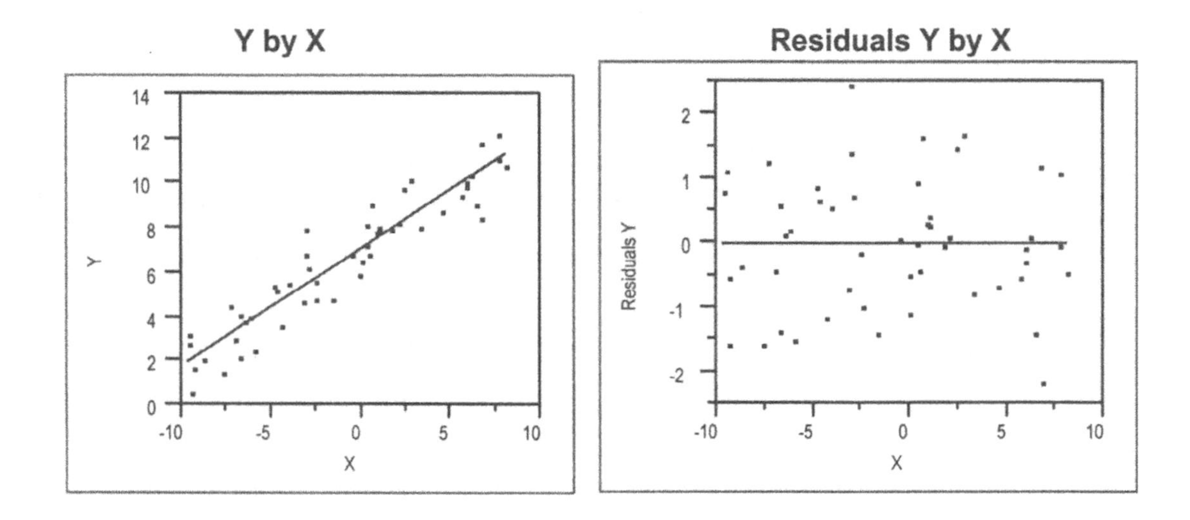

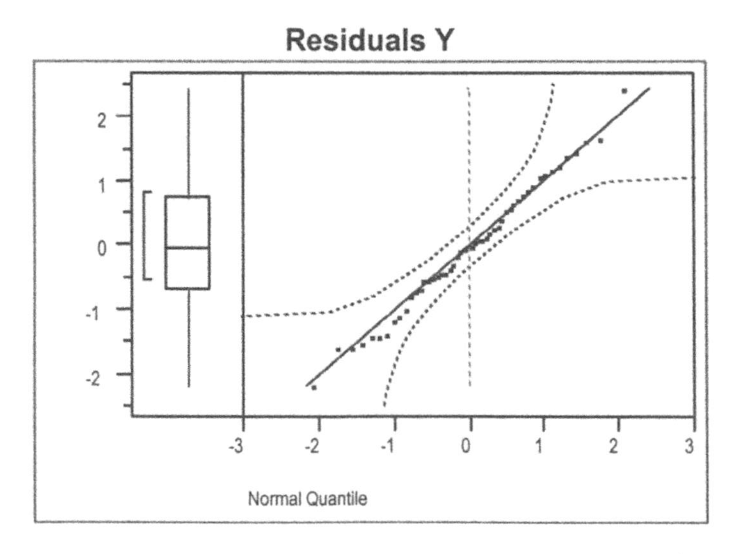

Residual Plots with Various Problems

Cellular - Correlated errors

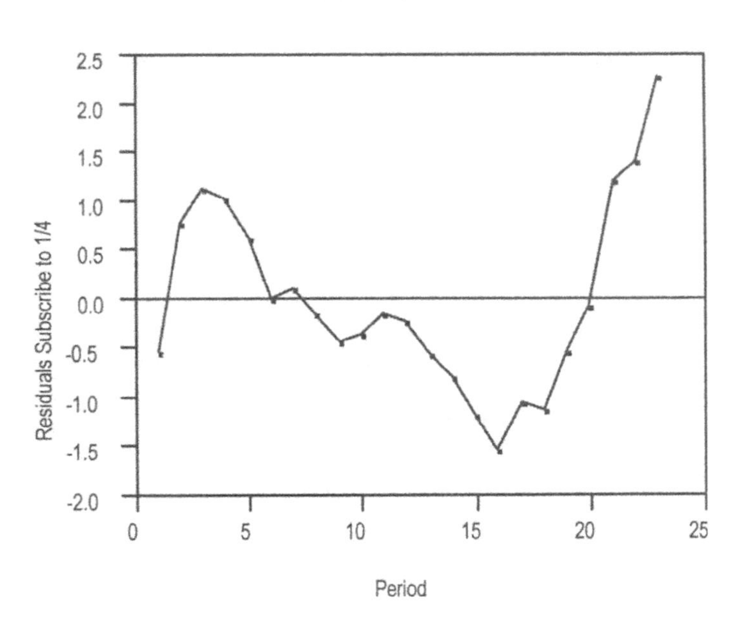

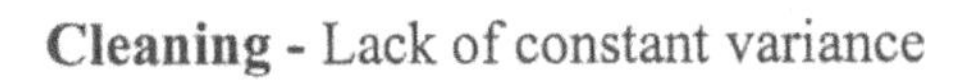

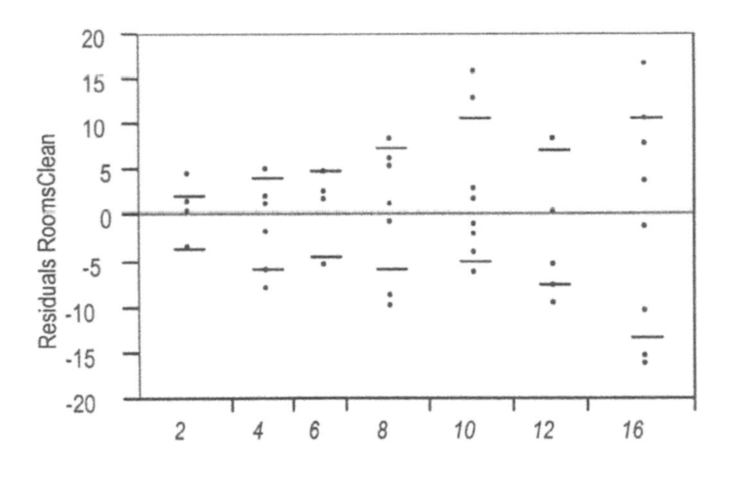

Outliers

Direct
Outlier not leveraged

Phila
Influential, leveraged outlier

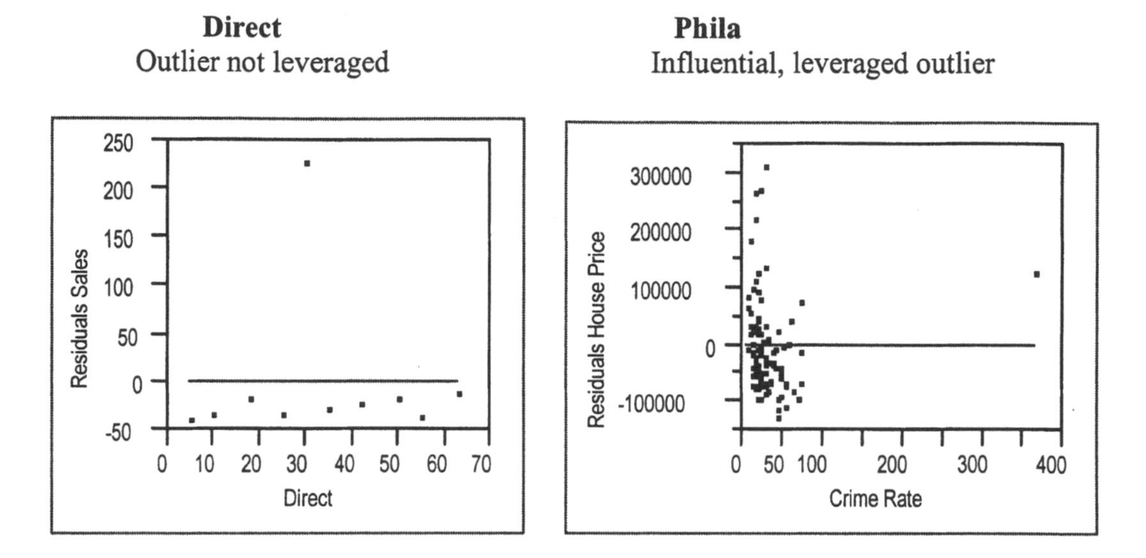

Cottages
Leveraged, but neither outlier nor influential

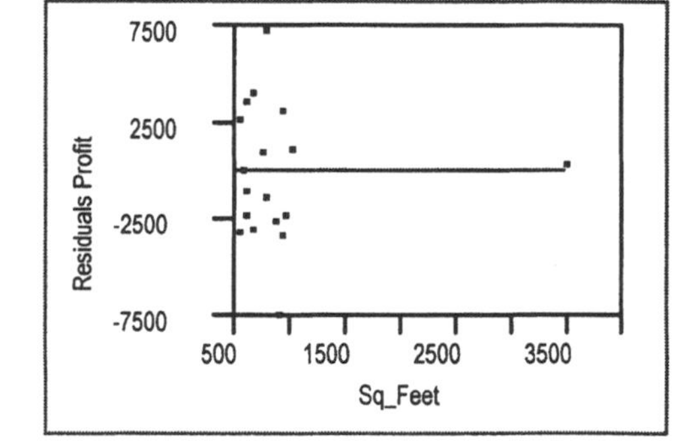

The Ideal Regression Model

Utopia.jmp

What do data look like that come from the idealized regression model?

The data for this example are simulated. We generated an artificial data set with 50 observations to illustrate the plots associated with data generated by the idealized regression model. Simulation is a powerful tool for experimenting with statistical tools and exploring their behavior under various conditions. Looking at the results for this small ideal data set give you a chance to "train your eyes" as to how the various regression plots appear when everything is working.

The initial scatterplot has a clearly defined linear structure, with the line capturing a large amount of the pattern in the data. The slope of the fitted line is large relative to the variability about the fitted line. Most of the points in the plot cluster about the line with large vacant areas apparent.

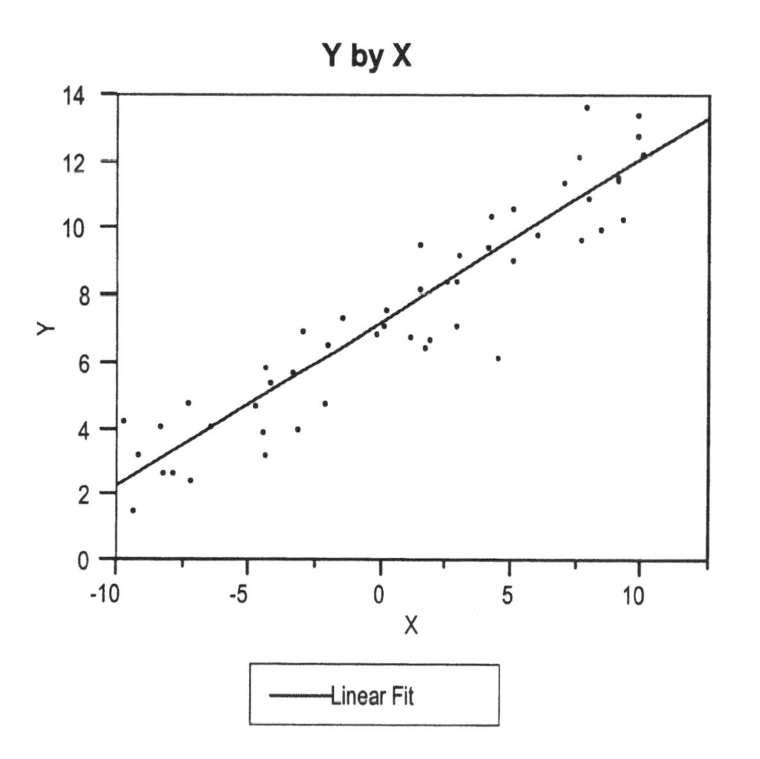

The associated numerical summary shows how close this model has come to what we are privileged to know is the true model (Y = 7 + 0.5 X + normal error).

Summary of Fit

... other stuff ...		
Root Mean Square Error	1.14	<- est error SD
Mean of Response	7.55	
Observations	50	

Parameter Estimates

Term	Estimate	Std Error	...more
Intercept	7.18	0.16	
X	0.49	0.03	

The standard errors of the estimates of the slope and intercept convey the same information about these statistical estimates as in the case of the standard error of a sample mean. The formulas used to obtain the standard error of a slope or intercept differ from that used for a mean, but the idea is the same. As before, the standard error measures how close we expect this sample estimate to lie to the population value.

For example, the standard error suggests that the slope estimate ought to be within about $2 \times 0.026 = 0.052$ of the true slope. Indeed, we know that it is. Notice that there is little reason to report more decimal digits in the estimates than the standard error suggests. JMP makes it easy to reformat the output tables. Double clicking on a table generates a dialog that lets you set the number of digits shown in the table.

So far, the initial plot and accompanying parameter summary indicate that the fitted model is doing well. The trend in the points in the plot is linear and the estimates are close to the true values. In order to assess the validity of a regression model, though, we have to examine the fit much more carefully than we have so far. We need a microscope to examine the fit, and the residuals are the key.

Regression residuals are the vertical deviations about the fitted line. These are the deviations that are squared, summed and then minimized in the defining least squares procedure. The plot on the next page shows the residuals from our regression line. By looking at deviations with the trend removed, this plot "zooms in" on the data near the line and expands the vertical scale to focus on what is happening in the data near the fit.

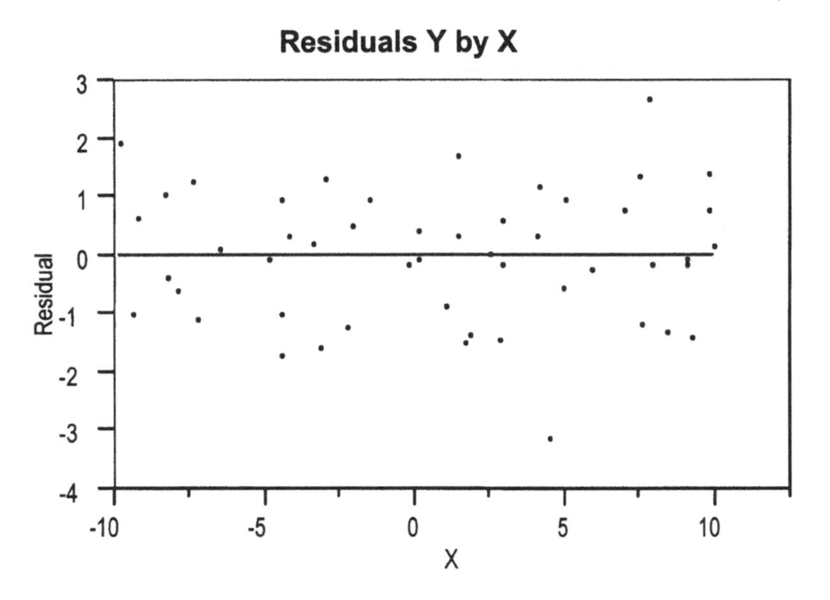

The residuals should appear as a "swarm" of randomly scattered points (such as these) about their mean (which is always zero because of the least squares criterion being minimized).

Plot linking and brushing make it easy to establish the correspondence between the points in the plot of the original data and the residual plot. With both plots visible on the screen, highlighting points in one view also highlights them in the other. JMP provides this plot linking automatically. Use the shift key with the brush tool to highlight the observations shown at the right of either plot on the next page (and then mark them using the menu item from the *Rows* menu)).

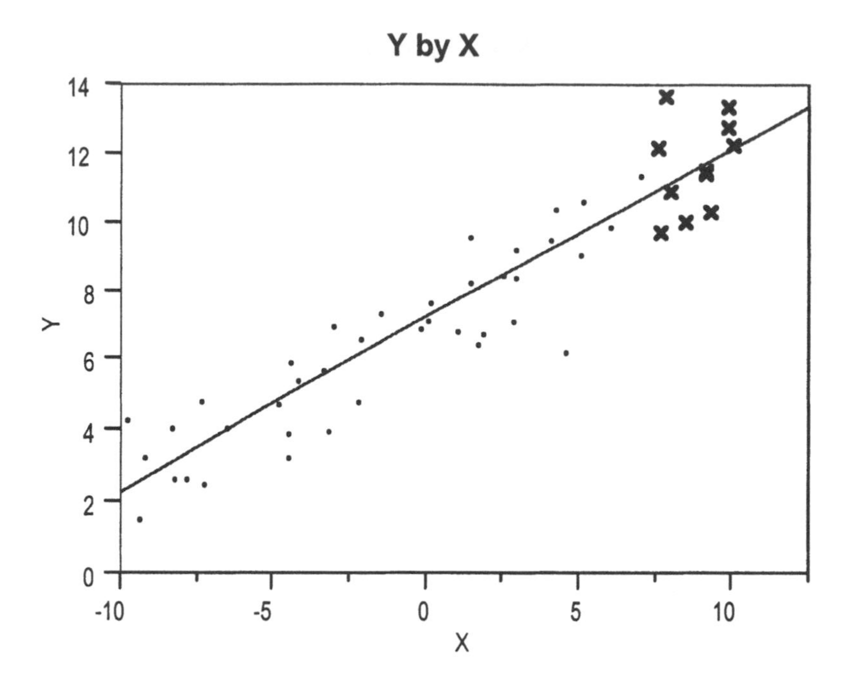
Y by X
Y
X
14
12
10
8
6
4
2
0
-10
-5
0
5
10

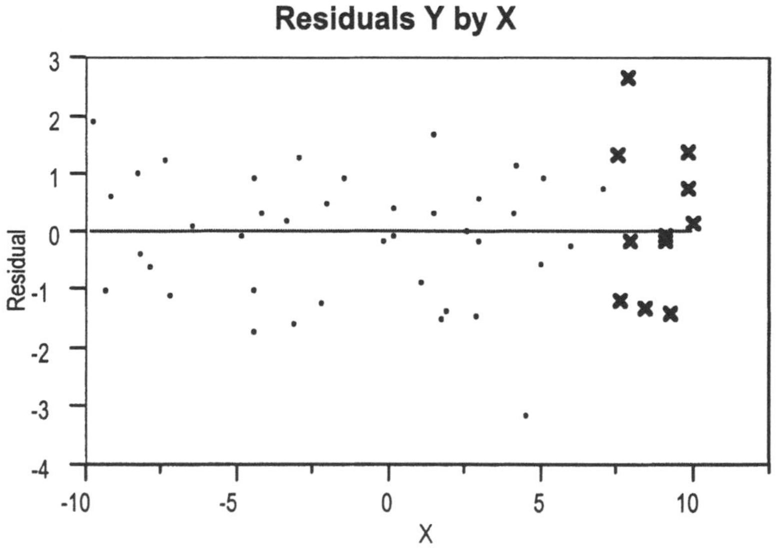

Residuals Y by X
Residual
X
3
2
1
0
-1
-2
-3
-4
-10
-5
0
5
10

Finally, we need to check the plausibility of the normality assumption. We use normal quantile plots, but of the residuals rather than the response variable itself. Remember, the underlying regression model only claims that the errors about the true fit ought to be close to normal. Neither the response nor predictor have to be normally distributed. Since residuals are estimates of the true errors, we use them to check the normality of the unobserved errors.

In this utopian setting, the normal quantile plot shows that the residuals are well within the limits appropriate to a sample from a normal population.

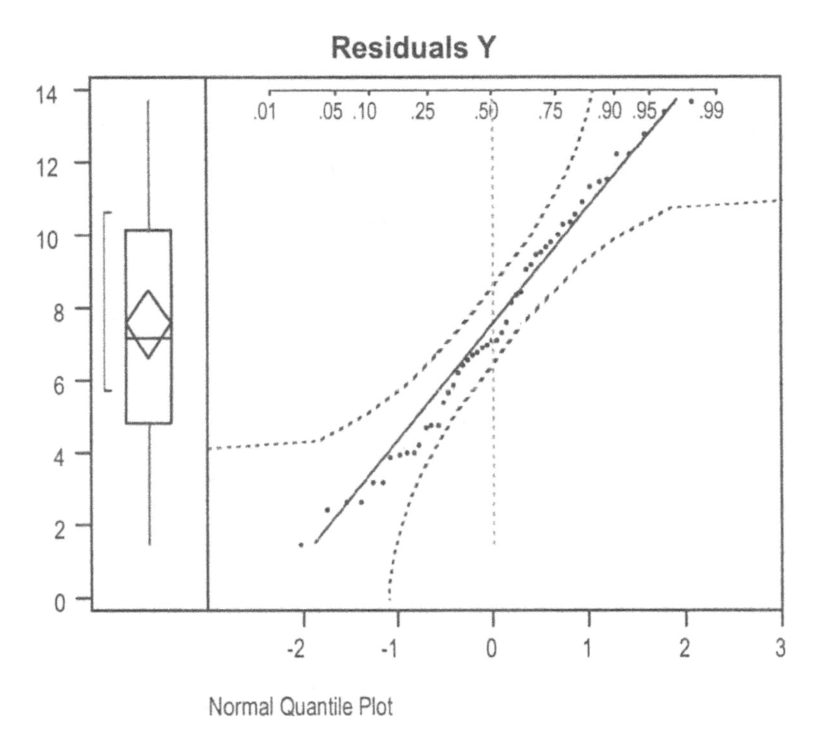

What do data look like that come from the idealized regression model?

These plots offer one example. You might want to build your own data set and experiment further. The random number generators and formula editor in JMP make these chores easy. Keep in mind that simulated data exist in a computer generated "virtual reality" which resembles an experimental laboratory, not the real world. It may be the last "perfect" regression model you ever see!

Another point regarding this illustration deserves comment. As part of a course like this, you will come to regard data like that in this example as "a good thing" to find. In later applications, you won't be so happy to find such normal, featureless residuals. Why? Once you have reduced the data to random variation, there isn't anything left for you to explain. You've reached the end of the modeling road, so to speak. Finding structure in the residuals, on the other hand, implies that you can explain more of the variation in the data. In the end, you want to reduce the data to random variation. You just hope that it does not come too soon.

A very useful exercise to do with this data set is to explore what happens when we add more observations to the data set. The results seen above with 50 were close to what we know the true model to be. What would happen if we had, say, 5000 observations rather than just 50? How would the regression model change?

For example, do you think that the slope would be larger or smaller? What about the RMSE, the estimated standard deviation of the errors around the underlying regression? Write down what you think will happen, then simply add more rows to the data set. JMP will automatically fill them in with more simulated values. Then just refit the model and compare. The plot alone is worth the exercise.

Predicting Cellular Phone Use, Revisited

Cellular.jmp

The use of cellular phones in the US grew dramatically, and the number of subscribers approached 35 million by the end of 1995.

Based on the data at that time, how many subscribers would you expect by the end of 1996? How many by the end of the 20th century?

This example appears first in Class 1. In this second examination of these data, we look more closely at the model that was fit previously with an eye on the assumptions. For regression predictions to be reliable, we need a model that comes close to meeting the assumptions posed as our idealized regression model.

Here is a plot of the raw data and fitted line. Clearly, a straight line does not fit well.

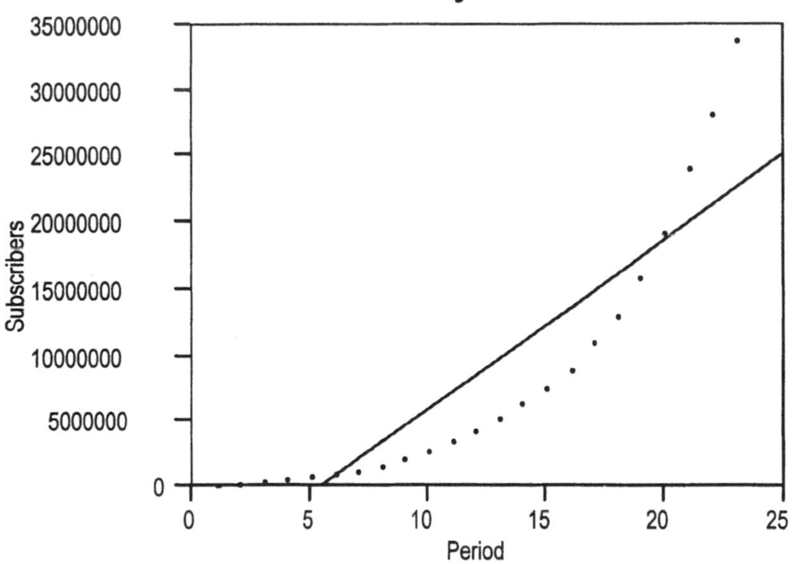

To obtain a better model, we transformed the response to obtain a linear relationship between the transformed response and *Period*. The quarter-power of the number of subscribers worked quite well in the sense of making the fitted data linear, and we used this model in the previous class to obtain predictions.

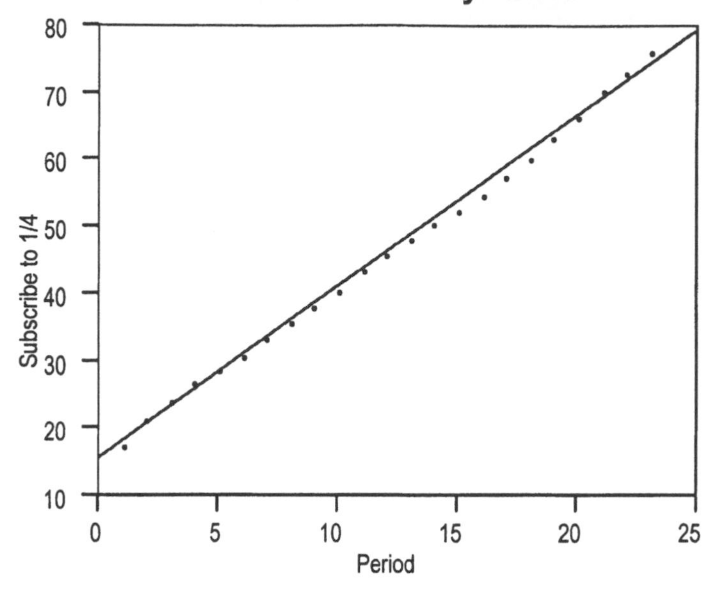

Subscribe to 1/4 = 15.38 + 2.55 Period

Summary of Fit

...

Root Mean Square Error	0.97
Mean of Response	45.95
Observations	23

Parameter Estimates

Term	Estimate	Std Error	More...
intercept	15.38	0.42	
Period	2.55	0.03	

Although the transformed model is a vast improvement over the original linear fit, it has a problem. The residuals appear to be correlated over time, tracking each other in the time series plot shown below. This type of correlation with time series is known as autocorrelation. (Again, use the button for the fitted model to generate the residual column in the JMP spreadsheet. Use the *Overlay plots* command of the *Graph* menu to get connected time plots like this one. Double click on the vertical axis to add the reference grid line at zero.)

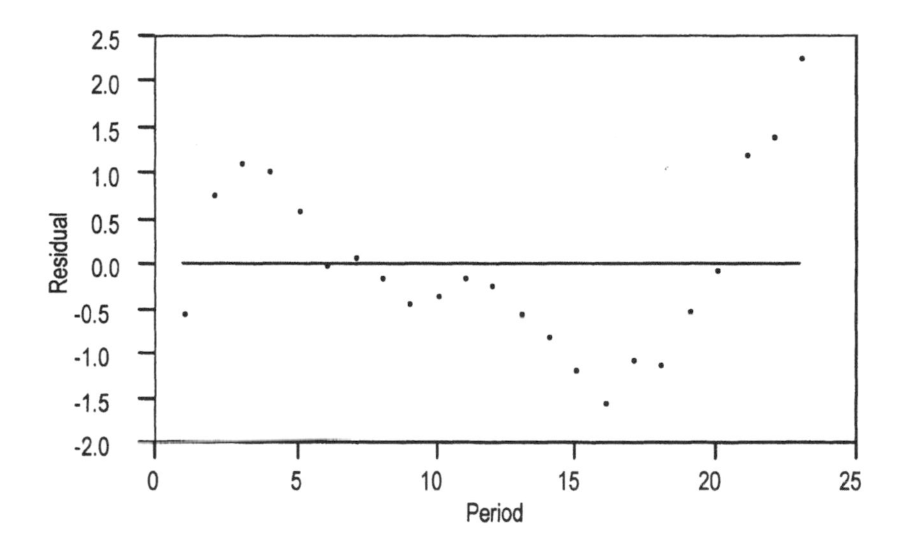

The pattern in the residuals suggests that predictions based on the quarter-power model are going to be too small. Perhaps the predictions of the alternative model (based on percentage growth) offered in that example are better. Recall that the prediction for the number of subscribers by the end of 1996 from the transformed model is **39.2 million**. In contrast, a model based on percentage growth predicts about 6 million higher, at **45.2 million**. The pattern in the residuals shown above suggests that the estimate based on the quarter-power could easily be 3 million too small if the recent trend continues. In fact, the prediction for the end of 1996 erred by more. The actual value for the end of 1996 was **44.0 million** – the short-term under-predictions suggested by the residual analysis happened.

The use of cellular phones in the US grew dramatically, and the number of subscribers approached 35 million by the end of 1995.

Based on the data at that time, how many subscribers would you expect by the end of 1996? How many by the end of the 20th century?

The residuals from the transformed model suggest that the unobserved errors about the regression fit are not independent as the idealized model assumes, but instead are dependent. Here, they appear autocorrelated and track over time. In this case, the pattern in the residuals suggests that the predictions from the transformed model are going to be too small. The lack of independence also leads to other, more complex problems that we will return to in Class 12.

Efficiency of Cleaning Crews, Revisited

Cleaning.jmp

A building maintenance company is planning to submit a bid on a contract to clean 40 corporate offices scattered throughout an office complex. The costs incurred by the maintenance company are proportional to the number of cleaning crews needed for this task.

The fitted equation determined by least squares indicates that about 10 crews are needed. Is this regression model reliable?

In Class 1, we noticed that while the trend seemed linear, the variation about the trend appears to grow.

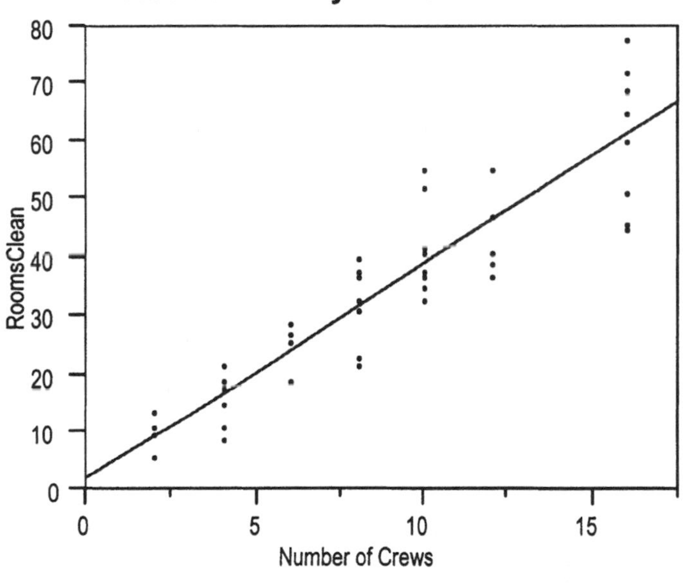

Parameter Estimates

Term	Estimate	Std Error
Intercept	1.78	2.10
Number of Crews	3.70	0.21

The trend to increasing variation is more evident when looking at the residuals. Residual plots focus on the deviations from the fitted equation and emphasize, in this case, the change in variation. With the linear trend removed, the change in variation stands out more clearly.

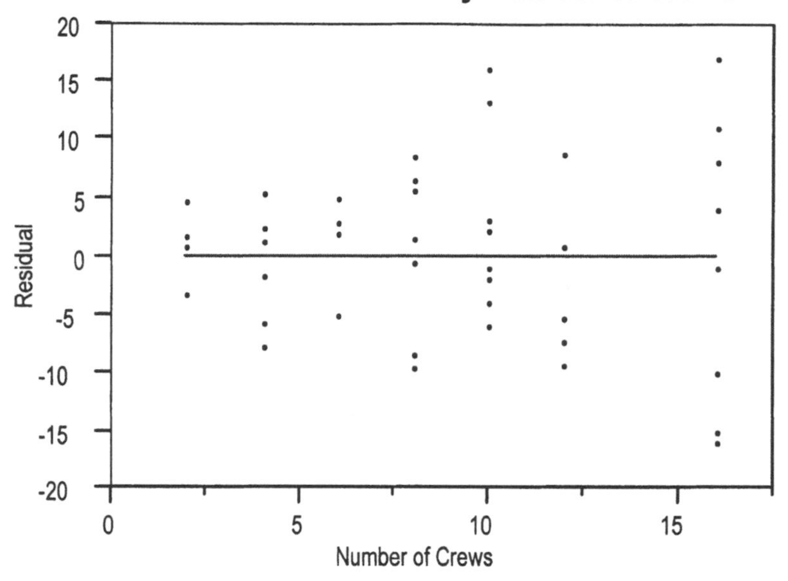

You can get the horizontal line in this graph by using the *Fit Y by X* command, and adding the mean to the plot. The average residual is zero by construction — it's got to be zero by the way that the least squares fitted line is calculated. Just because the average value of the residuals from a model that you have fit is zero does not mean that you have fit a good model. The average residual value *must* be zero if the least squares calculations have been done correctly, whether the model makes sense or not.

We can build a separate table which shows the standard deviation for each number of cleaning crews. The *Group/Summary* command of the *Tables* menu produces a table such as the following.

Cleaning.jmp by Number of Crews

Number of Crews	N Rows	Std Dev(RoomsClean)
2	9	3
4	6	4.967
6	5	4.690
8	8	6.643
10	8	7.927
12	7	7.290
16	10	12.00

Using this summary data set, we can plot the standard deviations directly. The growth of the SD with the increase in the number of cleaning crews is very clear. This pattern of increasing variation is reminiscent of the airline passenger data studied in *Basic Business Statistics*. Heteroscedasticity is the technical name for nonconstant variance in the residuals.

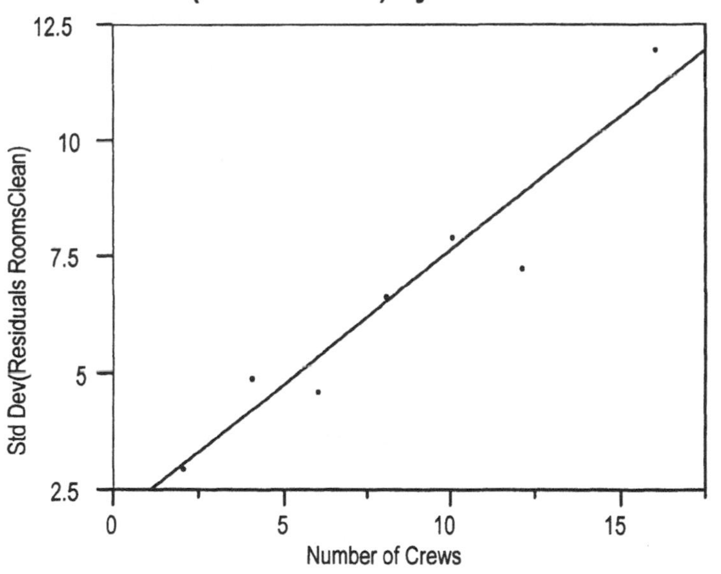

For this regression line, what do you think the intercept term should be?

A building maintenance company is planning to submit a bid on a contract to clean 40 corporate offices scattered throughout an office complex. The costs incurred by the maintenance company are proportional to the number of cleaning crews needed for this task. The fitted equation determined by least squares indicates that about 10 crews are needed. Is this regression model reliable?

The residuals from this fit suggest that the variation of the unobserved errors in the regression model increases with the number of crews sent out. Although the effects upon our conclusion in this example seem mild, we must think about the consequences for prediction. Unless our model incorporates the lack of constant variation, the prediction intervals developed next in Class 3 will not be reliable. Those intervals assume that the assumptions of the idealized model hold. Without constant variation, the prediction intervals may be too narrow or too wide.

Weighted Least Squares

Since the errors in this data appear to lack constant variance (heteroscedasticity), one should use a weighted least squares fit rather than an ordinary least squares fit. The idea is simple: weight the more precise observations more heavily than those that possess more error variation. All you need is a method to determine the weights. Since we have grouped data in this example, it's pretty easy to adjust the regression fit to accommodate the change in variation. We use the group variance estimates as determined by the group summary statistics. For this reason, it can be very useful in regression to have grouped data in which you have multiple values of the response Y for each value of the predictor X.

Ideally, weighted least squares weights the observations inversely to their variance. Looking back at the SD table provided in the text, we can create a column that has the variances and use this column as weights in the analysis. It's easiest if you sort the data by the number of crews; this make it easier to paste in the group variances from the summary table. Once you have pasted in the variances, remember to weight the analysis using the inverse variances so that observations with small variance get the most weight. The supplemental data set Clean2.jmp has this data. In the new fit summarized below, the slope estimate has increased a bit, and its standard error is just a bit smaller. The intercept is about half of its previous size with a sizable drop in its standard error as well.

Term	Estimate	Std Error
Intercept	0.81	1.12
Number of Crews	3.83	0.18

Summary of Outlier Examples

Outliers are observations that depart from the pattern followed by most of the data. Observations that are on the extremes as defined by the horizontal (or X) axis are said to be *leveraged*. Those that deviate from the fitted line along the vertical (or Y) axis are said to have a large *residual*. If an observation is *influential*, the regression slope changes if that point is removed from the fitting process. Typically, influential observations are leveraged with a substantial residual. Neither characteristic alone, however, determines whether an observation is influential; for example, an observation can be so leveraged that it is influential even though it has a relatively small residual, as in the following case of housing prices. As we shall see in the next three cases, influential observations require careful treatment.

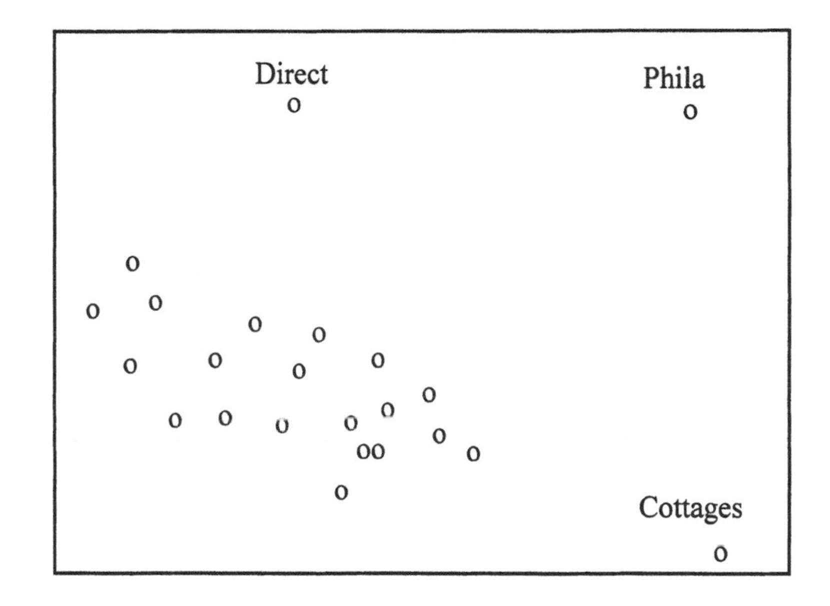

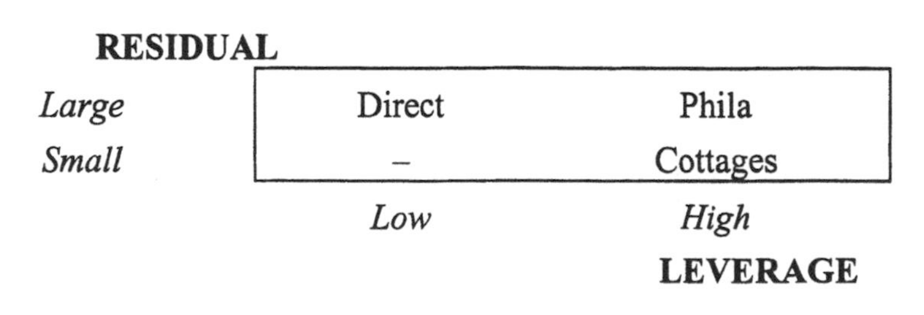

Housing Prices and Crime Rates

Phila.jmp

A community in the Philadelphia area is interested in how crime rates affect property values. If low crime rates increase property values, the community might be able to cover the costs of increased police protection by gains in tax revenues from higher property values.

If the community can cut its crime rate from 30 down to 20 incidents per 1000 population, what sort of change might it expect in the property values of a typical house?

To answer this question, the town council looked at a recent issue of *Philadelphia Magazine* (April 1996) and found data for itself and 109 other "communities" in Pennsylvania near Philadelphia. The names of the communities are part of the data set and make natural labels for points in plots. In addition, the data set includes

House Price	Average house price for sales during the most recent year.
Crime Rate	Rate of crimes per 1000 population.
Miles CC	Miles to Center City, Philadelphia.
PopChg	Change in population, as a percentage.

For this example, we will focus on the first two. The scatterplot below shows a clear outlier. How will this one outlier affect a regression analysis? Which community is this?

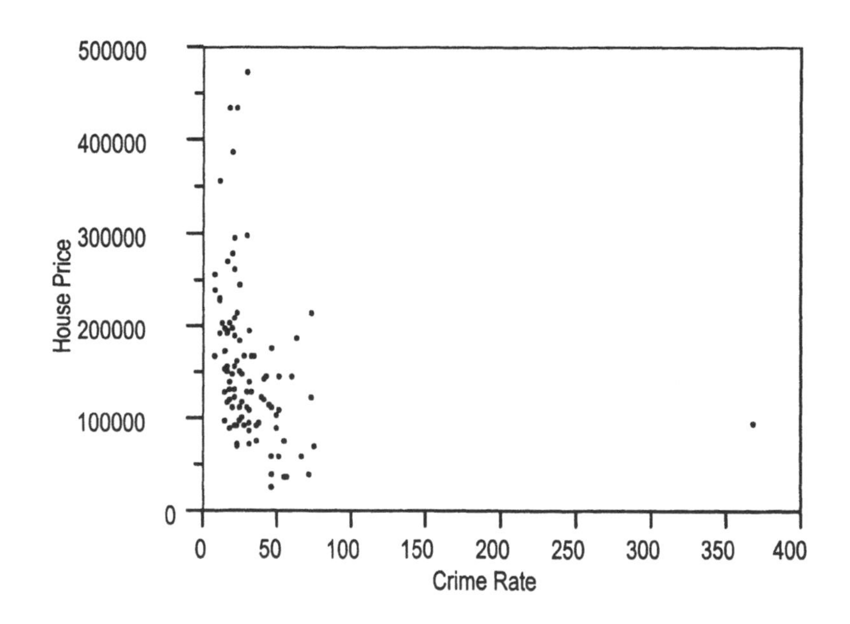

First use the name of the community as the plot label. (Select the column "Name" and then use the "Label/Unlabel" item from the "Cols" menu.) Now JMP will show the name of the community when an observation is highlighted in a plot. The accompanying boxplots show that this community (yes, it's Center City, Philadelphia) is an outlier on the horizontal (*Crime Rate*) axis, but is not unusual on the vertical (*House Price*) axis.

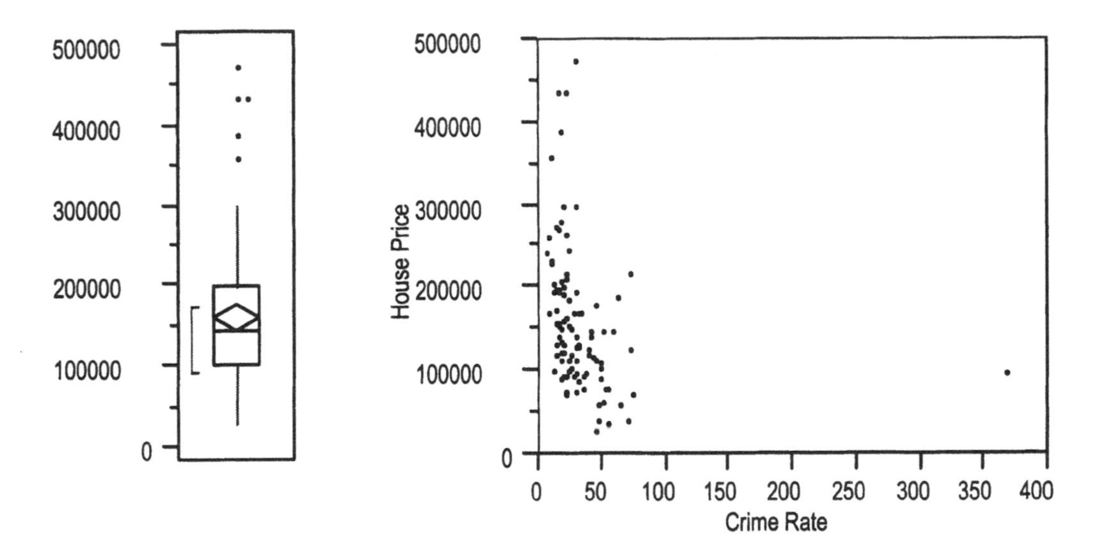

If we fit a least squares line to the data as shown, the single outlier pulls the regression fit toward itself. The least squares criterion minimizes the sum of squared deviations from the regression. Since the deviation is squared, observations that are far from the fit may have substantial influence on the location of the line. The size of the impact depends upon the leverage of the outlying value.

In this example, Center City is a highly *leveraged* observation. Leverage measures how unusual an observation is along the x-axis (here the *Crime Rate* axis). In bivariate regression, the leverage of the ith observation is computed as

$$\text{Leverage}_i = h_i = \frac{1}{n} + \frac{(X_i - \overline{X})^2}{SS_X}$$

where SS_X is the sum of squared deviations about the mean of X. The values of the leverage vary in the range $1/n \le h_i \le 1$.

For this regression, Center City is quite highly leveraged; its leverage is 0.82, approaching the maximum possible. (To compute the leverages, you need to use $ button from the output of the *Fit Model* platform rather than the *Fit Y by X* platform; leverages are labeled as *h(i)*.) With so much leverage, this one observation has substantially more impact on the location of the fitted regression line than others. The plot below shows the regression fit. Note that only 99 of the 110 communities have both of these variables in the data set.

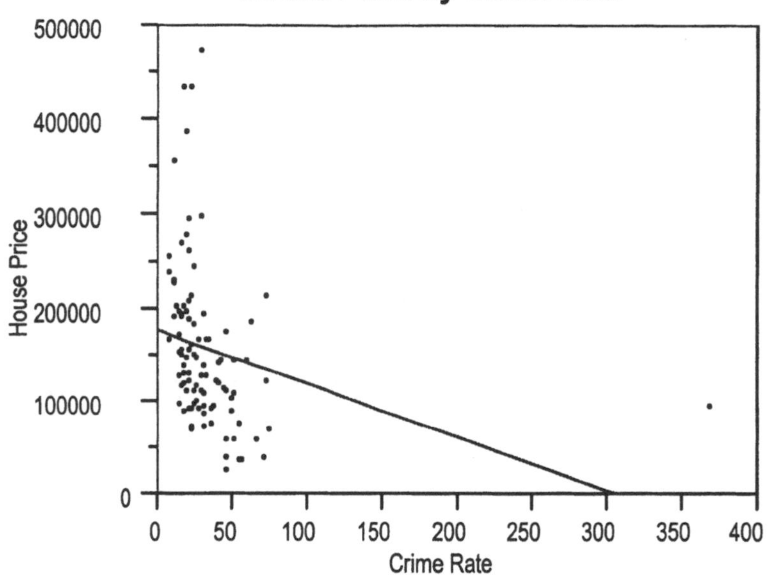

Linear Fit

House Price = 176629 – 576.908 Crime Rate

Observations	**99**

Term	**Estimate**	**Std Error**
Intercept	176629.4	11246
Crime Rate	-576.9	226.9

Residuals offer further evidence of the extreme impact of this one point on the model. The button shown beneath the scatterplot allows you to plot the residuals quickly. Here's the plot.

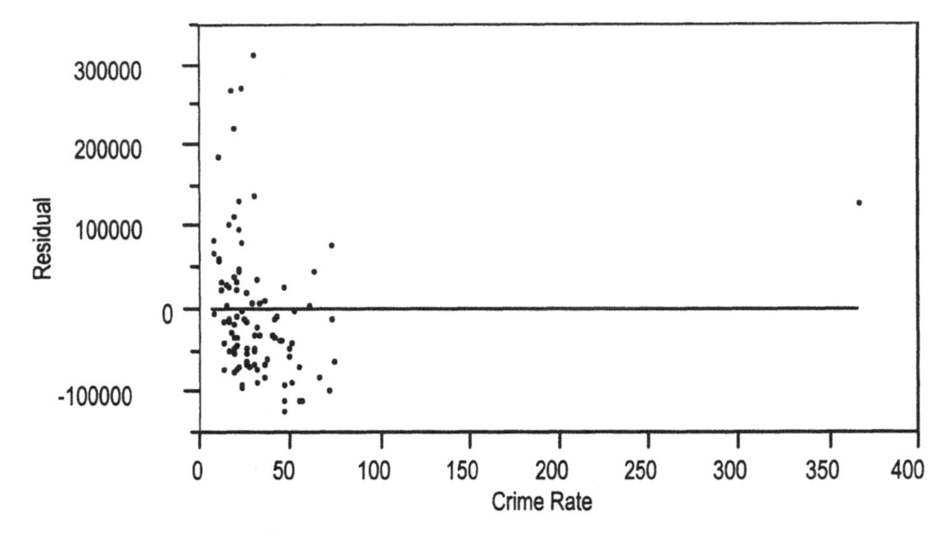

If we also save the residuals to the data spreadsheet, we can see the impact of the outlier on the normality of the residuals about the fit. Although Center City is extreme in terms of *Crime Rate*, its residual is in the middle of the pack. The full set of residuals is seen to be skewed, with a predominance of large positive residuals. Before we go further with any more residual analysis, we need to handle the outlier.

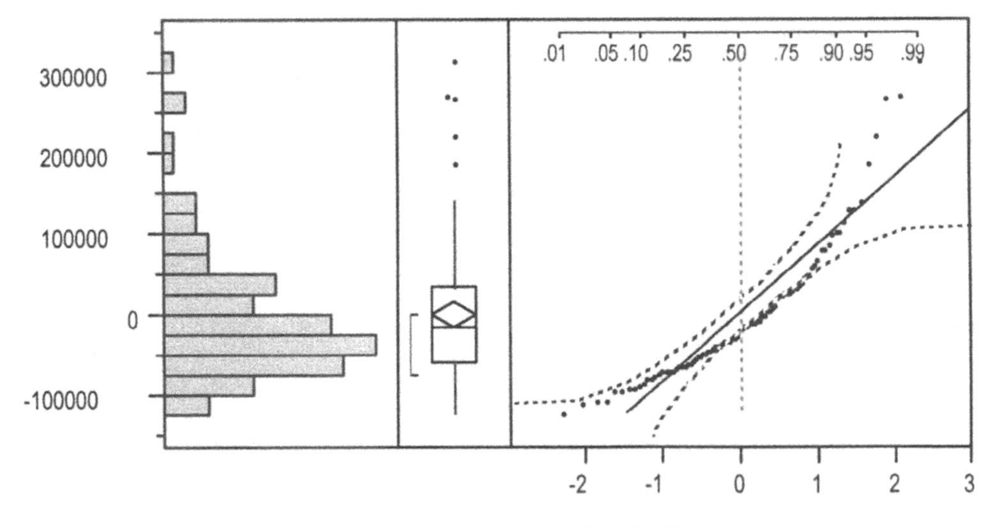

With Center City set aside, the fitted model changes dramatically. To exclude the outlier, select the point, and then use the *Exclude/Include* command from the *Rows* menu. Then fit a line again in the same display. The following plot shows both regression lines. The slope is about 4 times steeper, suggesting that crime rates have a much higher impact on house prices than would have been suggested by the fit including Center City.

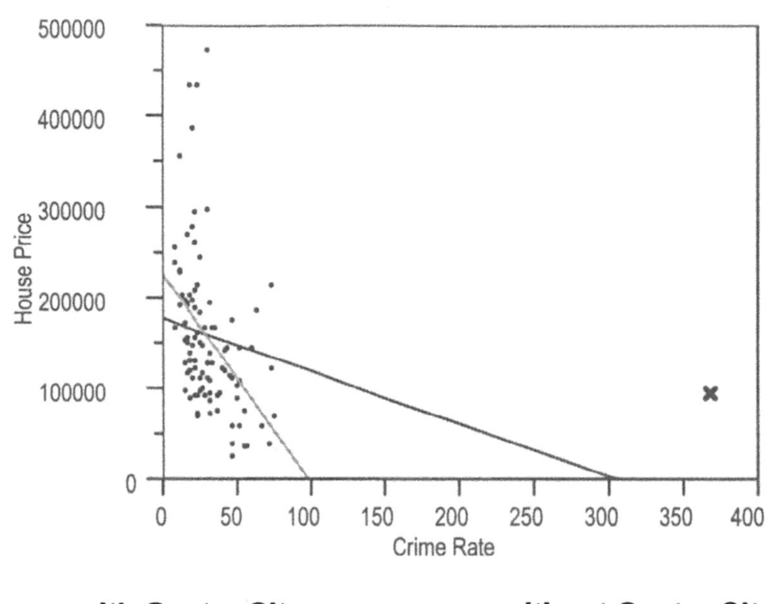

with Center City
House Price = 176629 – 576.908 Crime Rate

without Center City
House Price = 225234 – 2288.69 Crime Rate

With Center City excluded, plots reveal something previously hidden due to the scale compression caused by the outlier: the relationship is a bit nonlinear.

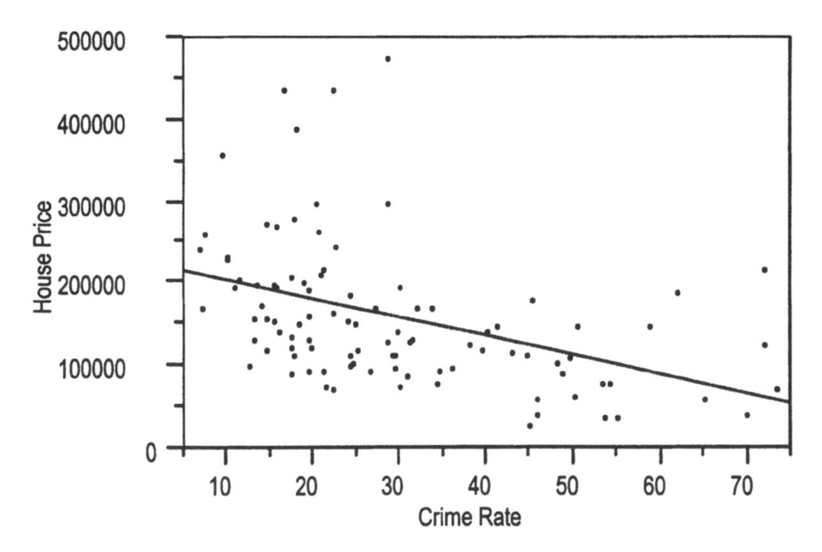

The residual plot from the linear fit without Center City also shows evidence of nonlinearity. Note the positive residuals at either side of the plot. What sort of transformation would help here?

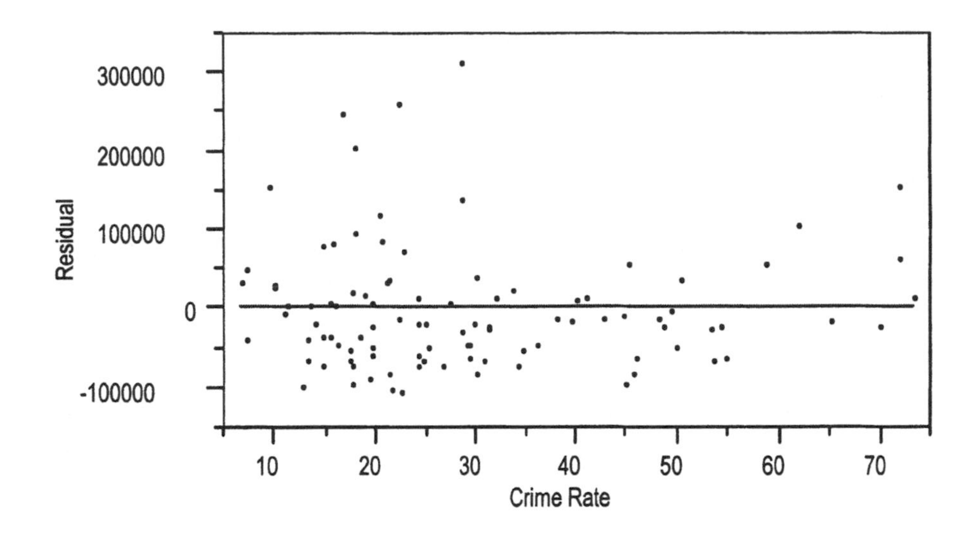

The skewness of the residuals confirms what we can see in the residual plot. Namely, there are many small negative residuals (points relatively close below the fitted line) and fewer, but larger, positive residuals.

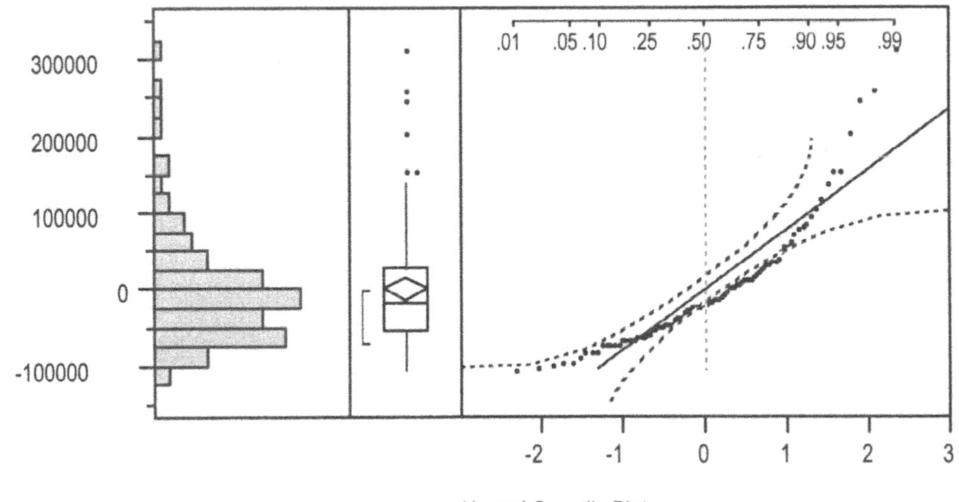

A community in the Philadelphia area is interested in how crime rates affect property values. If low crime rates increase property values, the community might be able to cover the costs of increased police protection by gains in tax revenues from higher property values.

If the community can cut its crime rate from 30 down to 20 incidents per 1000 population, what sort of change might it expect in the property values of typical house?

Ignoring issues of nonlinearity, the two linear fits imply very different changes in prices for a "ten-point" drop in the crime rate.

with Center City	**without Center City**
House Price = 176629 – 576.908 Crime Rate	House Price = 225234 – 2288.69 Crime Rate

With Center City, the increase is only 10 × 576.9, or about $5770 per house. Excluding Center City, the increase is about four times larger, at 10 × 2288.7 or $22,890 per house. There is a concealed catch in all of this analysis, however. The data for house prices are obtained from sales during the current year, not the full housing stock. If the sales are not representative of typical houses in the community, we are not getting a good impression of its housing stock and will not be able to say much about the net impact on the community.

It is worth spending a moment to think about how the possible nonlinear relationship might affect these conclusions. Below is a plot using the 1/X transformation.

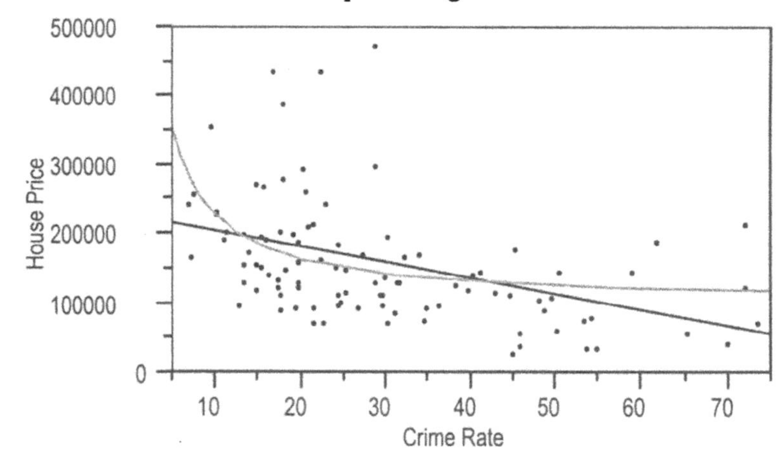

A log-log transformation works as well and gives an interesting interpretation that we'll come to in Class 5.

The $1/X$ transformation captures quite a bit of the nonlinearity. The residuals from the nonlinear fit don't "bow" as much as those from the linear model, though outliers remain for the communities with low-crime rate. Perhaps the variance is not constant.

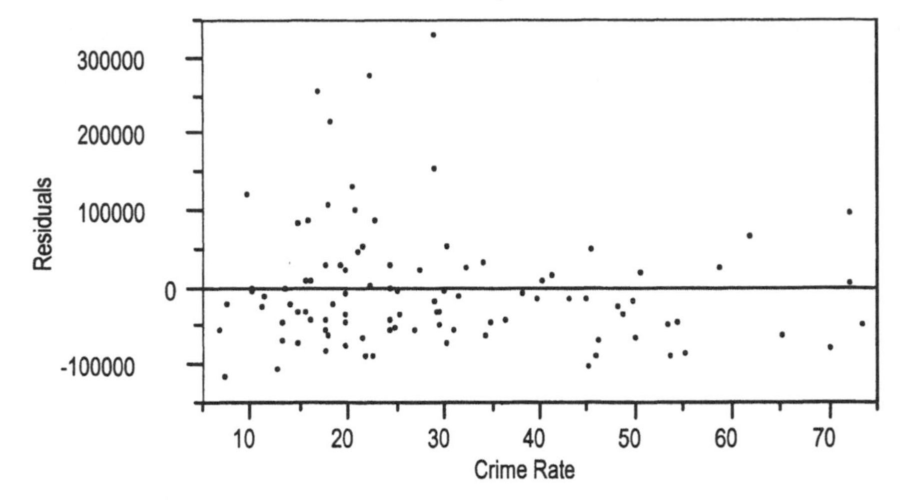

This nonlinear model is also interpretable. Rather than look at the number of crimes per 1000, count the number of people per crime (multiply 1/(*Crime Rate*) by 1000 so that, for example, a crime rate of 10 per 1000 becomes 100 people per crime).

House Price = 98120.1 + 1298243 Recip(Crime Rate)

Fit to the reciprocal of the crime rate, the fitted model flattens out as the crime rate increases. This nonlinear model suggests that an increase (or decrease) in the crime rate has a big effect on house prices when the crime rate is small, but only a slight effect on the prices when the crime rate is larger. For example, reducing the crime rate from 20 to 10 per 1000 raises the prices (according to this fit) by an impressive (the constants cancel)

$$\text{Change in fit(20 to 10)} = \frac{1298243}{10} - \frac{1298243}{20} = \$64{,}910\,.$$

A decrease of 30 down to 20 has a smaller impact that is close to what we got from the fitted line (at a crime rate of 20-30, the fitted line is parallel to a tangent to the curve)

$$\text{Change in fit(30 to 20)} = \frac{1298243}{20} - \frac{1298243}{30} = \$21{,}640\,.$$

Pushing the crime rate down to near zero is associated with very large increases in property values, and may help explain the popularity (and expense) of "walled" suburban communities.

Equally impressive is the effect of the nonlinear transformation on the outlier. For the plot below, we show all of the data, including Center City. The fitted nonlinear model, though, is the same one as above, the one based on the data without Center City. The fit is the same, but the plot shows all of the data. Center City does not seem like so much of an outlier now. Its crime rate is extreme, but its housing prices are consistent, in the sense of the fitted model, with the impact of crime rates elsewhere.

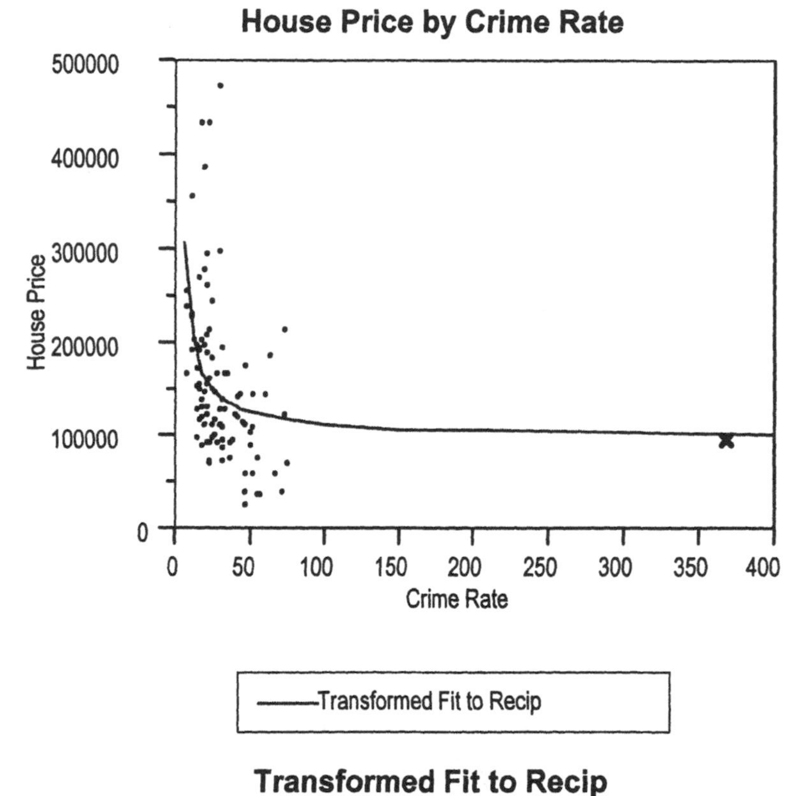

Transformed Fit to Recip

House Price = 98120.1 + 1298243 Recip(Crime Rate)

Often, as in this example, outliers are unusual only because we have not modeled the relationship between the factors correctly. Certainly, Center City has much higher crime rates than the other areas, but when the nonlinearity is taken into account, its housing prices seem right in line with what might be expected. In cases with highly leveraged observations like Center City, it's a good idea to fit the model with the outlier set aside, then use the outlier to see how well your model extrapolates.

Aggregation and Causation

This example also offers a setting for us to mention two other important issues that come up frequently in regression analysis. The first is aggregation. The data used in this example are observations of communities, not people. The data are aggregated. It would be inappropriate to infer a conclusion about the individuals living in these communities on the basis of such data, a mistake known in sociology as the "ecological fallacy." You cannot make statements about individuals on the basis of aggregate data.

The second comment regards causation. Do crime rates affect house prices, or does it work the other way around? Perhaps some third factor, like level of education, is affecting both. Might high average house prices imply an affluent community whose members are busy with other things? Here's the inverse regression fit for this relationship.

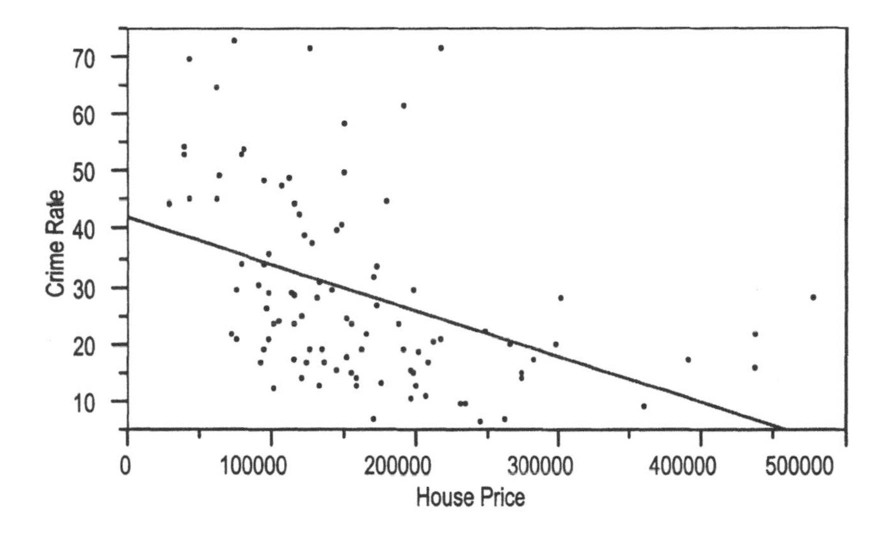

Again, nonlinearity is evident. Remember that regression, like correlation, is based on association. You cannot deduce cause and effect from these models (albeit, experimental models to be considered later come a lot closer). This issue is important for our community decision makers, for if indeed the causation works as shown here, their efforts to increase property values by lowering the crime rate will have no effect. It's housing prices that determine the crime rate, and not vice-versa.

Direct Mail Advertising and Sales

Direct.jmp

A catalog sales company would like to assess one segment of its mail order business. It knows the number of catalogs of a particular lineup mailed to customers over 10 previous mailings. Each catalog in these mailings costs the company $1.50 to prepare and mail. One of the mailings was for a clearance sale.

How much is the company profiting from this particular catalog?

A scatterplot of the level of sales (in multiples of $1,000) on the number of direct mail customers (1000's) reveals a very unusual observation. In the fifth month, the company ran a large sale to clear a backlog of inventory. Clicking on the outlier shows its point label and reveals that it is observation #8.

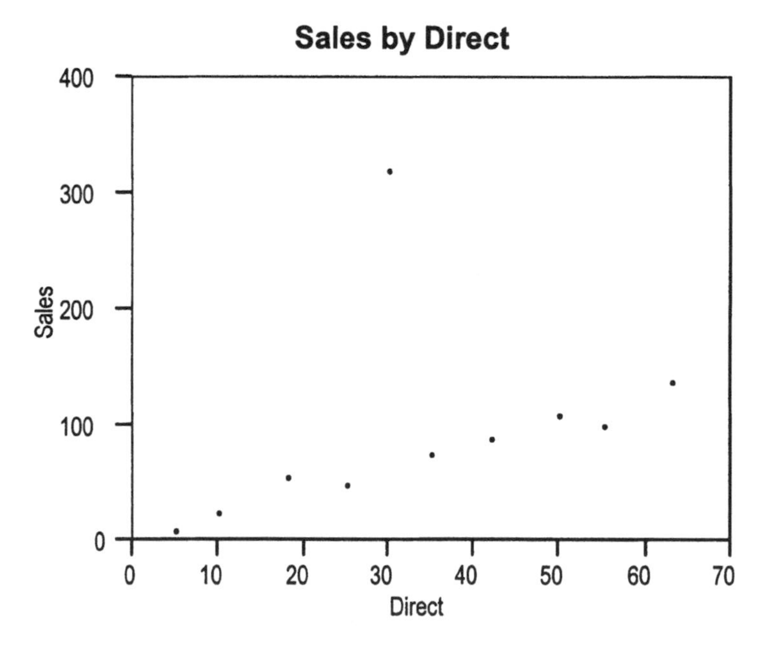

As the accompanying boxplots indicate, this observation is rather unusual on the *Sales* axis, though right in the center of the *Direct* axis.

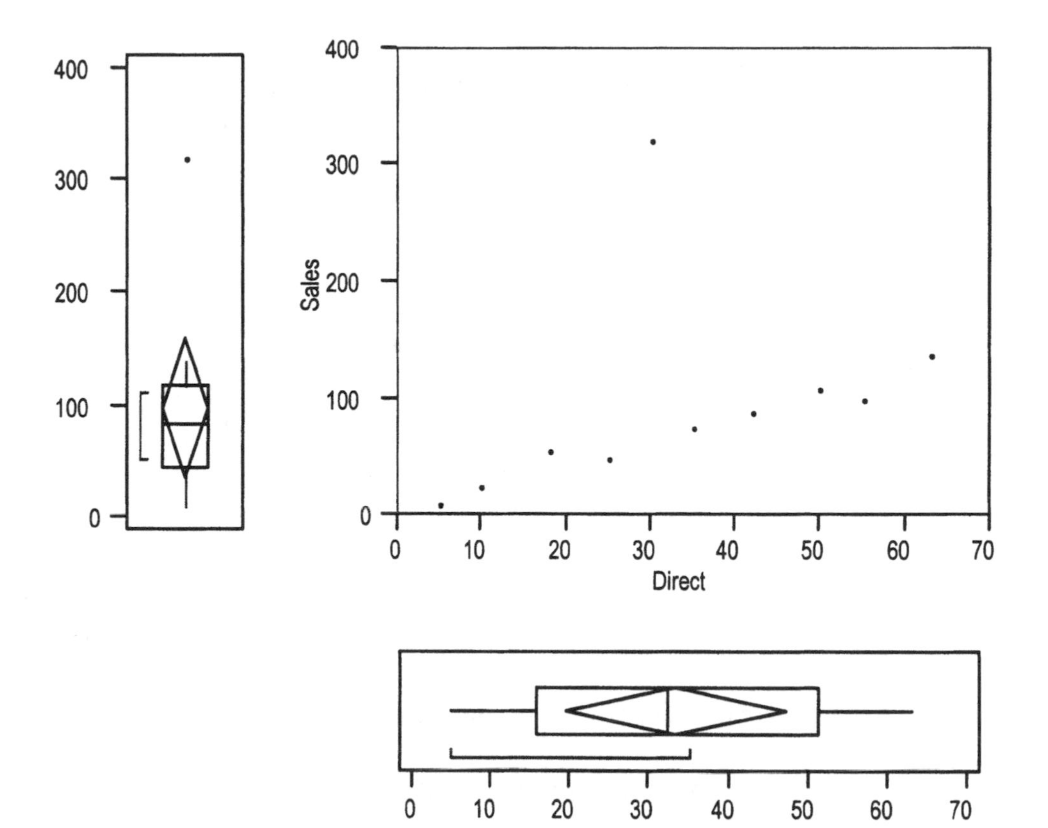

How does this one observation affect the regression equation? The fit of a linear equation seems quite reasonable but for this aberration. Plots of the fit show that the outlier has "lifted" the regression line, evidently not affecting the slope by very much, but pulling the intercept up.

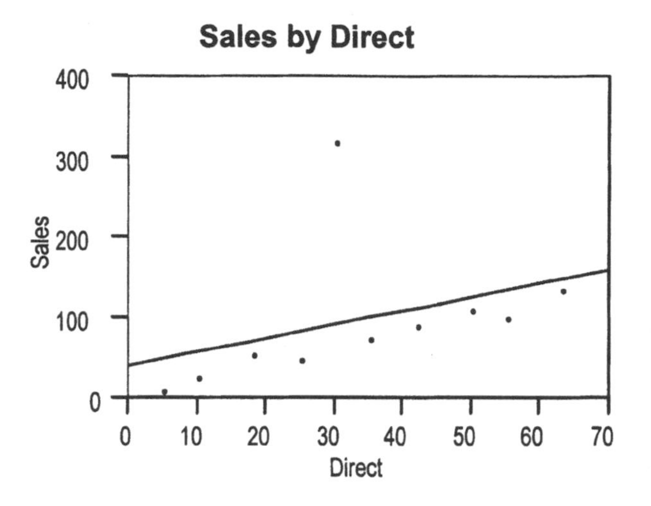

All but one of the residuals are negative (all but the outlier lie beneath the mean line).

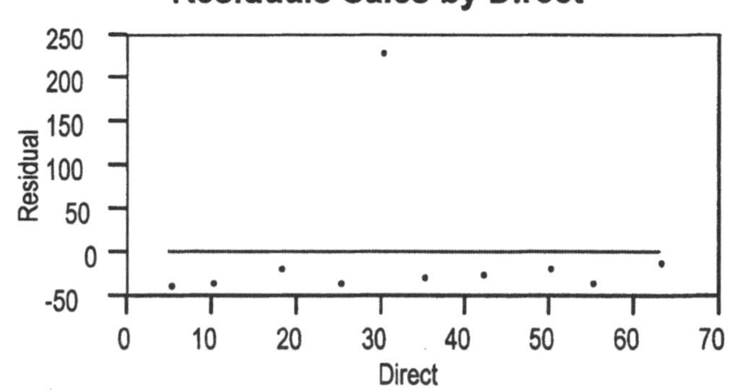

Since this month is very different from the others because of the clearance sale, we ought to set it aside for the time being and assess the trend in the rest of the data. To exclude this observation from further analyses, select this point in a plot and use the *Exclude/Include* command from the *Rows* menu.

Here is the scatterplot with fitted line, after having excluded the outlier in one row. The vertical scale has changed dramatically, and a linear fit captures the pattern in the data quite well. The residuals (not shown) appear random (though there are too few to be serious about checking for normality).

Here is the summary of the fit excluding the outlier.

Summary of Fit (excluding one)

Root Mean Square Error	8.82
Observations	9

Parameter Estimates

Term	Estimate	Std Error	...more
Intercept	5.80	5.9	
Direct	1.98	0.15	

The fit based on all 10 sales has a similar slope (1.98 versus 1.73 with the outlier). The revised fit, however, has a rather different intercept (5.8 versus 39.6) and very different estimates of the residual standard deviation, labeled root mean square error. The resulting standard errors of the estimated slope and intercept are also different.

Summary of Fit (using all)

Root Mean Square Error	85.68
Observations	10

Parameter Estimates (with outlier)

Term	Estimate	Std Error ...more
Intercept	39.58	56.1
Direct	1.73	1.5

A catalog sales company would like to assess the success of one segment of its mail order business. It knows the number of catalogs of a particular lineup mailed to customers over 10 previous mailings. Each catalog in these mailings costs the company $1.50 to prepare and mail. One of the mailings was for a clearance sale.

How much is the company profiting from this particular catalog?

With the outlying sale set aside, the sales per customer (the slope) have been averaging near $2.00, a profit of $0.50 per catalog mailed.

With the outlier included, the estimated profit is much smaller since then the estimate of sales per catalog is also smaller.

Housing Construction

Cottages.jmp

A construction company that builds lakefront vacation cottages is considering a move to construction of a line of larger vacation homes. It has been building cottages in the 500–1000 square foot range. It recently built one which was 3500 square feet.

Should the company pursue the market for the larger cottages?

The company has records on the profits earned on 18 prior buildings, including the large 3500 sq. ft. building. In this example, the large house stands out on both axes, being much larger than the rest as well as much more profitable.

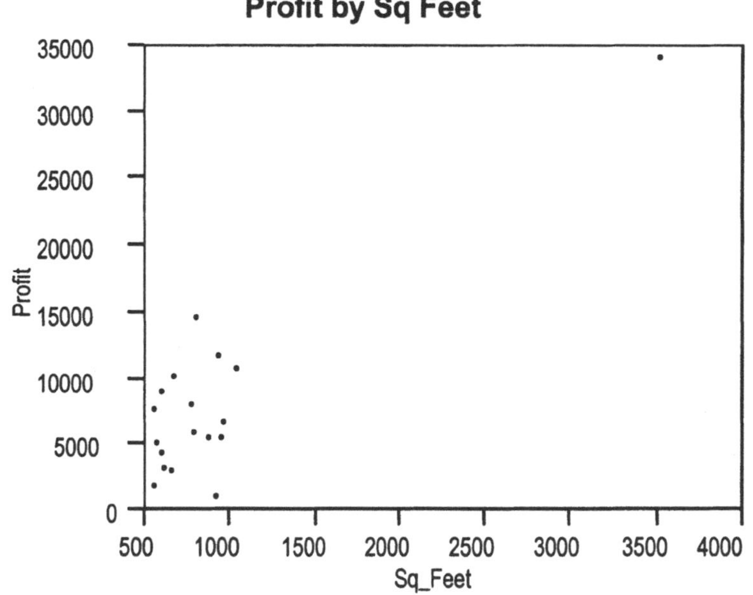

The marginal boxplots indicate (as if it were not already obvious) that this "cottage" is an outlier on both axes.

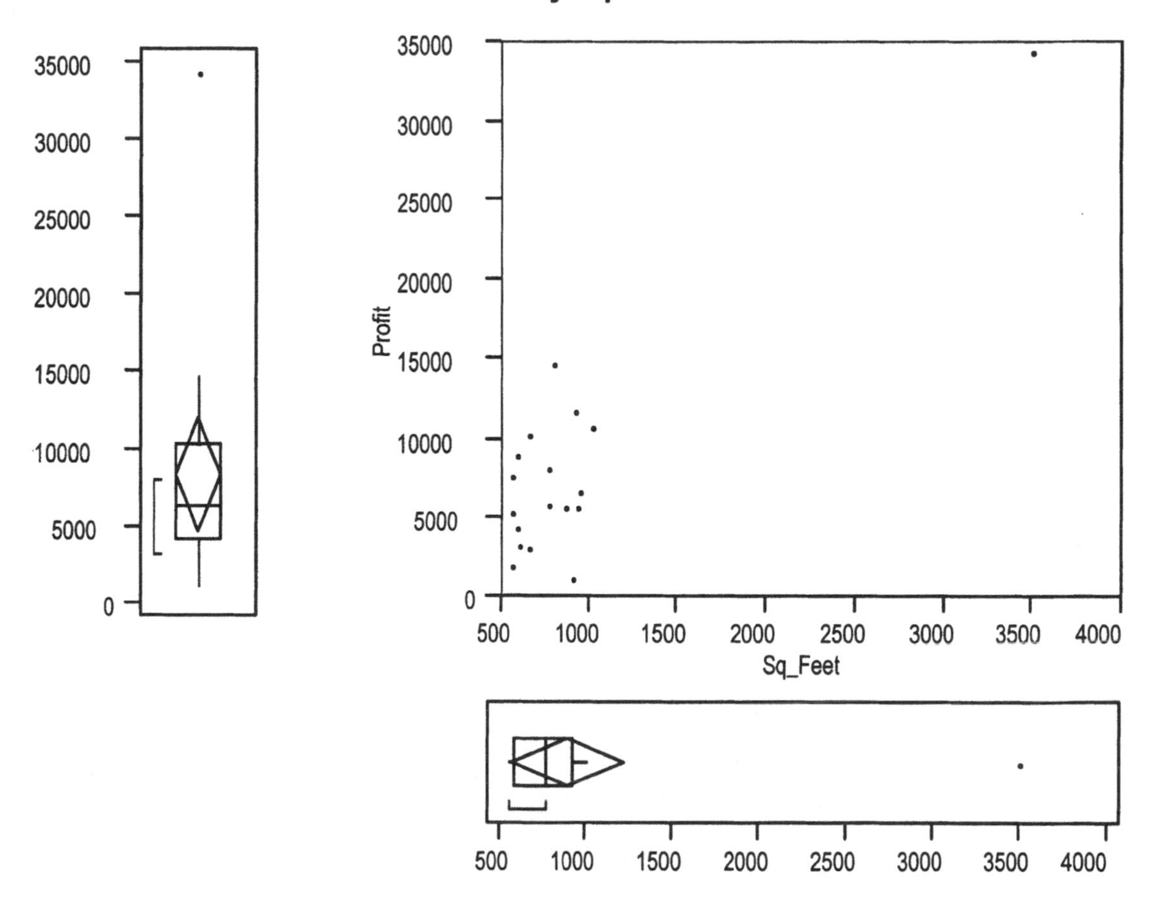

Observations that are unusual – distinct from the rest of the data – on both axes are going to have a big effect upon the regression model. They may not be "influential" in the sense of having a large effect upon the location of the fitted slope, but they will exert an effect on other features of the regression, such as our impression of how well the model has fit the data.

The linear regression analysis for these data seems reasonable (the point for the large cottage is near the fitted line), but one needs to be careful with such an extreme value present. The leverage of this outlier is $h(i) = 0.95$, near the upper limit of 1.

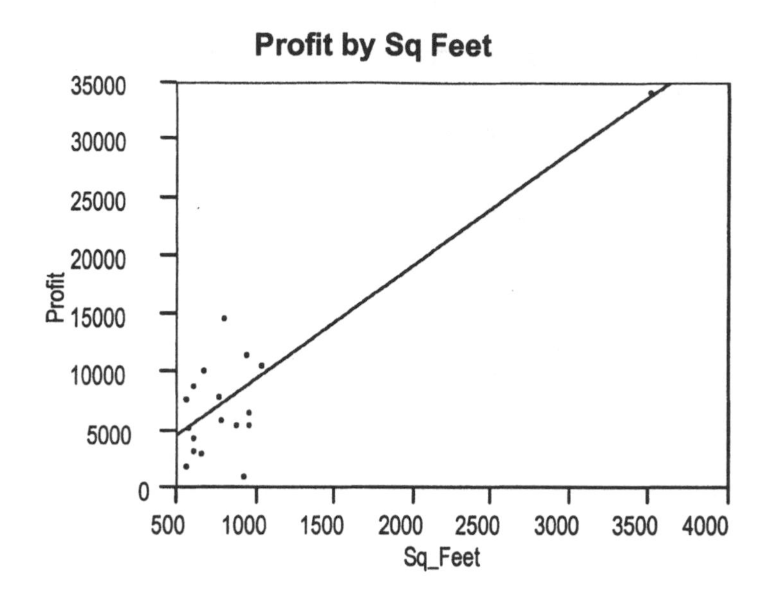

Although quite unusual in size, the large house does not stand out among the residuals from the model. The residual for the large house is near zero by comparison to the others in this fit. Evidently the builder seems to make a profit proportional to the size of the house with neither increasing nor decreasing returns to scale.

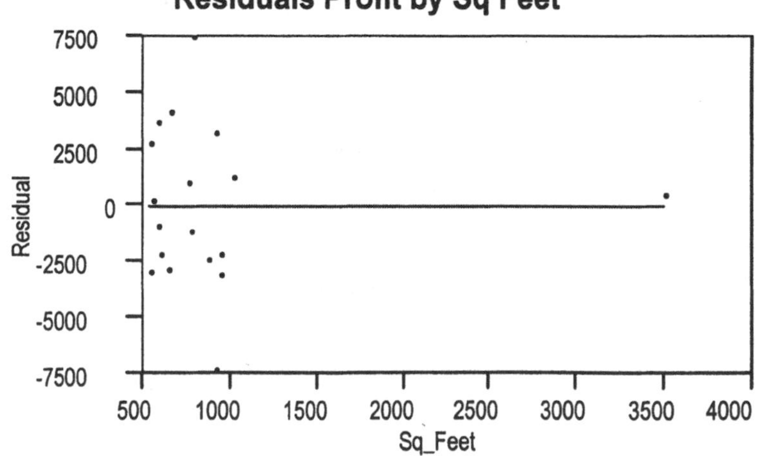

Even though the outlying house is highly leveraged, it has relatively little *influence* on the *slope* of the fitted regression. Using the *Exclude/Include* command from the *Rows* menu, we set this point aside and consider the fit without it.

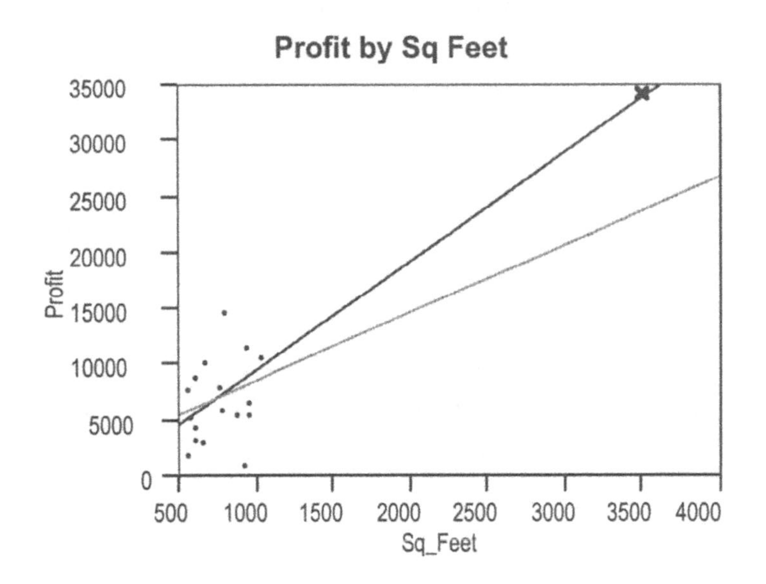

The trend is less evident with the plot focused on the smaller cottages. That is, by focusing our attention on the points in the lower left corner of the plot shown above, the trend seems much weaker.

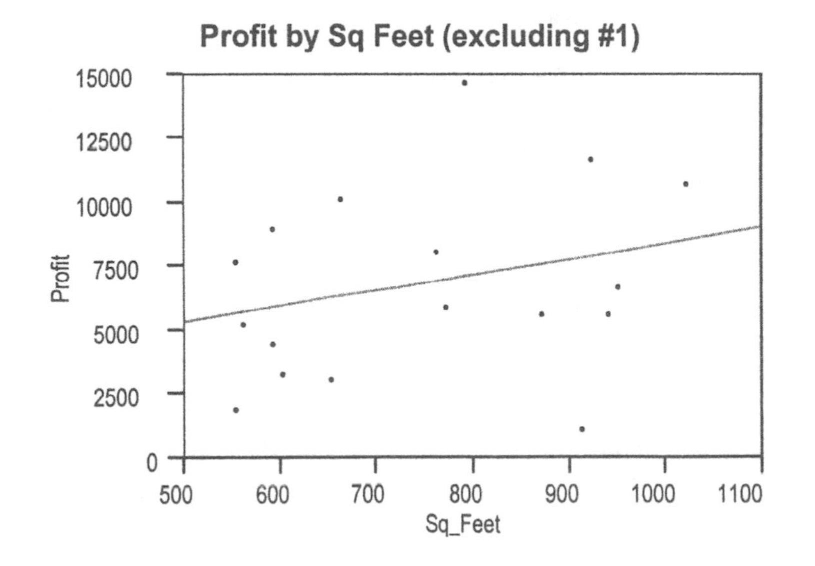

Summaries of the two fits appear below. The most dramatic differences appear in the estimates of the coefficient standard errors. Without the highly leveraged observation, the parameter estimates have larger standard errors. The variability around the fit (root mean square error or standard error of the regression) is similar.

With all data

Summary of Fit

...	
Root Mean Square Error	3570
Mean of Response	8348
Observations	18

Parameter Estimates

Term	Estimate	Std Error	... more
Intercept	-416	1437	
Sq Feet	9.75	1.3	

Without the large cottage

Summary of Fit

...	
Root Mean Square Error	3634
Mean of Response	6823
Observations	17

Parameter Estimates

Term	Estimate	Std Error	...more
Intercept	2245	4237	
Sq_Feet	6.14	5.6	

In a very narrow sense, the outlying value in this example is not influential. The slope of the fitted model without the outlier is "close" to the estimated value when the outlier is included. By "close" we mean that the difference between the two estimates is judged relative to the standard error of the fit without the outlier. More generally, though, it is clear that this large house has a dramatic effect upon our impression of the goodness of the model.

A construction company that builds lakefront vacation cottages is considering a move to construction of a line of larger vacation homes. It has been building cottages in the 500–1000 square foot range. It recently built one which was 3500 square feet.

Should the company pursue the market for the larger cottages?

Based on this one large house, it would seem that more profit is made on large houses than others. However, the analysis is quite contingent on this one observation. Without the one large cottage, the rest of the data suggest a less pronounced upward trend with considerable variation in profits.

Influence refers to how much the regression equation changes when a single observation is deleted. Although the slope decreases when the large house is removed, the change is small considering how distinctive this observation is. Further discussion of the size of the change in this slope appears in Class 3.

Class 3. Prediction and confidence intervals in regression

This class considers prediction intervals, confidence intervals, and the associated statistics for a regression model. As with confidence intervals for the mean, these intervals rely on an estimate of variability known as the standard error. Although the interpretation of confidence intervals in regression is similar to that of the confidence intervals developed in *Basic Business Statistics*, the regression model itself is more complex and complicates the interpretation of the intervals. In addition, similar intervals known as prediction intervals provide estimates of the accuracy of predictions from the fitted model.

Topics

Prediction intervals for new observations; extrapolation
Standard error of regression coefficients; confidence intervals; hypothesis tests
R^2 as a measure of goodness-of-fit; relationship to correlation
Correlation versus causation

Examples

1. Housing construction, revisited
2. Liquor sales and display space, revisited

Key application

Forecasting and prediction. Recall the introductory application of Class 2 in which the city was trying to determine the optimal issue size for its bonds. Once we have decided on the optimal issue size, it is natural to ask what average cost to expect.

One may want to do two types of prediction. The first is to predict a "typical" value of average cost, the second is to predict what the average cost of a single new issue will be. The best guesses for both types of prediction are, not surprisingly, the same. The range of feasible values, however, is very different. The bottom line is that single observations are harder to predict than averages. The uncertainty in predicting a new value needs to take into account the inherent variability of observations about the regression line.

Definitions

R squared (R^2, the coefficient of variation). The proportion of variation in the response variable Y that is explained by the regression line. We consider a regression successful if it explains a "useful amount" of the variability. In bivariate regression, it is also the square of the usual correlation between X and Y (lending us a new interpretation of the size of a correlation).

Prediction interval. A range of feasible values for a new observation at a given X-value.

Confidence interval for the regression line. A range for the value of the regression line at a given X-value.

Concepts

Two types of intervals for regression predictions. When we do regression there is an implicit understanding that the data we are analyzing is generated through a two-part process. Heuristically, the way we generate an (X, Y) pair, a row with two columns in the spreadsheet, is to go up to the regression line above the X-value and add some noise to the value on the regression line, that is, the underlying regression model implies that (in the notation of many text books with Greek letters for parameters)

$$Y_i = \beta_0 + \beta_1 X_i + \varepsilon_i.$$

In words, this assumes that

Observed Response = Signal + Noise

or that

Observed Response = Deterministic + Random .

If you focus on this two-part process, it becomes apparent that there are going to be two types of confidence statements to be made about the regression. First, you may only want to make a statement about the deterministic part of the model, that is estimate the value of the true regression line above a particular X-value. In JMP such predictions are obtained in simple regression via the "Confid Curves:Fit" command. Alternatively, you may want to forecast a single, new observation at a given X-value. In this case you need to take into account the error component in the model as well. JMP calls this "Confid Curves: Indiv".

The critical difference between the two statements is the range of feasible values provided for the estimate. You know much more about the regression line than you do about an unobserved individual. The uncertainty in the single, new observation includes the error component in the model, whereas the uncertainty in the regression line (the fitted line) does not. It is not surprising that the "Confid Curves: Indiv" intervals are wider than the "Confid Curves: Fit" intervals because they include this error component.

To make things more concrete, take the example of hurricane prediction. Over the last 40 years, there has been a slight downward trend in the number of Atlantic hurricanes that form each year. If we fit a regression line to a scatterplot of the number of hurricanes each year against year, it would slope down from top left to bottom right. There is also a lot of variability about the regression line. Now imagine making a statement about the number of hurricanes next year. We could make one statement about where the regression line is going to be next year. This is the "Confid Curves: Fit". Or we could make a statement about the actual number of hurricanes that will occur next year. This latter statement needs to reflect the considerable variability about the regression line and will produce a wider interval.

R squared (R^2). This ubiquitous statistic aims to summarize how well the regression model fits the data (goodness-of-fit). The bigger R^2 is, the more successful the regression is in the sense of explaining variability in the response. Another way of saying this is that the bigger R^2 is, the closer the points lie to a line. An oft asked question is "What's a good R^2?" Watch your statistics professor wince each time this is asked in class! The problem is that there is no single or simple answer. The usual response is "It depends on the context." If you walked around the school you would see different professors getting excited about very different values for R^2.

One way of feeling good about your R^2 is to find out how your competitors are doing. Say you are trying to predict aluminum futures. If all your competitors have R^2's of about 30% and you manage to get an R^2 of 50%, then clearly you are probably in a much better position to do forecasting than they are. Equally obviously, if all your competitors have R^2 of about 80% and you are at 50%, then this is not good. So once again, it is all relative.

Heuristics

Am I using the right confidence region?

It is easy to confuse the two types regions for prediction, but there is a simple way to tell them apart. The confidence region for the fitted model is like the confidence interval for a mean in that it becomes infinitely small with increasing sample size. As the sample size becomes larger, an interval for the location of the regression line narrows. The prediction region for a single new observation, however, does not. Rather, when using the 95% variety, this region should capture about 95% of the data.

Potential Confusers

Why do we see statistically significant regression coefficients even when R^2 is minuscule (not much larger than zero)?

This is in fact not a contradiction at all. All a regression coefficient is telling you is the slope of a line. The *p*-value associated with the regression coefficient tells you how certain you are of whether or not the true slope is zero. A small p-value tells you that you can be quite certain that the slope is not zero. As illustrated in *Basic Business Statistics*, we can get a very small *p*-value just by having a large sample size.

In contrast, R^2 measures the proportion of the variability in the response that is explained by the regression. It is quite possible to have a regression line with a small (but nonzero) slope but for the data to have considerable variability about the regression line. In this situation, if you collect enough data, you will get conclusive evidence that the slope of the regression line is not zero (i.e., small p-value), but R^2 will remain small. This might all sound a little abstract, but if you read the classic series of papers by Fama and French on testing the CAPM, you will see that this is precisely the case when they do regressions of returns versus the market. Eventually, the slope is significantly different from zero, but the R^2 remains tiny.

Housing Construction, Revisited

Cottages.jmp

> The construction company has decided to build another large cottage, this time with 3000 square feet.
>
> How much profit might the company expect from selling this new dwelling?

Using the same data as before, the company has records of the profits made on 18 prior "cottages," including the large 3500 sq. ft. mansion. The scatterplot of this data is repeated below.

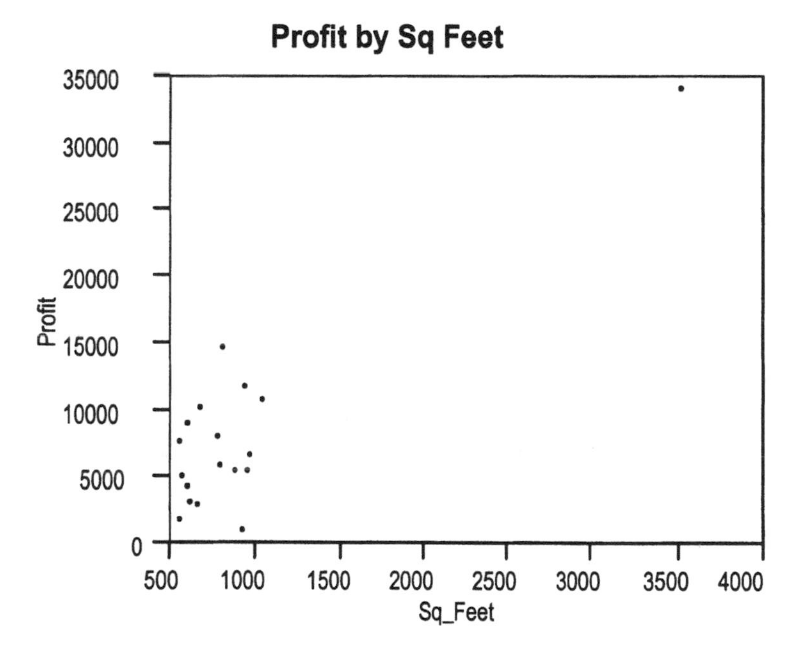

One approach used by the company is to retain the large house in the data used to predict the sale price of a 3000 sq. ft. home. The figure below shows the fitted regression line together with the associated 95% prediction intervals as a pair of dashed lines. (You get these via the *Confid Curves: Indiv* command accessed by the button associated with the linear fit). These intervals are designed to capture 95% of *new* observations.

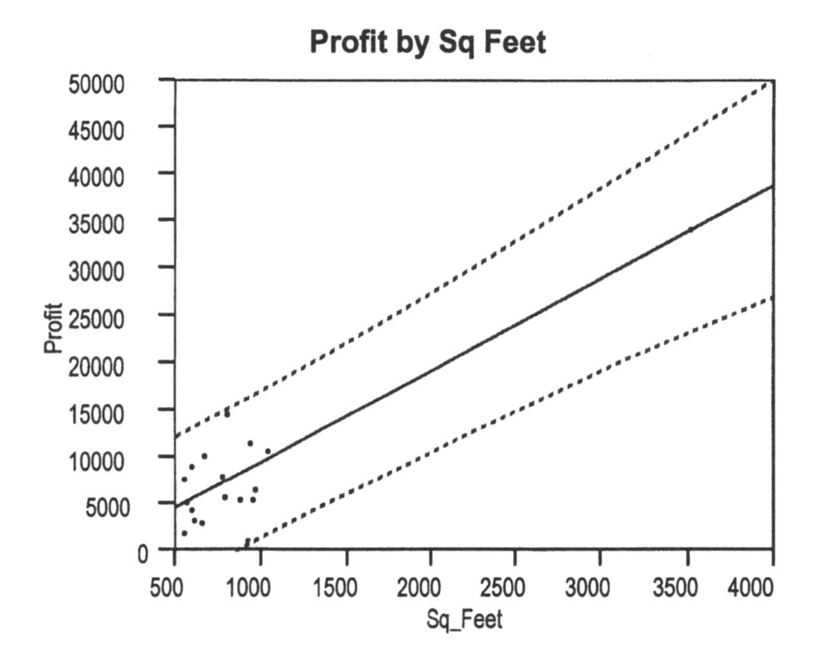

Summary of Fit

RSquare	0.78	
RSquare Adj	0.77	
Root Mean Square Error	3570.4	est SD about line
Mean of Response	8347.8	
Observations	18	

Parameter Estimates

Term	Estimate	Std Error	t Ratio	Prob>|t|
Intercept	-416.86	1437.02	-0.29	0.78
Sq Feet	9.75	1.30	7.52	<.0001

Some details of the regression output.

The standard error shown in the regression summary is analogous to the standard error of the sample mean, though computed by a different formula. The standard error of the sample average is an estimate of the accuracy of the sample average as an estimator of the true population mean. In *Basic Business Statistics*, we showed that the standard error of the mean is the standard deviation of one observation divided by the square root of the sample size, denoted n:

$$\text{SE(average of } n) = \frac{\text{SD(one observation)}}{\sqrt{n}} \approx \frac{s}{\sqrt{n}}$$

The standard errors shown in the regression summary are computed analogously, but by different formulas. The standard error for the slope is not hard to understand and gives some insight into the properties of regression models. The standard error of a slope estimate is the standard deviation of one observation (about the regression line) divided by a measure of the spread of the points on the X axis:

$$\text{SE(estimated slope)} = \frac{\text{SD(error about line)}}{\sqrt{SS_x}} \approx \frac{s}{\sqrt{n}} \times \frac{1}{SD(X)},$$

where SS_x is the sum of squared deviations about the mean,

$$SS_x = \Sigma\,(X_i - \bar{X})^2 \ .$$

From this formula, we can see the *three factors* that determine the standard error for the slope:

(1) as n increases with more data, SE(estimated slope) gets smaller;

(2) as the predictors become more dispersed, SE(estimated slope) gets smaller; and

(3) as the observations cluster more closely around the regression line, the SD(error about line) and resulting SE(estimated slope) both get smaller.

The R^2 statistic is an overall summary of the goodness-of-fit of the regression model. One way to think of R^2 is to interpret this value as the "proportion of explained variation." To get a sense of this interpretation, suppose in this example you had to predict the profit obtained on a cottage *without* knowing the size of the cottage. About the best (at least in some sense) that you could predict would be to use the average of the available profits. The sum of squared errors made by this prediction in the observed data is known as the total sum of squares,

$$Total\ SS = \Sigma\,(Y_i - \bar{Y})^2.$$

Now suppose that you know the size of the cottage and can use the regression line to predict the profits. The total squared error made by this predictor when applied to the available data is just the sum of the squared residuals,

$$Residual\ SS = \Sigma\ \text{residual}_i^2\ .$$

Has regression helped? If the regression has worked to our advantage, *Residual SS* will be smaller than *Total SS*. (It cannot be larger.) The R^2 statistic is a measure of this gain in predictive accuracy and is defined as

$$\begin{aligned} R^2 &= 1 - \frac{Residual\ SS}{Total\ SS} \\ &= \frac{Total\ SS - Residual\ SS}{Total\ SS} \\ &= \frac{Regression\ SS}{Total\ SS}, \end{aligned}$$

which is the proportion of the total variation (or total sum of squares) "captured" by the regression model. Thus R^2 is confined to the interval from zero to one,

$$0 \le R^2 \le 1\ .$$

The anova table in the output of the fitted regression model summarizes these sums of squares. Finally, R^2 is the square of the correlation between the predictor and response.

The prediction limits for a cottage with 3000 square feet are read from the plot as shown below.

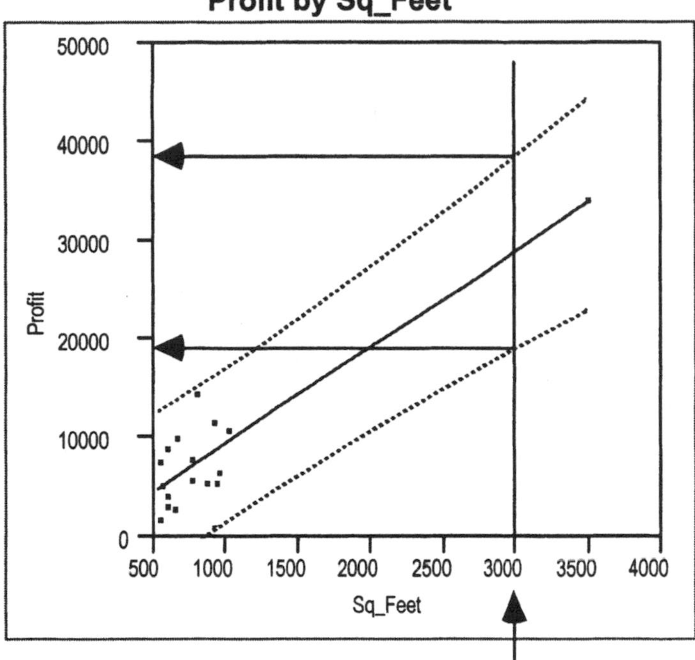

For now, this approximate, graphical approach is fine. Later, we will see how to get JMP to compute the precise limits for us (as part of our analysis in multiple regression). For the curious, the prediction is $28,830 and the limits are $19,150 to $38,520. Without this large "cottage," the prediction limits become far wider.

Setting aside this large house also changes the prediction. The prediction decreases when this observation is excluded. By excluding the very large cottage and recomputing the linear fit, we obtain the plot shown below.

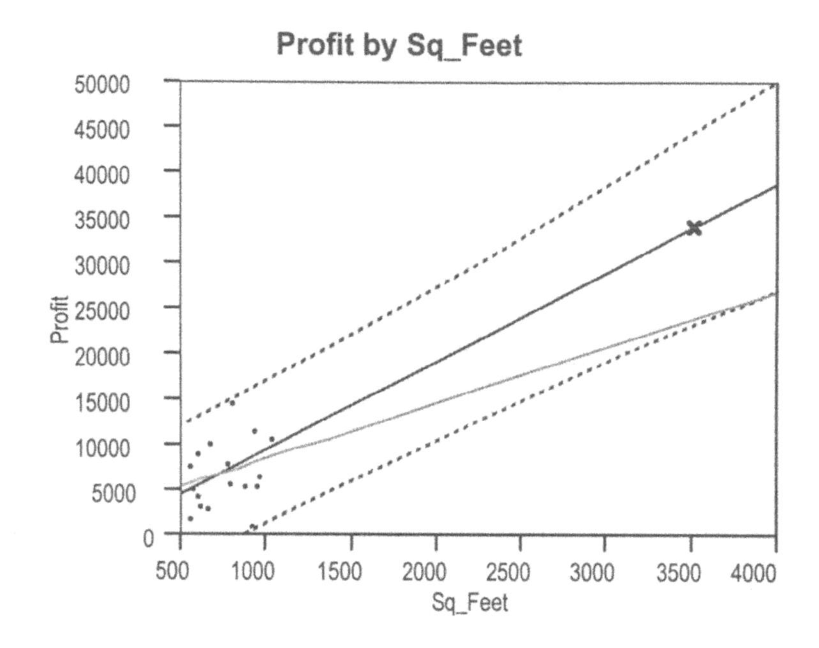

Here is the summary of the fit without the large cottage. Not only has the slope decreased, but the R^2 has dropped to near zero, and the standard errors have grown much larger.

Summary of Fit

RSquare	0.075
RSquare Adj	0.013
Root Mean Square Error	3634
Mean of Response	6823
Observations	17

Parameter Estimates

Term	Estimate	Std Error	t Ratio	Prob>|t|
Intercept	2245.40	4237.25	0.53	0.60
Sq Feet	6.14	5.56	1.10	0.29

This next plot shows both fitted lines (with and without the large cottage), but now the prediction intervals from the model exclude the large cottage. The intervals from this model are vastly wider about a slightly smaller prediction. The prediction is $20,660 with an interval ranging from –$7,200 to $48,500.

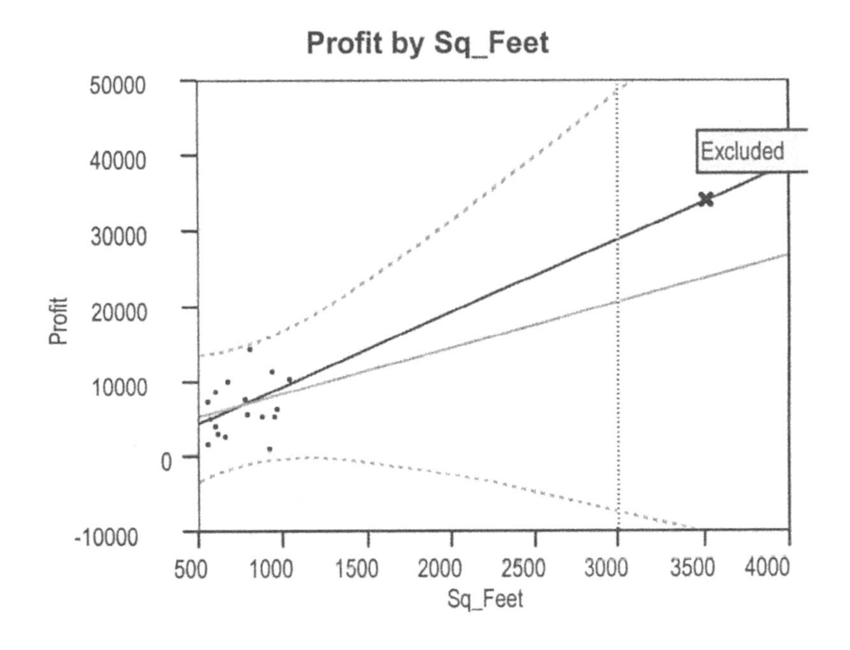

Notice that the prediction intervals become much wider as the value of the predictor increases. Without the large cottage, predicting what will happen to profits when constructing cottages larger than, say, 2000 feet represents quite an extrapolation from the rest of the data. This effect is not so pronounced when the large cottage is included since we have information on what happens to profits for at least one occasion.

The construction company has decided to build another large cottage, this time one with 3000 square feet.

How much profit might the company expect from selling this new dwelling?

If we believe that the new house will reap a profit similar to that of the other large home that was built, we anticipate a profit in the range of $20,000 to $40,000. This is a wide range and is heavily dependent on the one previous house of such large size. Without the experience of the previous large house, we have virtually no information upon which to base prediction. From the analysis without the highly leveraged previous large house, the linear interval ranges from a loss of $7,000 to a profit of nearly $50,000.

This example also illustrates that the effect of removing an outlier from an analysis depends upon where the outlying value is located. Without the outlier, there is little relationship between footage and profit. Sometimes removing an outlier reveals a better fit (as in the Philadelphia housing example); sometimes, such as here, the fit almost disappears when the outlier is set aside.

Note that the validity of a regression model is crucial if one is to trust the offered prediction intervals. Of the assumptions associated with the regression model, the assumptions of

- constant variation and
- normality

are usually less important than independence and capturing nonlinearities. However, if one is doing prediction (particularly when it involves extrapolation), *these assumptions are crucial.* Also, recognize that the best regression is not necessarily the fitted model with the highest R^2. It's easy to get a large R^2 by fitting one or two outlying values well. Such models can explain a lot of variation, but often miss the pattern in most of the data by fitting the outliers.

Prediction limits and confidence intervals for the slope.

Prediction intervals also describe the precision of the slope estimate. If the prediction intervals completely include a horizontal line drawn at the mean of the response, the fitted slope is "not significantly different from zero." This means that the 95% confidence interval for the slope includes zero.

Returning to this example, here is the initial model using all 18 cottages. The summary table shown here includes optional confidence intervals. The prediction limits *do not contain* the horizontal mean line, and the confidence interval for the slope *does not contain* zero. The slope is significantly different from zero.

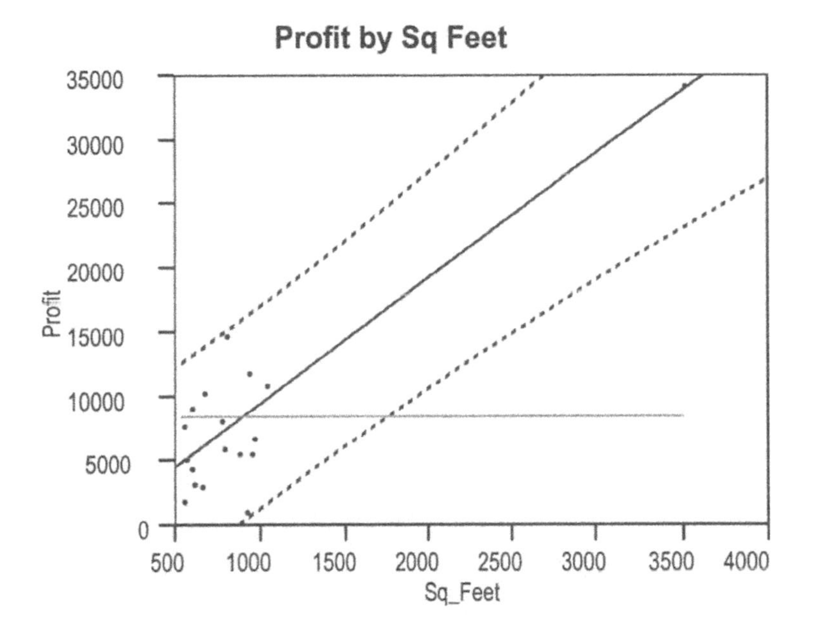

Summary of Fit

RSquare	0.780
RSquare Adj	0.766
Root Mean Square Error	3570
Mean of Response	8348
Observations	18

Parameter Estimates

Term	Estimate	Std Error	t Ratio	Prob>|t|	Lower 95%	Upper 95%
Intercept	-416.86	1437.02	-0.29	0.78	-3463.18	2629.5
Sq_Feet	9.75	1.30	7.52	0.00	***7.00***	***12.5***

In contrast, omitting the large cottage from the analysis gives the following results. The prediction limits *do contain* the horizontal line for the mean profit, and the confidence interval for the slope *does contain* zero. The slope is *not* significantly different from zero. To make the case even more sobering, the wide statistical extrapolation penalty is in fact optimistic because it presumes that the model continues to hold even as we extrapolate far outside the observed range. The true uncertainty in this problem is larger still.

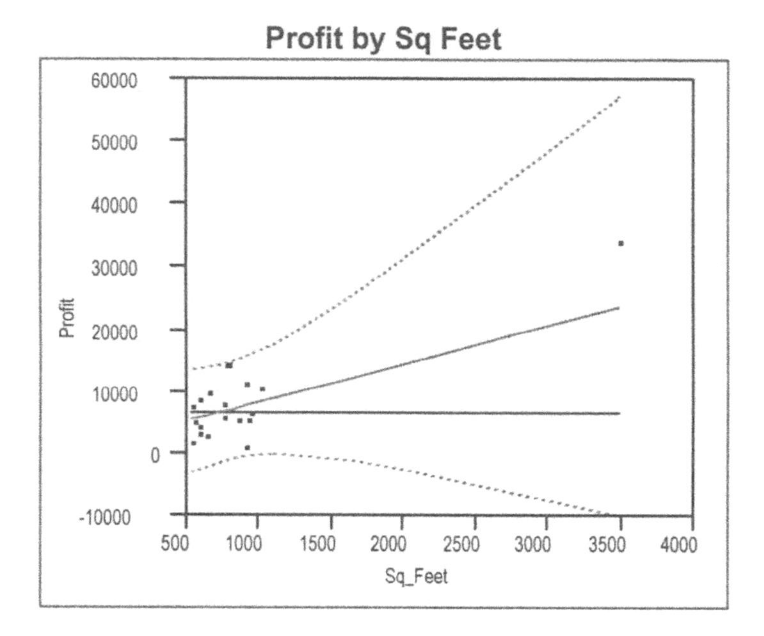

Summary of Fit

RSquare	0.075
RSquare Adj	0.014
Root Mean Square Error	3634
Mean of Response	6823
Observations	17

Parameter Estimates

Term	Estimate	Std Error	t Ratio	Prob>|t|	Lower 95%	Upper 95%
Intercept	2245.40	4237.25	0.53	0.60	-6786.05	11276.85
Sq_Feet	6.14	5.56	1.10	0.29	***-5.71***	***17.98***

Liquor Sales and Display Space, Revisited

Display.jmp

> (1) How precise is our estimate of the number of display feet yielding the highest level of profit?
>
> (2) Can the chain of liquor stores use the fitted model to predict the level of sales anticipated per display foot when this new wine is featured in an upcoming one-store promotion that uses 20 feet of display space?

The data collected by the chain were obtained from 47 of its stores that do comparable levels of business. The following scatterplot shows the data with the model using the log of the number of display feet as the predictor. From the scale of the plot, it is clear that the intended special display requires extrapolation far from the range of past experience.

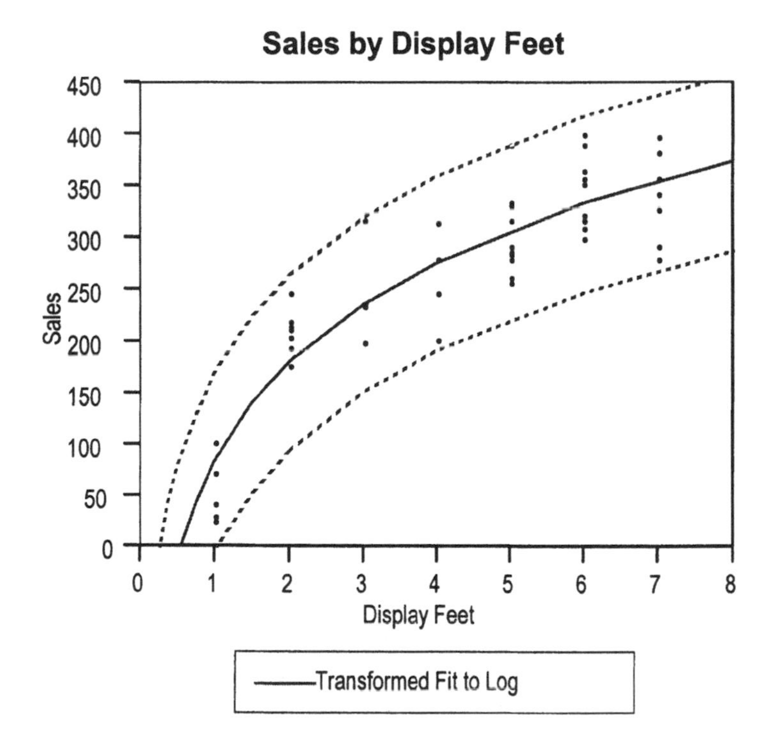

Here is a summary of the fitted model. We can use the results summarized here to check the precision of our estimate of the optimal amount of display feet.

Transformed Fit to Log

Summary of Fit

RSquare	0.82
RSquare Adj	0.81
Root Mean Square Error	41.31
Mean of Response	268.13
Observations	47

Parameter Estimates

Term	Estimate	Std Error	t Ratio	Prob>|t|	Lower 95%	Upper 95%
Intercept	83.56	14.41	5.80	<.0001	54.5	112.6
Log(Display Feet)	138.62	9.83	14.10	<.0001	118.8	158.4

The number of display feet for the optimal profit was the solution of the equation where the derivative of the "gain" function equals zero:

$$0 = \frac{138.6}{x} - 50 \qquad \Rightarrow \qquad x = \frac{138.6}{50} = \frac{\text{estimated slope}}{50} \ .$$

Plugging the endpoints of the 95% confidence interval for the slope into this same expression, we obtain the associated interval for the optimal amount of shelf space:

$$\text{95\% interval for optimal shelf space} = \frac{[118.8,\ 158.4]}{50}$$

$$= [\ 2.38,\ \ 3.17\ ] \text{ feet of shelf space} .$$

This interval implies that the stores ought to display about two and a half or three feet of this item. If they stock more, they will not likely gain enough from this item to make up for other things that could have been shown. Displaying less misses the opportunity. This conclusion is also not too sensitive to the choice of the transformation; we get about the same interval whether we use logs or reciprocals. The predictions at 20 feet are different in this regard.

As to the question of predicting sales at 20 feet, we need to extend the scale of the plots to show the new prediction intervals. Unfortunately, rescaling the plot is not enough since JMP will not expand the prediction limits to the new range. The easiest way to see the intervals on the expanded scale is to add a single, artificial row to the data set. From the *Rows* menu, choose the *Add rows...* command and direct JMP to extend the data set by one row. In this additional row, enter a value of 20 for the number of display feet. Leave the other column missing (denoted by the • symbol).

With this partially missing observation added, JMP does the analysis on the extended scale, and we obtain the plot shown next. (The useful grid lines at 250 and 500 are obtained by double clicking on the *Sales* axis and using the resulting dialog.) The prediction from the logarithmic model is $499 in sales, with a 95% interval of
[409, 589].

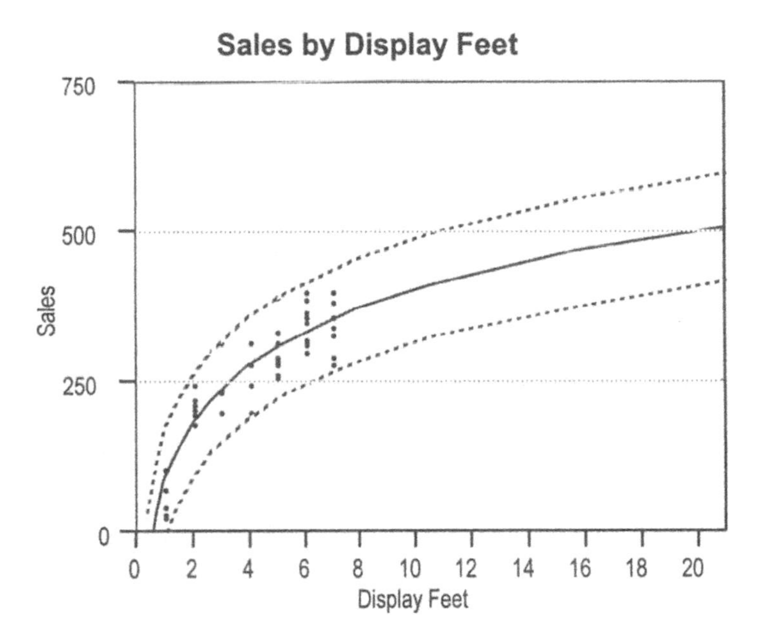

Although this prediction interval is rather wide, it does not approach the actual uncertainty in this extreme extrapolation. Recall that we entertained an alternative model for this data based on the reciprocal transformation. A summary of this other model follows.

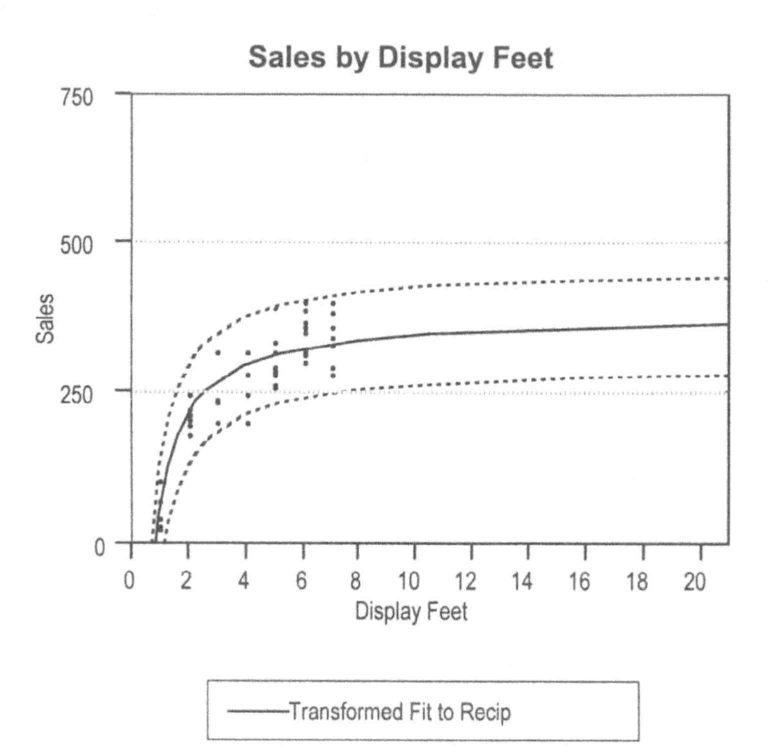

Transformed Fit to Recip
Summary of Fit

RSquare	0.83
RSquare Adj	0.82
Root Mean Square Error	40.04
Mean of Response	268.13
Observations	47.00

Parameter Estimates

Term	Estimate	Std Error	t Ratio	Prob>|t|	Lower 95%	Upper 95%
Intercept	376.70	9.44	39.91	<.0001	357.7	395.7
Recip(Display Feet)	-329.70	22.52	-14.64	<.0001	-375.1	-284.3

Although these two models are comparable over the observed range of data (the R^2's are virtually identical), they offer very different views of the sales at 20 feet. The farther out we extrapolate these two models, the greater is the difference in predictions.

The prediction at 20 display feet from the model that uses reciprocals ($360) lies outside the prediction interval from the model based on logs.

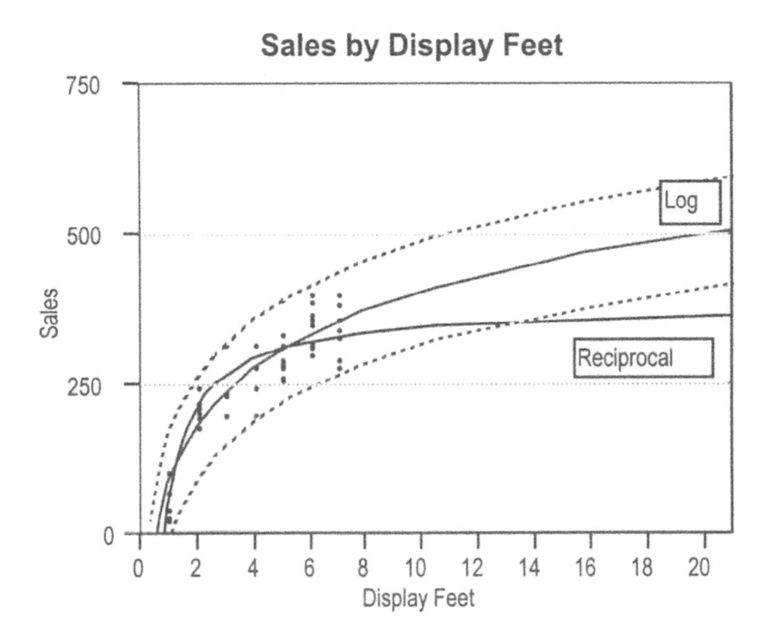

Given our difficulty in choosing between these two models, the true range of uncertainty in the prediction is likely to be closer to the union of the associated intervals rather than to either one. Certainly, the actual prediction interval ought to be wider than either one shown here. As noted at the end of the previous example, statistical prediction intervals penalize only for statistical error, and assume that the fitted model continues to hold when extrapolated.

Suppose sales were to drop as display feet increase, when customers become over-exposed to a product. Would either range be useful?

(1) How precise is our estimate of the number of display feet yielding the highest level of profit?

(2) Can the chain of liquor stores use the fitted model to predict the level of sales anticipated per display foot when this new wine is featured in an upcoming one-store promotion that uses 20 feet of display space?

(1) Confidence intervals for the coefficient provide a simple method for assessing the uncertainty in a derived formula such as that for the optimal shelf space.

(2) In problems of prediction, regression prediction intervals are not reliable far away from the observed data. Models that seem similar over the observed range of the data may offer very different prediction intervals. Since the standard prediction interval does not include a penalty for this so-called *model selection error* or *extrapolation penalty*, treat such extrapolations as an optimistic hope, even when accompanied by wide intervals.

Class 4. Multiple regression

Multiple regression extends the ideas from bivariate regression by allowing one to include several explanatory factors in the equation of the model. As in simple regression, a regression coefficient measures the impact of changes in each explanatory variable on the response. One uses the equation for prediction by again filling in values for the explanatory variables. The confidence intervals and tests for these coefficients closely resemble their counterparts covered in Class 3, and the multiple correlation R^2 and residual standard error again measure the goodness-of-fit of the equation.

The various diagnostics from bivariate regression also generalize to multiple regression. As in bivariate regression, residuals are the key to the diagnostics. For example, residuals and leverage again quantify the size of outlying values. In addition, special graphical methods offer a sequence of diagnostic scatterplots that assist one in judging the validity of assumptions. These leverage plots let one diagnose multiple regression as a sequence of bivariate regressions.

Topics

- Scatterplot matrix and correlation matrix
- Interpretation of multiple regression coefficients: partial and marginal slopes
- t-test, R^2 in multiple regression
- Comparison of regression models via the partial F-test (with details in Class 5)
- Leverage, influence, and outliers in multiple regression
- Leverage plots (partial regression plots) and residual plots in multiple regression
- Prediction intervals in multiple regression.

Example

- Automobile design.

Key application

Understanding what drives sales of sports team products. The major sports leagues, the NHL, NFL, MLB, and NBA, collect a royalty every time someone purchases a team product with the league logo on it. For example, if you buy a Bulls shirt and it's got the NBA logo on it somewhere, the NBA collects money (assuming you buy your clothes from a legit outlet and not some street vendor). It makes sense to try and understand what drives purchases of these products and hence determines royalties. Two potential predictors (independent variables) that might help to explain royalty income generated by a team are share of the local TV audience and the number of games a team has won in a season. Understanding this relationship could help to identify teams that are not marketing their products effectively.

Up until this moment in the course, all of the analyses we have done have involved a single explanatory variable (X-variable, independent variable, predictor), but the problem described above has two potential explanatory variables, TV audience and number of wins. The question is "How do we cope with the two variables?" One approach would be to perform two simple regressions. The crucial flaw in this approach is that the two independent variables are themselves related: TV share is likely to increase as a team becomes more successful. This one-at-a-time approach does not tell the whole story about the association between all the variables. A better way is to treat the two predictors simultaneously. The technique that does this is called multiple regression and is the topic for most of the remaining classes.

Definitions

Leverage plot. A scatterplot that shows the data associated with a fitted multiple regression coefficient. (See the Heuristics.)

Partial regression slope. The change in the response for every one unit change in the particular predictor under consideration, holding other predictors constant.

Concepts

Multiple regression. Multiple regression allows you to build more complicated models than does simple regression. It is conceptually very similar to simple regression. Whereas in simple regression we find the "best" line through the points, in a multiple regression with two explanatory variables, we find the best "plane" through a cloud of points in 3-D space. An essential difference between simple and multiple regression lies in the fact that in multiple regression the X-variables may themselves be related.

The fact that there is more than one X-variable in multiple regression leads to a different interpretation of the regression coefficients. In simple regression, a regression coefficient is a slope. For every one unit change in X, the slope tells you how much Y has changed. In multiple regression, each coefficient is still a slope, but now it is a "partial" slope. It tells you the change in Y for every one unit change in a particular X, "holding all the other X-variables constant". The "holding all other X-variables constant" statement is deceptively important. If the other predictors were not held constant or controlled for (i.e., if they were ignored), we may get a very different slope. A slope calculated ignoring all the other variables and looked at a single predictor at a time, is called a "marginal slope." This is what we get from simple bivariate regression. It tells you the impact of one particular X-variable ignoring the other X-variables. The partial and marginal slopes can be very different.

Multiple regression also gives a t-statistic for each predictor, and the interpretation of these statistics is also different. A t-test for a single X-variable in multiple regression tells you whether or not this particular variable is important in the regression (has a nonzero partial slope), after taking into account all the other X-variables in the fit. It determines whether the X-variable adds anything to a model that already has all the other X-variables in it. Yet another way is to think of it as a "last in" test: if this variable were the last to be entered in the multiple regression, do we really need it? Does it add anything more to what we already know based on all of the other X-variables?

Leverage plot. When doing multiple regression with many X-variables, it becomes impossible to draw a scatterplot as was done in simple regression. The lack of a simple plot makes it harder to pick out influential and high-leverage points. However, these concepts are still important since a multiple regression can be highly sensitive to a single observation. We need special tools to identify these points. Fortunately, the residual plots we have seen generalize easily to multiple regression so all that you have learned so far about them has an immediate carryover effect. The new tool that is available for multiple regression is the leverage plot. This plot allows you to essentially look at a multiple regression, one X-variable at a time. It does the calculations that control for all the other variables and presents you with a scatterplot that shows the association between each X-variable and the Y-variable one at a time. All that you have learned about scatterplots in simple regression is applicable to these leverage plots, so you can pick out outlying and influential observations.

Heuristic

Judging changes in R^2.

What's a big change in R^2? It depends on many things (such as sample size). Don't think in terms of R^2 itself, but rather the proportional change in $1-R^2$. A change in R^2 from 0.98 to 0.99 is quite large, but from 0.50 to 0.51 is not. In both cases R^2 changes by 0.01, but in the first case going from 0.98 to 0.99 implies that the extended model explains half of the remaining variation missed by the first model. That's a significant improvement.

Potential Confusers

Don't get excited that R^2 has increased when you put more variables into a model.

R^2 has to increase, even if you added the distance from the earth to the closest stars to your model R^2 would increase. R^2 measures the amount of explained variability in the response (Y-variable). Adding another X-variable to the model can never reduce the amount of variability explained, only increase it. The right question to ask is not whether or not R^2 has increased (it has to), but rather whether or not R^2 has increased by a useful amount. The t-statistic for each coefficient answers this question.

What does "holding all other variables constant" mean in a polynomial regression?

This heuristic interpretation is often inadequate, particularly in polynomial regression in which the predictors are X and X^2. There's no way to change X and not also change X^2. To more fully appreciate this phrase, we have to understand how multiple regression controls for the other predictors $X_2, ..., X_p$ when it judges the effect of X_1. In regression, to "control" for $X_2, ..., X_p$ means to remove their influence from Y and X_1. How does regression do this? It uses regression! That is, it removes their influence by regressing both Y and X_1 on the other predictors. Then it obtains the multiple regression coefficient for X_1 as the simple regression of the *Y residuals* on the *X_1 residuals*. In fact, a leverage plot shows these two sets of residuals; that's how you can get a scatterplot that shows a multiple regression coefficient. We'll discuss this further in Class 5.

Automobile Design

Car89.jmp

A team charged with designing a new automobile is concerned about the gasoline mileage that can be achieved. The team is worried that the car's mileage will result in violations of Corporate Average Fuel Economy (CAFE) regulations for vehicle efficiency, generating bad publicity and fines. Because of the anticipated weight of the car, the mileage attained in city driving is of particular concern.

The design team has a good idea of the characteristics of the car, right down to the type of leather to be used for the seats. However, the team does not know how these characteristics will affect the mileage.

The goal of this analysis is twofold. First, we need to learn which characteristics of the design are likely to affect mileage. The engineers want an equation. Second, given the current design, we need to predict the associated mileage.

The new car is planned to have the following characteristics:

Cargo	18 cu. ft.	
Cylinders	6	
Displacement	250 cu. in.	(61 cu. in. ≈ 1 liter)
Headroom	40 in.	
Horsepower	200	
Length	200 in.	
Leg room	43 in.	
Price	$38,000	
Seating	5 adults	
Turning diameter	39 ft.	
Weight	4000 lb.	
Width	69 in.	

An observation with these characteristics forms the last row of the data set. The mileage values for this observation are missing so that JMP will not use this row in fitting a regression model. It is useful later in getting JMP to do the calculations needed to predict the mileage of the new car.

How should we begin an analysis with so many possible predictors? It is quite hard to know where to start if you do not have some knowledge of what the variables measure. Fortunately, most of us are somewhat familiar with cars and recognize that the bigger the vehicle, the more fuel it will tend to consume. So, let's start there using weight to predict fuel consumption.

The following plot shows the mileage (miles per gallon) plotted on the weights of the cars (in pounds), along with two fits. The relationship appears to be slightly nonlinear, particularly for relatively light or heavy cars. The linear fit under-predicts miles per gallon at both extremes. Since the contemplated car design has a weight of 4,000 lb., putting it at the right of this collection of cars, the linear model will probably under-predict its fuel consumption as well. The alternative model is only moderately nonlinear, but seems to capture the curvature at the extremes of this range. The particular transformation shown here is obtained by taking the reciprocal of the response. The choice of this transformation is only partly driven by the data, however, and is also motivated by ease of explanation: the reciprocal of miles per gallon is gallons per mile (GPM). Note that the slopes of these models have different signs. Why does this happen?

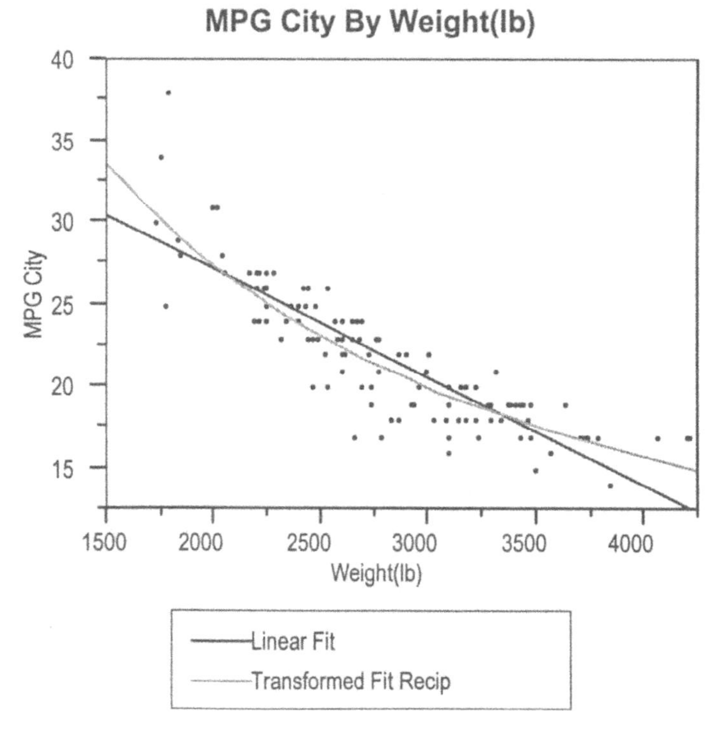

MPG City = 40.118279 - 0.0065504 Weight(lb)

Recip(MPG City) = 0.0094323 + 0.0000136 Weight(lb)

Since we'll be adding other predictors of fuel consumption, we will use the explicitly transformed response on the transformed scale of gallons per mile. The following plot shows the relationship between fuel consumption and weight. In keeping with the positive sign observed in the fit of the nonlinear model on the previous page, the relationship positive: heavier cars burn more fuel per mile.

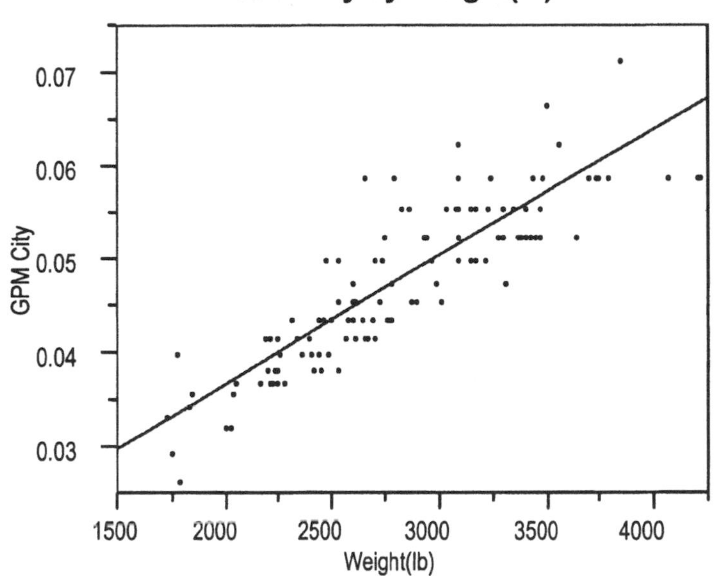

GPM City = 0.00943 + 0.00001 Weight(lb)

Summary of Fit

RSquare	0.765
Root Mean Square Error	0.004
Mean of Response	0.048
Observations	112

Analysis of Variance

Source	DF	Sum of Squares	Mean Square	F Ratio
Model	1	0.006	0.006426	358.62
Error	110	0.002	0.000018	**Prob>F**
C Total	111	0.008		<.0001

Parameter Estimates

Term	Estimate	Std Error	t Ratio	Prob>|t|
Intercept	0.0094323	0.002055	4.59	<.0001
Weight(lb)	0.0000136	7.19e-7	18.94	<.0001

The units of gallons per mile produce a very small slope estimate since each added pound of weight causes only a very small increase in fuel consumption per mile. We can obtain a "friendlier" and perhaps more impressive set of results by rescaling the response as gallons per 1000 miles. The results follow. Little has changed other than relabeling the Y-axis of the plot. In keeping with this change, though, the slope, intercept, and RMSE are 1000 times larger. The goodness-of-fit measure R^2 is the same.

GP1000M City by Weight(lb)

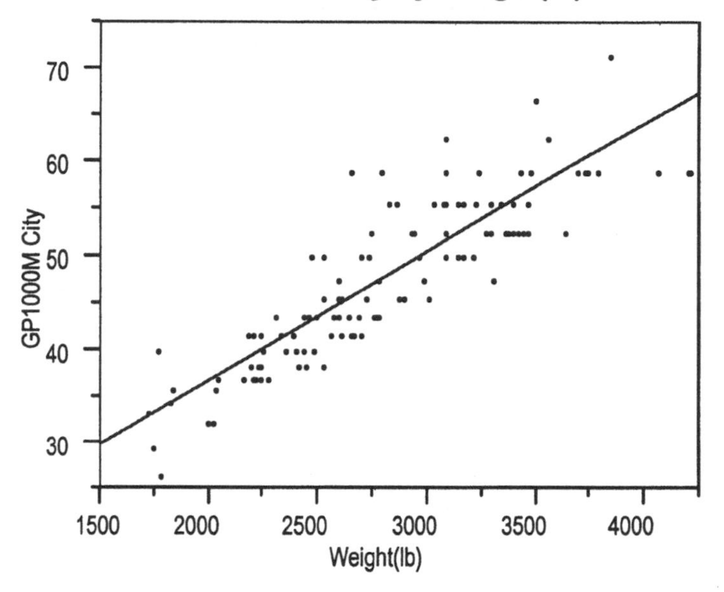

Linear Fit

GP1000M City = 9.43234 + 0.01362 Weight(lb)

Summary of Fit

RSquare	0.765
Root Mean Square Error	4.233
Mean of Response	47.595
Observations	112

Analysis of Variance

Source	DF	Sum of Squares	Mean Square	F Ratio
Model	1	6426.4	6426.44	358.6195
Error	110	1971.2	17.92	**Prob>F**
C Total	111	8397.6		<.0001

Parameter Estimates

Term	Estimate	Std Error	t Ratio	Prob>|t|
Intercept	9.4323	2.0545	4.59	<.0001
Weight(lb)	0.0136	0.0007	18.94	<.0001

Before we turn to the task of prediction, we need to check the usual diagnostics. The residuals in the previous plot do not appear symmetrically distributed about the line. Notice that several high-performance vehicles have much higher than predicted fuel consumption.

Saving the residuals lets us view them more carefully. The skewness stands out in the following normal quantile plot. How does this apparent skewness affect predictions and prediction intervals from the model?

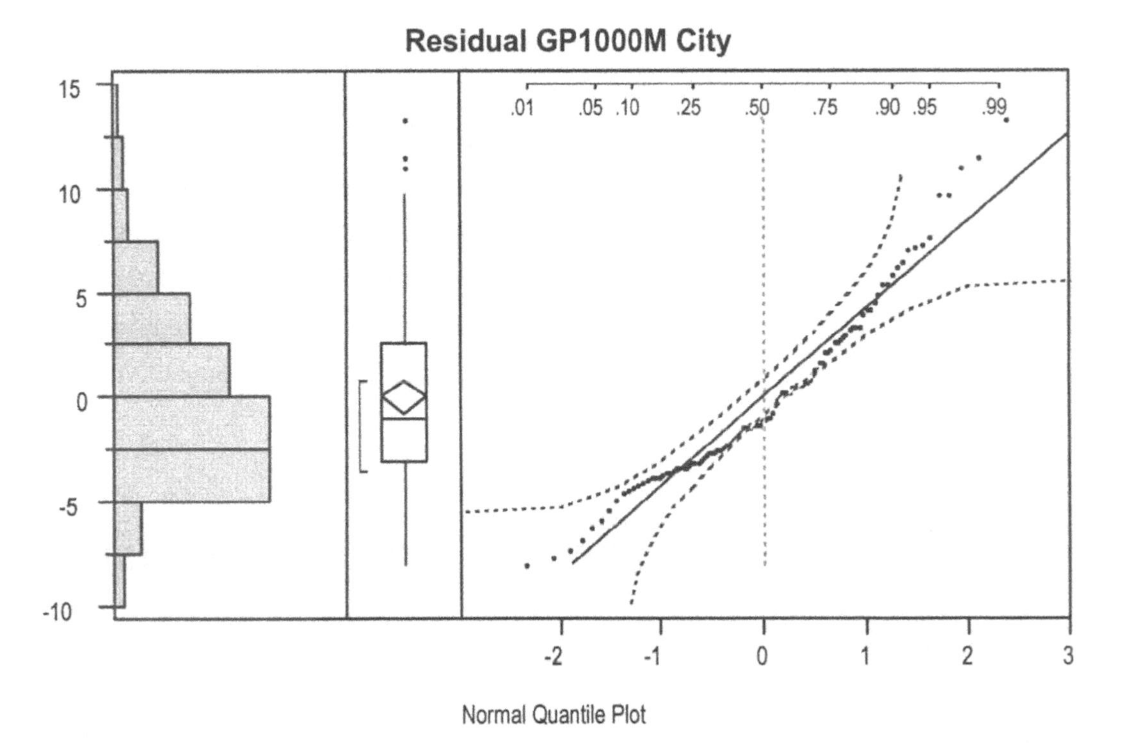

It is perhaps too early to be concerned with prediction; we have only used one predictor in our model so far. Other factors will perhaps be useful and help to explain the lack of normality visible in this quantile plot of the residuals. Just the same, for the sake of comparison, we will find the predicted fuel consumption of the new design.

From the summary of the regression of gallons per 1000 miles on weight, we can obtain a prediction of the mileage of the new car being designed. From the output shown, the equation for the fitted line is

Fitted GP1000M = 9.43 + 0.0136 *Weight*.

If we substitute the design weight of 4000 lb., we obtain the prediction

Predicted GP1000M for new design = 9.43 + 0.0136 (4000) = 63.8 *GP1000M*

This prediction agrees with fitted line shown two pages back.

We also need to determine the associated prediction interval. We can either estimate the interval endpoints from the plot or, more accurately, use the *Fit Model* platform to compute both the prediction and interval for us. The *Fit Model* platform lets us save the predictions and the prediction intervals. (Use the $ button to save the predictions and prediction intervals, labeled *Save Indiv Confidence.*) From the last row of the spreadsheet, we find that using weight to model the *GP1000M* of the vehicle leads to the prediction and interval:

	Predicted *GP1000M*	95% Prediction Interval *GP1000M*
Weight = 4000	63.9	[55.3 – 72.5] ,

which implies an interval of [13.8, 18.1] miles per gallon. (Note: The predicted *GP1000M* found by JMP differs slightly from the value obtained above due to rounding in our earlier calculation.) Since confidence intervals transform in the obvious way, we can also find intervals for related quantities like the gasoline operating cost per 1000 miles. At $1.20 per gallon, the cost interval is

[55.3 – 72.5] × 1.2 = [66.36 – 87.00] $/1000M.

The search for other factors that are able to improve this prediction (make it more accurate with a shorter prediction interval) begins by returning to the problem. *Weight* aside, what other factors ought to be relevant to mileage? Right away, the power of the engine (horsepower) comes to mind. Some other factors might be the size of the engine (the engine displacement is measured in cubic inches; 61 $in^3 \approx 1$ liter) or the amount of space in the vehicle, such as the passenger capacity or the cargo space available. Correlations show that the latter two have a slight relationship with the response. Horsepower, like weight, has a substantial correlation. Horsepower and displacement are highly correlated with the response, with each other, and with the weight of the car.

Correlations

Variable	GP1000M City	Weight(lb)	Horsepower	Cargo	Seating
GP1000M City	1.00	0.88	0.83	0.17	0.16
Weight(lb)	0.88	1.00	0.75	0.18	0.35
Horsepower	0.83	0.75	1.00	-0.05	-0.09
Cargo	0.17	0.18	-0.05	1.00	0.49
Seating	0.16	0.35	-0.09	0.49	1.00

7 rows not used due to missing values.

We have seen that outlying values and nonlinearity can " fool" simple regression into misleading summarizes of the relationship between two variables. Correlations are similarly sensitive to outliers. Plots tell a more complete story. The scatterplot matrix on the next page (with variables arranged as in the correlation matrix above) shows the data that go into each of the 10 distinct correlations in the previous table. The cargo variable essentially captures a few unusual vehicles. (Use point-labeling to identify these cars.)

The ellipses in the scatterplot matrix graphically convey the size of each correlation: the more narrow the ellipse and the more it is tilted toward the 45° line, the higher the correlation. If an ellipse looks like a circle, the correlation between that pair is near zero.

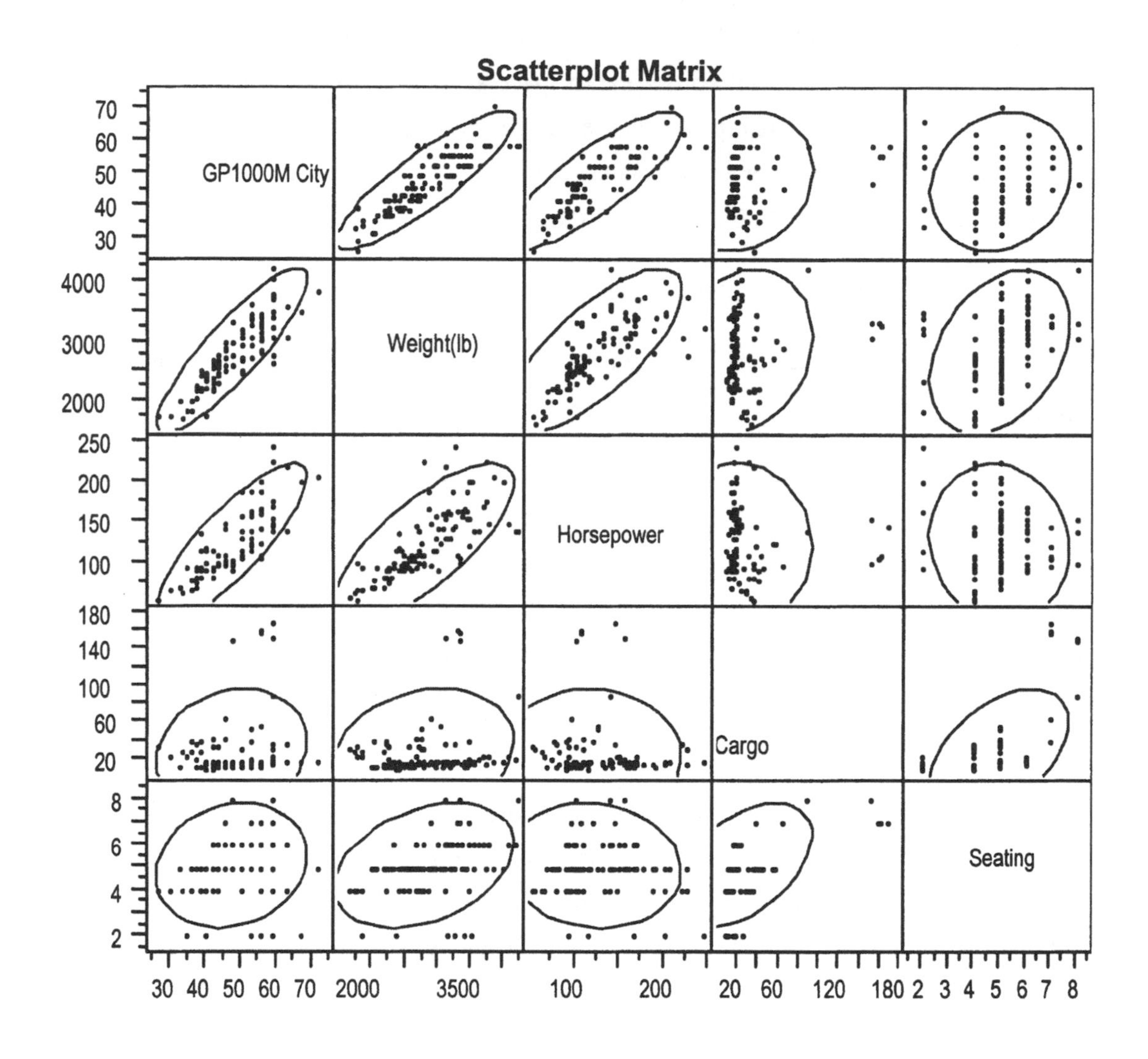

Outliers are also important in multiple regression, only it becomes harder to spot them with so many more variables. The *Correlation* platform in JMP produces a useful summary plot that helps spot overall outliers, but does not suggest how they will affect the multiple regression equation. Other plots will show the how individual points contribute to a multiple regression model. This plot shows a measure of how far each observation lies from the center of the selected variables, plotted on the row number from the associated spreadsheet. The dashed line is a reference line; only a small fraction of the data should exceed this threshold under some typical assumptions.

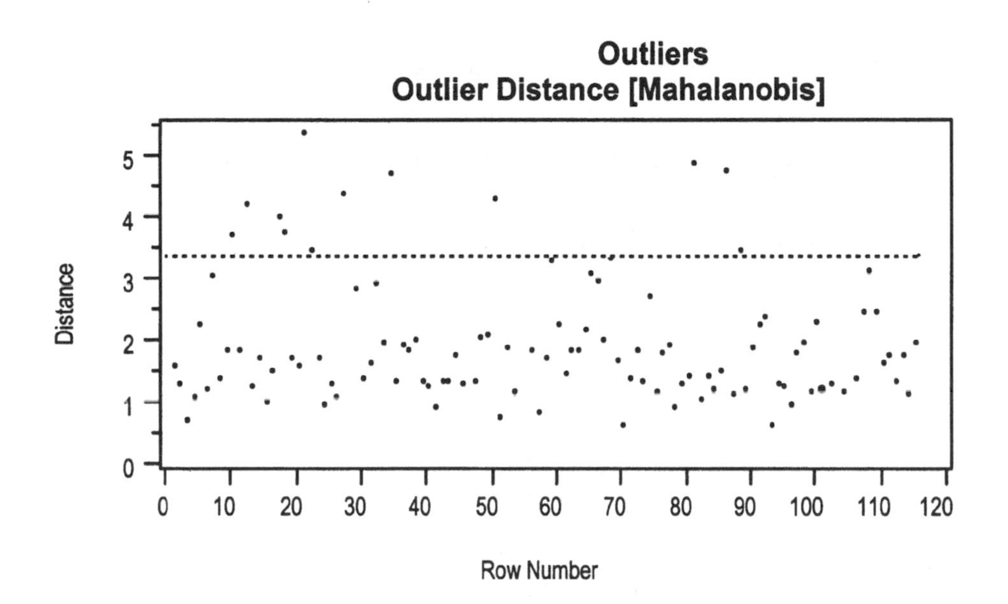

The outliers in this view are by-and-large the same as we can see in the scatterplot matrix: point-labeling identifies them as mostly vans and sports cars. The problem with this plot is that without linking it to the scatterplot matrix, it alone does not offer an explanation for why the observation is an outlier. Perhaps the outlier has an unusual value for the predictor, or perhaps it has moderately unusual values for several of the predictors. We cannot tell which explanation is the right one from just this one view – with it linked to the scatterplot matrix, we have a much more useful tool.

With the addition of horsepower to our model, the regression equation using both weight and horsepower is

Fitted *GP1000M* = 11.7 + 0.0089 *Weight* + 0.088 *Horsepower*.

The addition of horsepower improves the explanatory power of the initial regression (R^2 is higher, rising from 77% to 84%) by a significant amount (the *t*-statistic for the added variable is $t = 7.21$). The addition of horsepower captures about a third of the residual variation remaining from the regression using weight alone.

The coefficient for weight, however, is smaller than when considered initially in the bivariate regression (also with a *larger* SE and hence *smaller* *t*-statistic in the multiple regression). The *t*-statistic for weight was 18.9 in the previous simple regression.

Response: GP1000M City

RSquare	0.841
Root Mean Square Error	3.50
Mean of Response	47.6
Observations	112

Parameter Estimates

Term	Estimate	Std Error	t Ratio	Prob>|t|
Intercept	11.6843	1.7270	6.77	<.0001
Weight(lb)	0.0089	0.0009	10.11	<.0001
Horsepower	0.0884	0.0123	7.21	<.0001

Focusing on the difference between marginal and partial slopes, consider this question. For a typical car in this data, how much more gas will it use to carry an additional 200 pound passenger for 1000 miles? Using the marginal slope suggests an increase of 0.0136×200 = 2.72 gallons. By comparison, the partial slope suggests that the fuel consumption will rise by only 0.0089×200 = 1.78 gallons. Which is right? Well, did adding the weight change the horsepower? No. The horsepower of the car is the same, with or without the added 200 pounds, and the partial slope gives the better estimate (1.78 gallons).

Some "stripes" are visible in the residual plot. Looking back at the original scatterplots, we can see that these are caused by discreteness in the response.

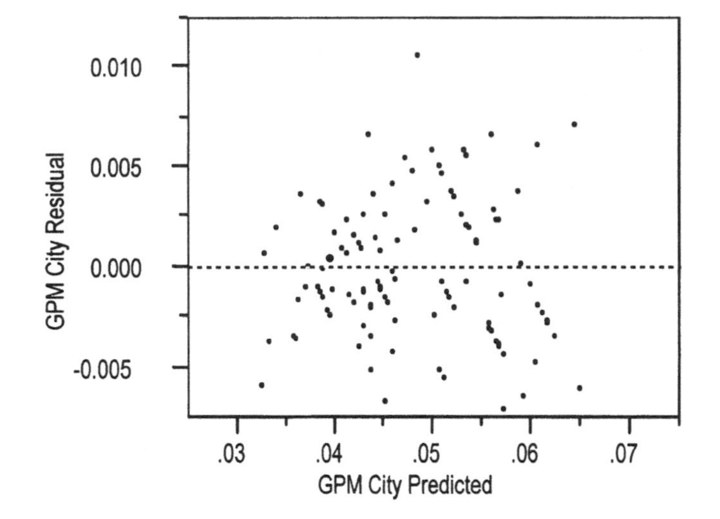

The residual plot displayed below shows that the addition of horsepower to the equation has reduced the amount of skewness in the residuals. The large outlier is the Mazda RX-7, the only car with a rotary engine.

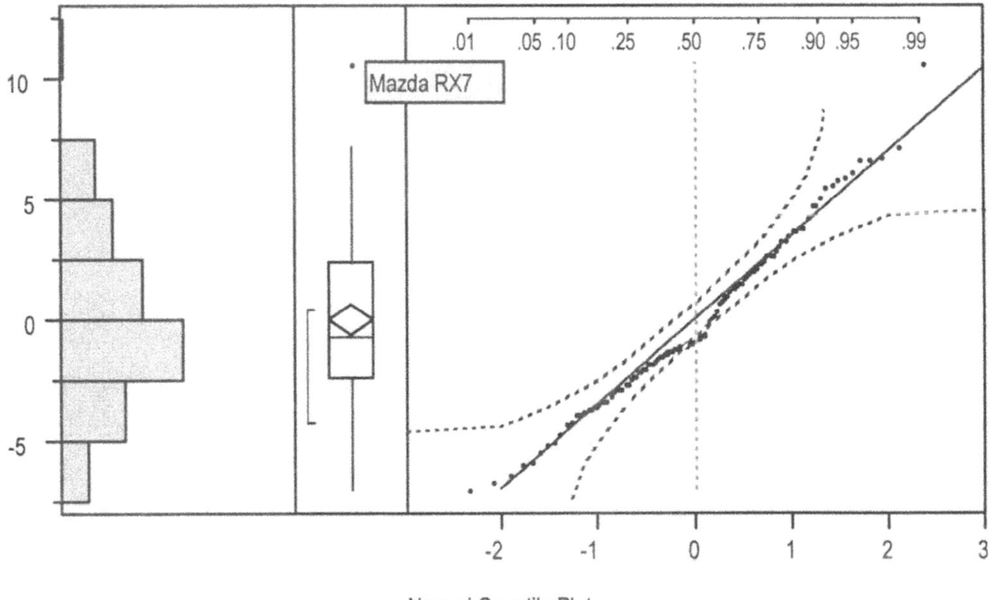

The addition of horsepower to the model has produced a better fit with narrower prediction limits. However, the increase in the standard error for the coefficient of weight is notable: the overall fit is better, but the estimate is different and its SE is larger. Why has the estimate changed? Why has our fitted slope also become less precise, even when we have "explained" more of the variation in the fuel consumption?

From the original correlation matrix or scatterplot matrix, notice that the two predictors used here, weight and horsepower, are highly related. The correlation is 0.75 between these two factors. A plot of the two appears below.

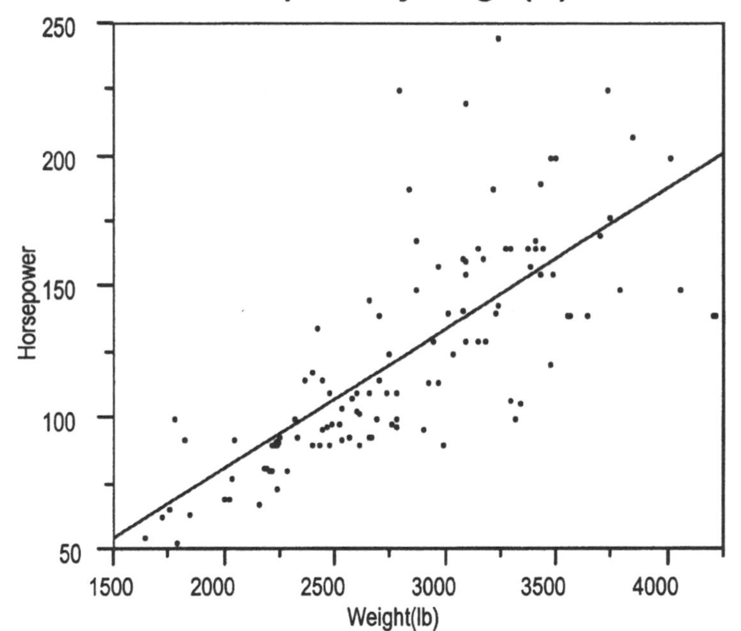

When used together in a regression model, these two factors interact with each other as both describe similar features of cars – both are related to the size of the car. This relationship also affects how well the multiple regression slopes are estimated from the data. The effect of the correlation between *Weight* and *Horsepower* shows in the standard error of the two slope estimates.

Recall that the SE of a slope estimate in a simple regression is determined by three factors:

(1) error variation around the fitted line (residual variation),
(2) number of observations, and
(3) variation in the predictor.

These same three factors apply in multiple regression, with one important exception. The third factor is actually

(3) "unique" variation in the predictor.

The effect of the correlation between the two predictors is to reduce the effective range of weight, as suggested in the plot below. Without *Horsepower* in the equation, the full variation of *Weight* is available for estimating the coefficient for *Weight*. Restricted to a specific horsepower rating, much less variation is available. As a result, even though the model fits better, the SE for *Weight* has increased.

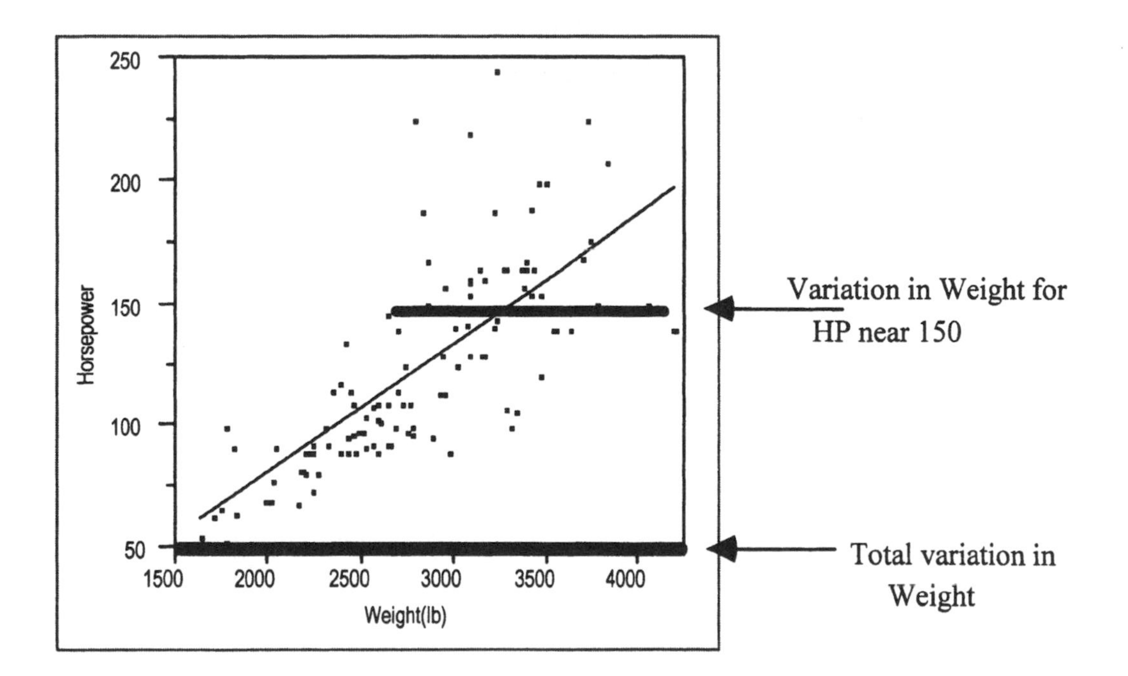

A spinning 3-D plot of *GP1000M* , *Weight*, and *Horsepower* helps to visualize the problem. The initial view shows the clear association between *GP1000M* and *Weight*

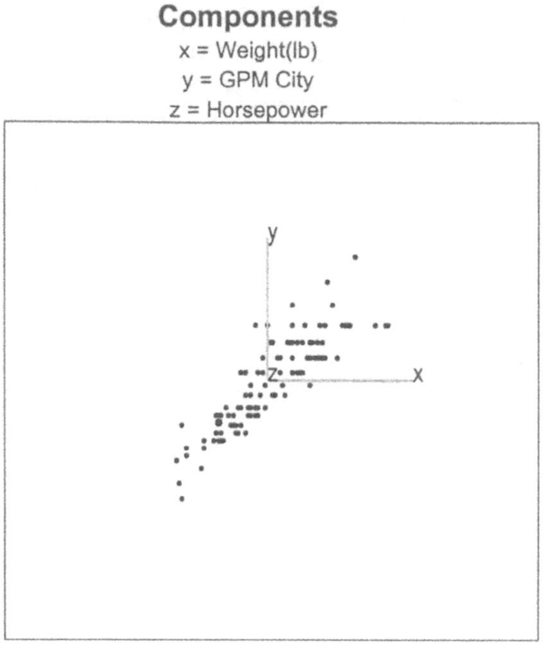

Components

x = Weight(lb)

y = GPM City

z = Horsepower

Rotating the plot shows, however, that most of the data fall into a cylindrical region. *Weight* and *Horsepower* are by and large redundant, with only a few points to identify the best fitting multiple regression surface.

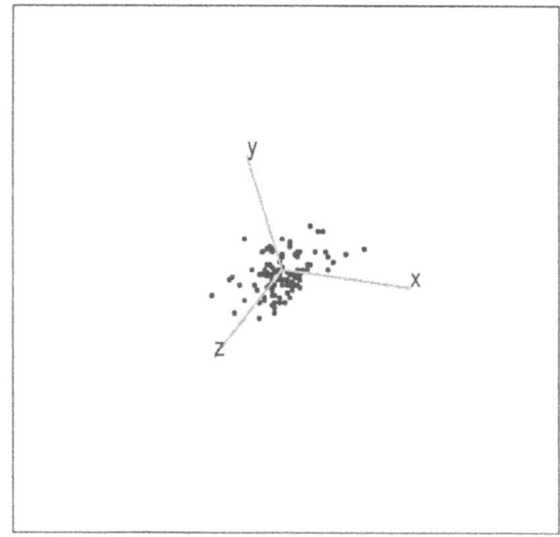

The next type of plot or a regression model, called a *leverage plot* in JMP, focuses on a single multiple regression coefficient. There is one leverage plot for each predictor in the model. Each leverage plot shows the contribution of each predictor to the multiple regression, exclusive of the variation explained by others. It also reveals to us how the observations affect the multiple regression slope and is particularly effective for finding leveraged outliers.

• The *slope* of the line shown in the leverage plot is equal to the coefficient for that variable in the multiple regression. In this sense, a leverage plot resembles the familiar scatterplot that makes regression with a single predictor so easy to understand.

• The *distances* from the points to the line in the leverage plot are the multiple regression residuals. The distance of a point to the horizontal line is the residual that would occur if this factor were not in the model. Thus, the data shown in the leverage plot are not the original variables in the model, but rather the data adjusted to show how the multiple regression is affected by each factor.

Avoiding the details of how a leverage plot is created, you should treat it as a "simple regression view" of a multiple regression coefficient. A variety of things make multiple regression more difficult to understand than bivariate regression. Chief among these is that there is no simple way to look at a multiple regression. Leverage plots offer a sequence of simple regressions that attempt to show how each predictor enters the model. For example, as in simple regression, the slope for a variable is significant if the horizontal line ever lies outside the indicated confidence bands. More useful to us, though, the ability of a leverage plot to show how individual observations affect the fitted model. The leverage plots for this model indicate how sports cars like the Corvette affect the fitted slope for horsepower. Performance cars are leveraged in this model because they have unusually high horsepower for their weight. A car with large horsepower in this model is not automatically leveraged – it could also be rather heavy so that its power is not unusual for its weight. Sports cars are leveraged in this multiple regression because they have as much horsepower as any of the other cars but are typically much lighter.

Leverage plots for *Weight* and *HP/Pound* suggest the high precision of the slope estimates in our revised model. Performance cars stand out on the right side in the second plot. Plot linking and brushing using the car names for labels is particularly useful when sorting out these plots since the variables shown in the plot are constructed from the multiple regression and are not in the initial data set.

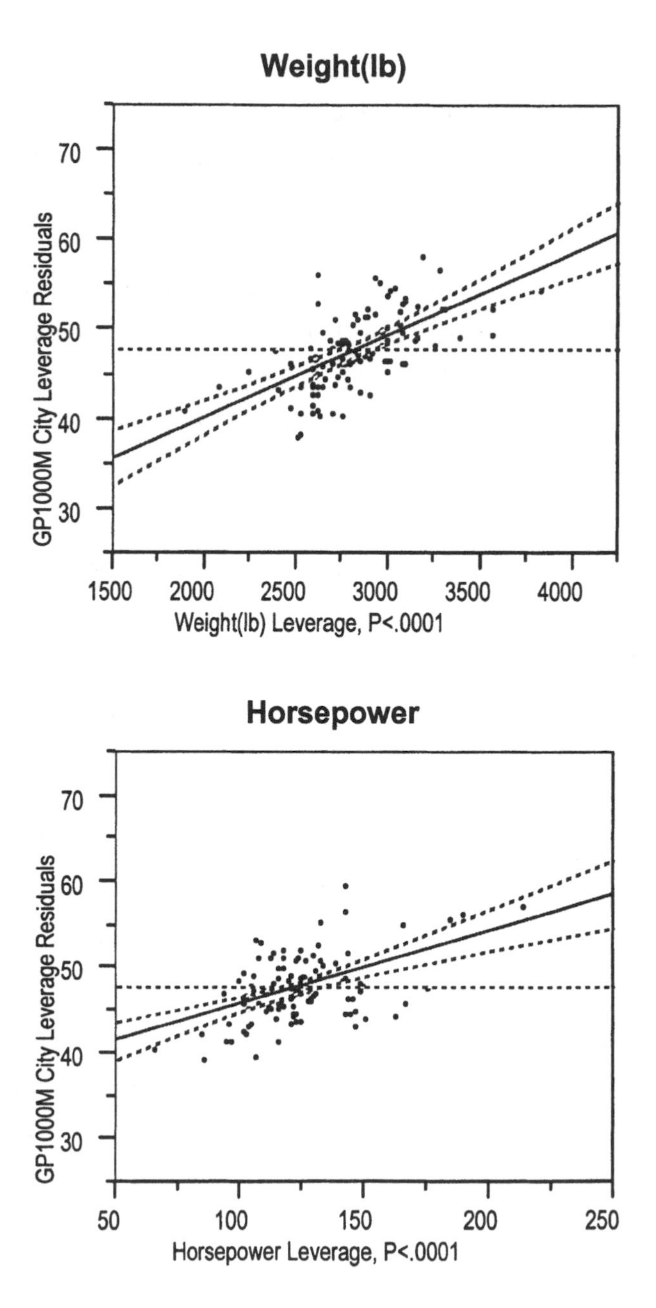

Returning to the problem of predicting the mileage of the proposed car, this multiple regression equation provides narrower prediction intervals than a model using weight alone. The prediction interval using both variables shifts upward (lower mileage) relative to the interval from simple regression. The higher than typical horsepower of the anticipated car leads to a slightly higher estimate of gasoline consumption. With more variation explained, the prediction interval is also narrower than that from the model with *Weight* alone.

Model	Predicted *GP1000M*	95% Prediction Interval
Weight alone	63.9	[55.3 – 72.5]
Weight & Horsepower	65.0	[57.9 – 72.1]

Other factors might also be useful in explaining the cars' mileage, but the analysis at this point is guided less by reasonable theory and becomes more exploratory. For example, adding both cargo capacity and number of seats indicates that the cargo space affects mileage, even controlling for weight and horsepower. Seating has little effect (small t-statistic, p-value much larger than 0.05).[1]

Response: GP1000M City

RSquare	0.852
Root Mean Square Error	3.412
Mean of Response	47.675
Observations	109

Term	Estimate	Std Error	t Ratio	Prob>|t|
Intercept	12.9305	2.0208	6.40	<.0001
Weight(lb)	0.0091	0.0012	7.88	<.0001
Horsepower	0.0858	0.0151	5.68	<.0001
Cargo	0.0346	0.0133	2.61	0.0104
Seating	-0.4765	0.4124	-1.16	0.2506

[1] The addition of these two predictors also has another surprising effect: the data set gets smaller, dropping from 112 to 109. Three of the cars have missing values for either the cargo or seating measurement and hence get dropped. Missing data complicates analyses because the data set changes as we consider other predictors.

Leverage plots make it clear, though, that the only reason that *Cargo* is significant is the presence of several vans in the data (here marked with x's). Seating capacity is not relevant, even though the leverage plot below is dominated by the two-seaters shown at the left.

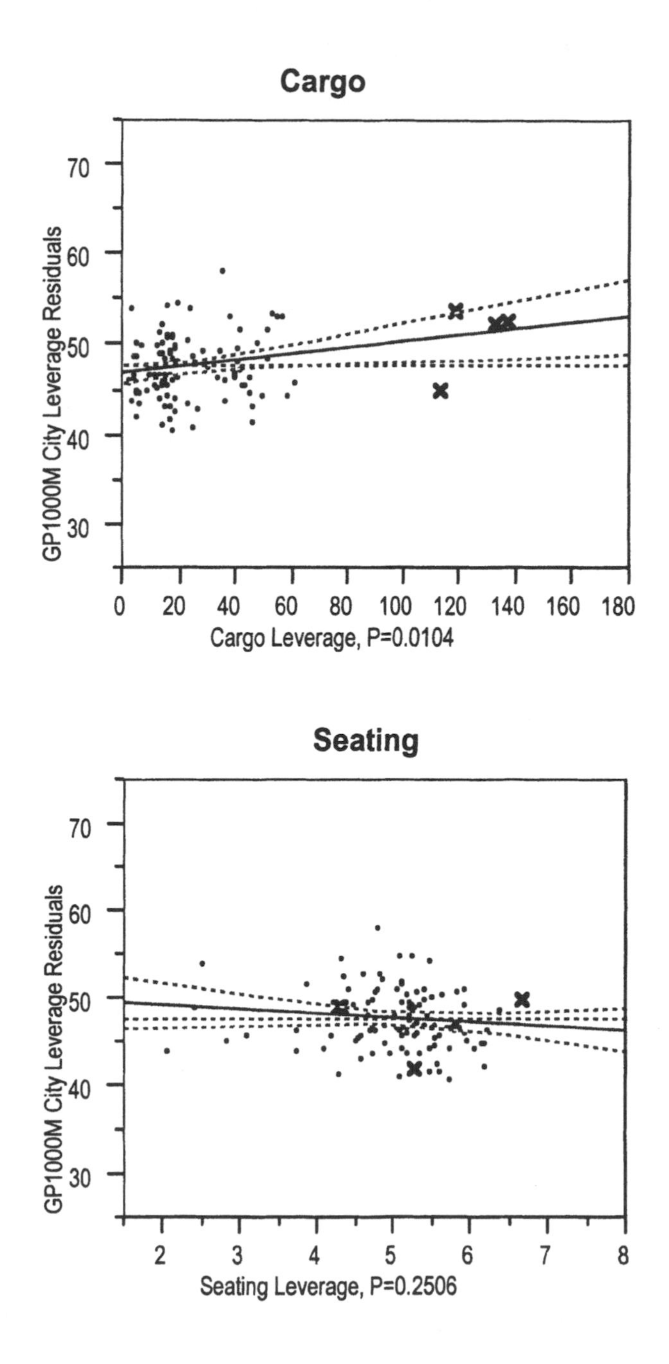

Further exploration suggests that price is a significant predictor that improves the regression fit (keeping the vans in the data set). But why should it be included in the model?

Response: GP1000M City

RSquare	0.861
Root Mean Square Error	3.335
Mean of Response	47.889
Observations	106

Parameter Estimates

Term	Estimate	Std Error	t Ratio	Prob>|t|
Intercept	14.2801	2.0387	7.00	<.0001
Weight(lb)	0.0079	0.0013	6.16	<.0001
Horsepower	0.0807	0.0153	5.29	<.0001
Cargo	0.0351	0.0130	2.69	0.0084
Seating	-0.2670	0.4301	-0.62	0.5362
Price	0.0001	0.0001	1.84	0.0682

Has the addition of these three new predictors significantly improved the fit of our model? To answer this question, we need to go outside the realm of what JMP provides automatically and compute the partial F statistic. The idea is to see how much of the residual remaining after *Weight* and *Horsepower* has been explained by the other three.

$$F = \frac{\text{Change in explained variation per added slope}}{\text{Remaining variation per residual d.f.}}$$

$$= \frac{(0.861 - 0.841)/3}{(1 - 0.861)/(107 - 6)} = 4.84$$

Each added coefficient explains about five times the variation remaining in each residual. This is significant, as you can check from JMP's calculator. As a rough "rule of thumb", an F-ratio is significant if its larger than 4. (It might be significant when it is less than 4 in some cases, but once F>4, you don't need to look it up unless you have very, very little data.)

Why should price be significant? Are more expensive cars better engineered and have a more efficient design? Why would such things would lead to higher fuel consumption, especially since we have taken into account the power and weight of the car?

Outliers are another possible explanation. Just as we found for *Cargo*, a small subset of the data produces the significant effect for *Price*. To reveal this subset, look at the leverage plot for *Price*. The leverage plot for *Price* shows that three cars (marked with o's) dominate this coefficient, just as four vans dominate the leverage plot for *Cargo*. Which models produce the leveraged observations for *Price*? Perhaps this regression model reaches too far, building on special features of this data set rather than those that might apply to our new car.

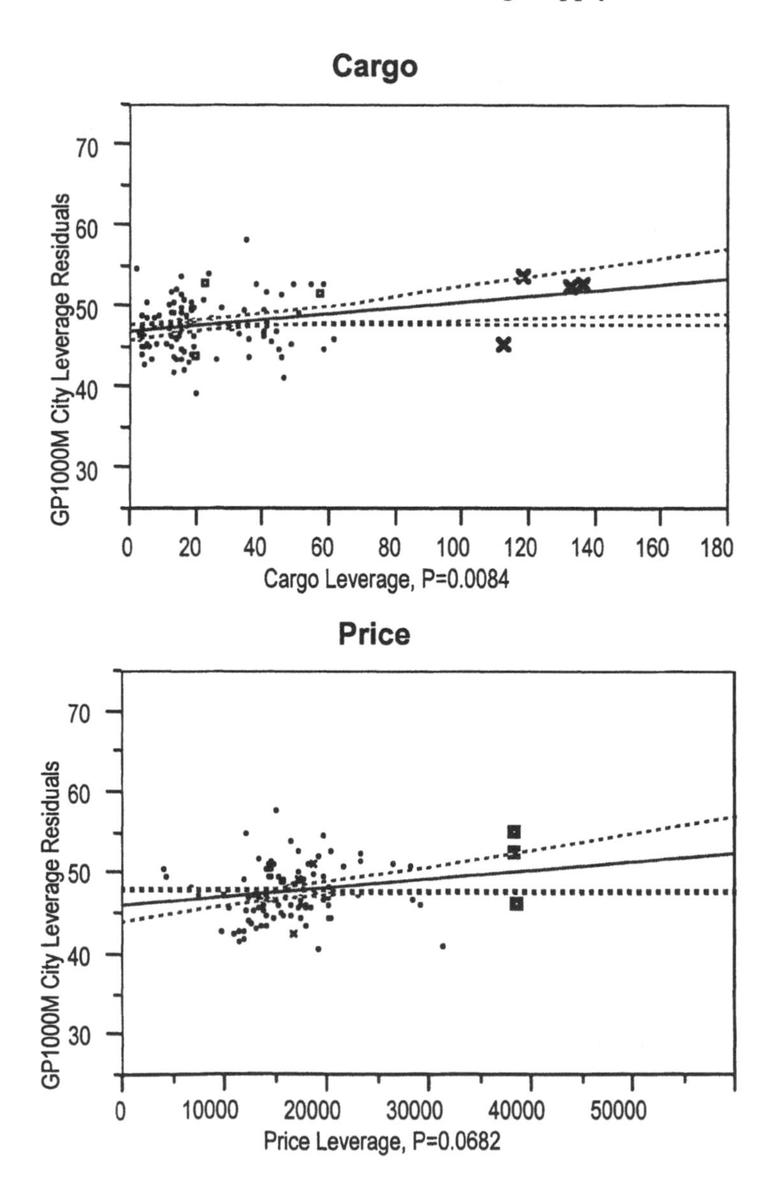

From looking at the leverage plots, it seems that just a small subset of the vehicles determines the value of these added variables that look significant, or at least close to significant. If we set aside the four vans and the three expensive cars on the right of the leverage plot for *Price* (BMW-735i, Cadillac Allante, Mercedes S), the regression coefficients for both *Cargo* and *Price* are no longer significant. The diminished size of these slopes when we remove the influential cases suggests that the original model over states the significance for these two factors, relying too much on a small subset of the available data.

Response: GP1000M City

RSquare	0.848
Root Mean Square Error	3.303
Mean of Response	47.088
Observations	99

Parameter Estimates

Term	Estimate	Std Error	t Ratio	Prob>|t|
Intercept	13.8122	2.1089	6.55	<.0001
Weight(lb)	0.0086	0.0014	6.35	<.0001
Horsepower	0.0782	0.0162	4.83	<.0001
Cargo	0.0288	0.0272	1.06	0.2931
Seating	-0.2755	0.4518	-0.61	0.5436
Price	0.0000	0.0001	0.52	0.6027

Here are the leverage plots for *Cargo* and *Price*, with the seven outlying or leveraged points excluded. Neither slope estimate differs significantly from zero based on the reduced data set.

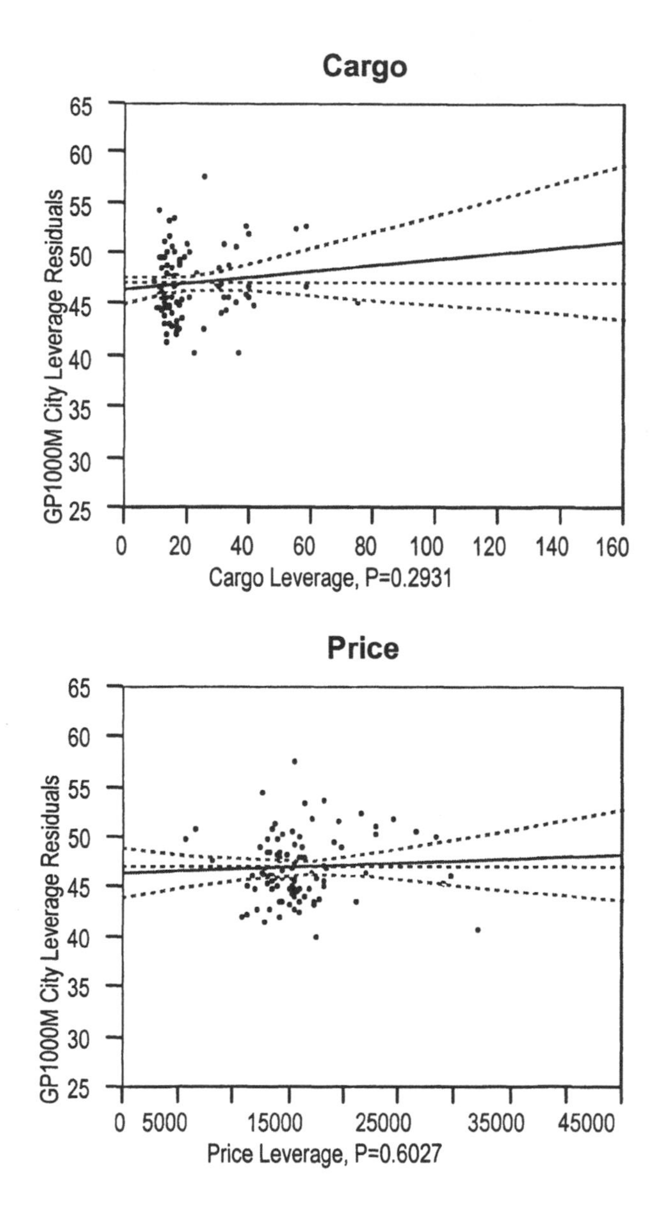

A caveat is in order. One can easily get carried away with this approach, removing any leveraged observation. This approach will fail, as we learned in the cottage example. Leveraged observations tell us the most about the slope, so long as they are consistent with the rest of the data. We have to decide if the leveraged points should be retained, and this choice is seldom straightforward.

The design team has a good idea of the characteristics of the car, right down to the type of leather to be used for the seats. However, the team does not know how these characteristics will affect the mileage. The goal of this analysis is twofold. First, we need to learn which characteristics of the design are likely to affect mileage. The engineers want an equation. Second, given the current design, we need to predict the associated mileage.

The analysis is done on a scale of gallons per 1000 miles rather than miles per gallon. With regard to the questions of this analysis:

(1) As expected, both weight and horsepower are important factors that affect vehicle mileage. Adding the power-to-weight ratio to a simple regression equation leads to a better fit and more accurate (as well as shifted) predictions. The inclusion of other factors that are less interpretable, however, ought to be treated with some skepticism and examined carefully. Often such factors appear significant in a regression because of narrow features of the data being used to fit the model; such features are not likely to generalize to new data and their use in prediction is to be avoided.

(2) Using the equation with just the weight and the horsepower of the car as predictors, we estimate the mileage of the car to be in the range

$$[57.9 - 72.1]\ GP1000M \quad \Rightarrow \quad [13.9 - 17.3]\ \text{MPG}$$

Prediction intervals (like confidence intervals) easily handle a transformation: just transform the endpoints of the interval (here, by taking the reciprocal and multiplying by 1000).

Other examples will show that regression models are more easily interpreted when the predictors are at least close to being uncorrelated. In this example, we can do a few things to make the predictors less correlated, but still predict the response. For example, *based on knowing something about cars*, consider using the power-to-weight ratio, *HP/Pound*, rather than *Horsepower* itself as a predictor. This ratio is not so correlated with *Weight*. The correlation between *Weight* and *HP/Pound* is 0.26, whereas that between *Weight* and *Horsepower* itself is 0.75. Typically, whenever companies make a heavier car, they also increase its horsepower. However, they do not tend to increase the power-to-weight ratio, and so the correlation is smaller. Using the power-to-weight ratio in place of horsepower alone yields a multiple regression with a comparable R^2=0.85. The *t*-statistic for the coefficient of *Weight* is much higher.

Term	Estimate	Std Error	t Ratio	Prob>|t|
Intercept	0.6703	2.0472	0.33	0.7440
Weight(lb)	0.0125	0.0006	20.71	<.0001
HP/Pound	270.7381	36.2257	7.47	<.0001

As a supplement to this analysis, we have included a somewhat newer version of this data set with information about cars produced in the 1993 model year. To see how well you are following along and get some practice, compare an analysis based on that data (car93.jmp) to what we found here. For example, what is the relationship between miles per gallon and weight? Is it nonlinear in that model year as well? Here is the plot for the 1993 cars along with summaries of the linear model and transformed model. Compare these to the fits for the 1989 models.

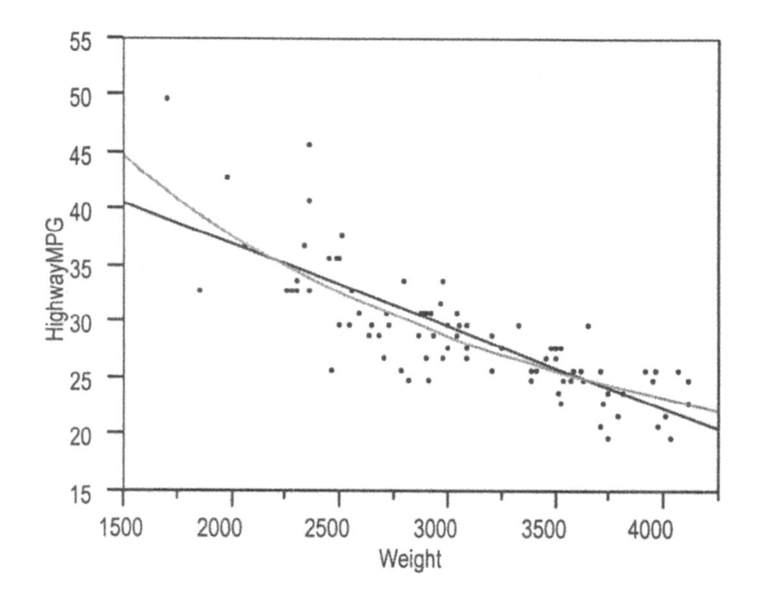

Parameter Estimates, Linear

Term	Estimate	Std Error	t Ratio	Prob>|t|
Intercept	51.601365	1.73555	29.73	<.0001
Weight	-0.007327	0.000555	-13.21	<.0001

Parameter Estimates, Reciprocal

Term	Estimate	Std Error	t Ratio	Prob>|t|
Intercept	0.0099398	0.001845	5.39	<.0001
Weight	0.0000083	5.9e-7	14.06	<.0001

If you consider other models, notice that the more recent data file includes some predictors not used here, but most of the factors are common to the two data sets.

Class 5. Collinearity.

Correlation among the covariates complicates the interpretation of coefficients in multiple regression. The size of the coefficient for one variable depends upon which other correlated variables also appear in the regression model. This class considers methods for understanding how the correlations among the explanatory variables affect the regression coefficients.

This class also introduces certain tests useful in multiple regression. Since multiple regression features several predictors, one requires additional testing methods beyond those introduced in Class 3 for bivariate regression. These new tests answer important questions such as "Does this collection of explanatory factors really explain anything?" and "Is this model better than that model?"

Topics

Anova table for the regression; F test
Interpretation of t-test in multiple regression
Comparison of t and F tests
Comparison of regression models via the partial F-test
Collinearity and variance inflation factor (VIF)

Examples

1. Stock prices and market indices
2. Improving parcel handling

Key application

Identifying the factors that determine profitability. Regression is often used to try to learn why some companies are more profitable than others. Data is gathered for numerous variables on a cross section of companies and then put into regression. Typically, many predictors in such models (such as the debt-to-equity ratio, modernization measures, productivity scores, etc.) are highly correlated. This correlation among the predictors makes it difficult (some would say impossible) to sort out which factors are most important. Collinearity is the source of this confusion and the topic for this class.

Definitions

Collinearity (multicollinearity). Correlation among predictors in a regression model.

Variance inflation factor (VIF). An index of the effect of collinearity upon the variance of coefficient estimates in regression. A VIF of 9 implies that the standard error of the associated coefficient estimate is 3 times larger than it would be were this predictor unrelated to the other predictors. Similarly, a VIF of nine implies that the confidence interval for that coefficient is 3 times longer than it would be were this predictor uncorrelated with the others.

Concepts

Collinearity. Collinearity is an aspect of regression that can only occur when you have two or more independent variables (predictors, covariates, explanatory variables, X's) in the regression. Collinearity simply means that two or more of your predictors are themselves correlated. It is a "column-based" problem: one predictor column in the spreadsheet looks from the regression perspective to be very similar to another or to a combination of others. The net consequence is that the regression coefficient values are unstable, meaning that a very small change in the data could lead to a substantially different appearing regression equation. The main diagnostics for collinearity are

(1) the scatterplot matrix, where you would see strong linear relationships between some of the predictors,

(2) counterintuitive signs on the regression coefficients,

(3) large standard errors on some of the regression coefficients caused by the fact that there is very little information to estimate them,

(4) narrow leverage plots in which all the observations tend to fall in a narrow band in the middle of the plot,

(5) high variance inflation factors (VIF) which tell you how much the variability of the regression coefficient has been inflated due to the presence of collinearity, and

(6) low values for t-statistics despite a significant overall fit, as measured by the F-statistic.

Recall that low variability of an estimator is desirable because that indicates high precision in the estimator itself.

Collinearity need not be a disaster. If your sole objective is to use your model for prediction and you are trying to predict in the range or the observed data (interpolation) then these predictions are *not* ruined by collinearity. The predictions are OK because, even though the regression model is unstable as a whole, it is in fact stable in the region where you have collected data. A note of caution: if you extrapolate from your data in the presence of collinearity, then you are courting disaster because now you are trying to predict in the unstable portion of the regression plane. The model does warn you by giving very wide prediction intervals.

There are a variety of approaches for dealing with collinearity. First, as stated above, it may not even be a problem if all you want to do is predict in the range of your data. Second, you may want to combine collinear variables into a single index. If you do this, then make sure that the new variable has a meaningful interpretation. A third and somewhat drastic alternative is simply to remove some collinear variables from the model.

The anova summary and partial F-tests. A multiple regression has at least two predictors. Questions concerning the values of regression coefficients, the partial slopes, can take many forms. The main question is to ask what is the relationship between that particular predictor and the response (Y). If you ask a question about *one* predictor then it is answered by the t-statistic. If you want to ask a question about *all* the predictors at once (whether all the slopes are zero so that there is no useful explained variation), then it is answered by the F-test from the anova summary. A significant F-test says that *some combination of predictors is useful* in the model but it doesn't tell you which. Clearly, many intermediate questions could be asked about subsets of predictors to learn if some subset is important in the regression. Such questions are answered by the partial F-test, the mechanics of which will be shown to you in the cases that come and also appeared at the end of the previous case involving cars (it is one of the few tests that JMP will not do for you).

There is a continuum of tests depending on how many variables you are testing:

Number being tested	*Test procedure*
One	t-test
Subset	Partial F-test
All	F-test

All of these tests are variations on the partial F-test, since one coefficient alone is a subset, and all of the coefficients also form a subset. The traditional regression output shows these special cases because the associated questions are so common.

Heuristics

What is a degree of freedom (as found in the anova table)? These counts are used in the numerator and denominator of F-statistics. In *Basic Business Statistics*, we measured the variation in a sample by computing the average squared deviation from the mean. Such a sample-based estimate is said to have "n-1 degrees of freedom." Why not n? We "lose" a degree of freedom because we had to first compute the sample mean in order to center the data to compute the sample variance. Fixing the sample mean "ties down" one observation in the sense that given $\bar{X}$ and any n-1 values, you can figure out the missing observation. In regression, estimating the constant and slopes cost one degree of freedom for each.

What's going on in the anova table? The process of computing slopes in a regression can be thought of as turning observations into slopes. Each time we estimate a slope, we "tie down" one observation in the sense of losing a degree of freedom.

Each observation is responsible for some of the variation in our data. When we add a variable to a regression, we explain some of this variation. Unless the variation explained by adding variables is more than the variation "explained" by the observations themselves, why bother turning them into slopes? The data themselves are simpler. The F-ratio (or test) in the anova table measures the "efficiency" of this conversion process. Large F-ratios imply that each slope explains much more variation than a single observation.

Potential Confusers

What's the difference between collinearity and autocorrelation? Collinearity is caused by a relationship between columns in the spreadsheet whereas autocorrelation is a relationship between rows.

Why bother with all the F-tests? Why not do a whole bunch of t-tests? The reason why F-tests are useful is that they are honest, simultaneous tests. "Honest" in this context means that you can credibly analyze more than one variable at a time. This is not so with doing many individual t-tests. The problem is that it is not clear how to combine the p-values for all the individual tests you have done into a single overall p-value. If each t-test has a 5% chance for error, then what is the probability for making an error in any of 10 t-tests? The Bonferroni inequality described in *Basic Business Statistics* suggests that it could be as large as 50%. Thus, to be safe, we can use the *F*-test or use a smaller threshold for the *p*-values. This issue will be dealt with in much more depth when we do the analysis of variance in Class 9.

Stock Prices and Market Indices

Stocks.jmp

How is the return on Walmart stock related to the return on various stock market indices, such as the S&P 500?

This data set consists of monthly returns on several stocks (IBM, Pacific Gas and Electric, and Walmart) along with two market indices, the S&P 500 and the value-weighted index (VW) of the full market. The returns are monthly, spanning the 25 years 1975-1999. The marginal regression coefficient of a stock regressed on the market is known in Finance as the beta for that stock. Beta is used to assess the performance and risk of a stock. Stocks with beta less than one vary less than the market, whereas those with large beta magnify the ups and downs of the market.

The analysis begins with correlations and a scatterplot matrix, using *Date* as the last variable so that the last column of plots in the scatterplot matrix shows each series versus time. The correlation matrix offers a numerical summary of this matrix of plots. The extreme correlation between the returns on the value-weighted index and the S&P 500 index is apparent. A clear outlier is also present (use the date as a point label to easily find the date of this event).

Correlations

	Wal Mart	VW Return	SP Return	Date
Wal Mart	1.000	0.597	0.578	-0.080
VW Return	0.597	1.000	0.987	0.012
SP Return	0.578	0.987	1.000	0.055
Date	-0.080	0.012	0.055	1.000

Since we put the date as the last variable in the correlations, the last column of plots in the associated scatterplot matrix shows a sequence plot of each variable. A quick glance down the last column shows a time series plot for each of the returns.

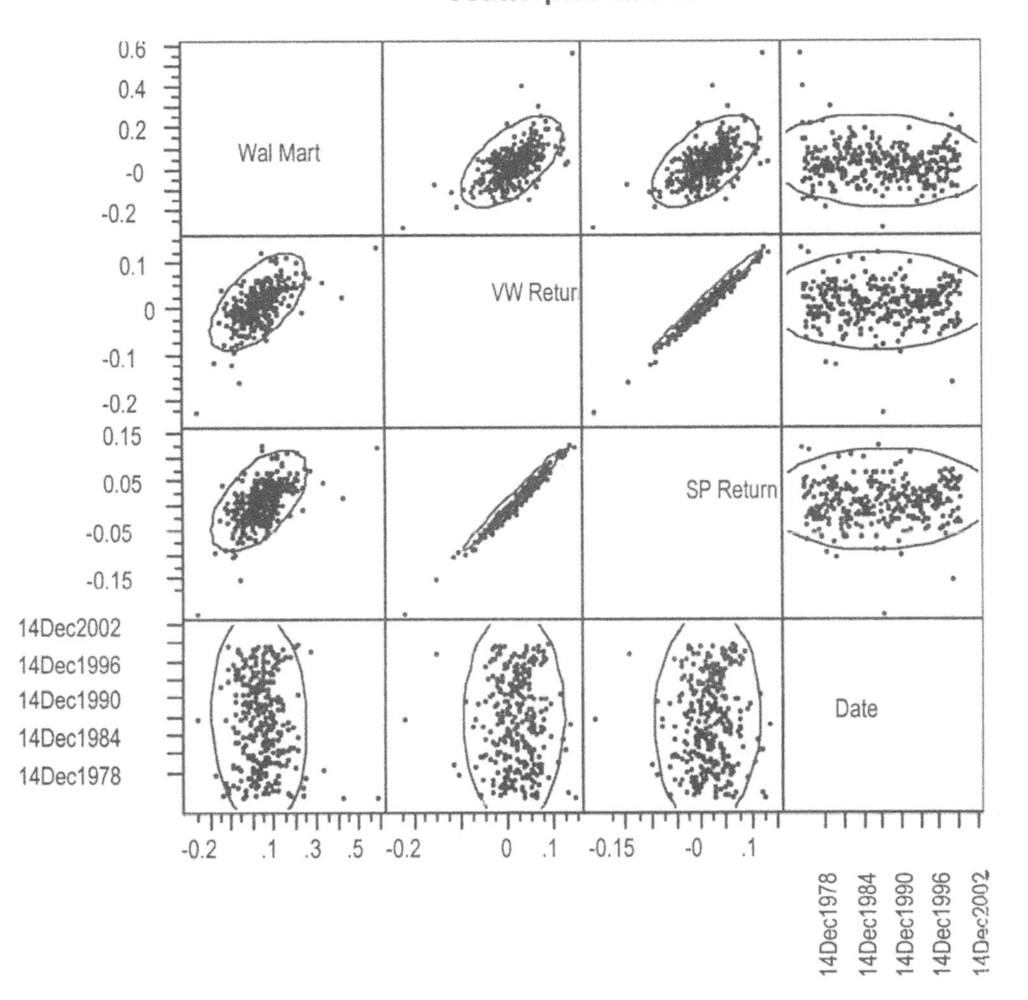

Time Series Plots

The multivariate view of data available from the *Correlation of Y's* platform also allows us to check for outlying periods. Here we have omitted the sequence number since we do not want early or late observations flagged as "unusual". The score for each observation is shown plotted on the row index, which in this case gives a time series plot of the outlier distance of the stock market data. This distance measures how distinct the returns in a given month are from those in other months, considering the relationship among all of the variables simultaneously. The returns for October, 1987, are unusual, but not the most unusual.

Correlations

	VW Return	SP Return	Wal Mart
VW Return	1.000	0.987	0.597
SP Return	0.987	1.000	0.578
Wal Mart	0.597	0.578	1.000

Outliers

Outlier Distance [Mahalanobis]

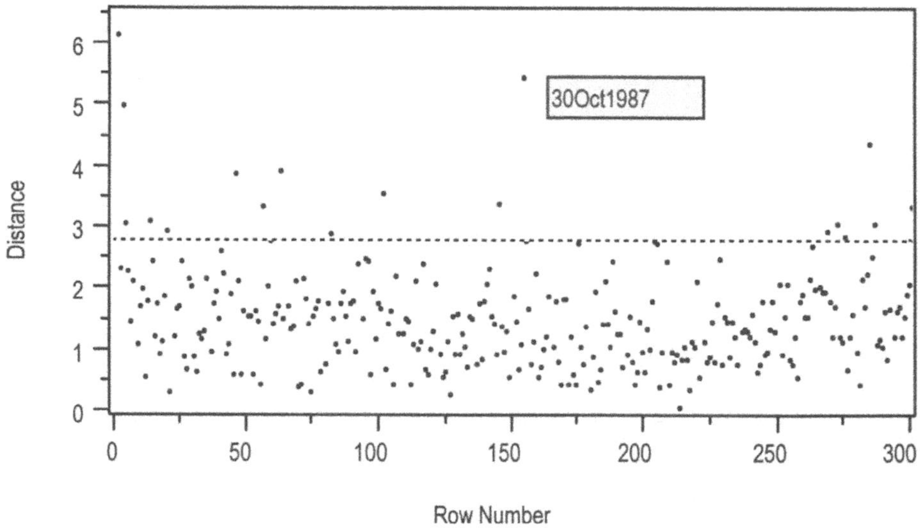

An initial regression analysis of the monthly returns for Walmart on those of the S&P500 shows a strong, positive relationship between the two. The fitted slope, known as the beta coefficient for Walmart, is very significant, falling over 12 standard errors above zero. This output is from the *Fit Model* platform; this tool makes it easy to extend the model to a multiple regression. The F-test in this model is equivalent to the t-test for the single slope ($t^2 = F$, with the same p-value).

Response: WALMART

RSquare	0.334
Root Mean Square Error	0.074
Mean of Response	0.032
Observations	300

Analysis of Variance

Source	DF	Sum of Squares	Mean Square	F Ratio
Model	1	0.8297	0.8297	149.4896
Error	298	1.6539	0.0056	Prob > F
C. Total	299	2.4836		<.0001

Parameter Estimates

Term	Estimate	Std Error	t Ratio	Prob>|t|
Intercept	0.018	0.004	4.15	<.0001
SP Return	1.228	0.100	12.23	<.0001

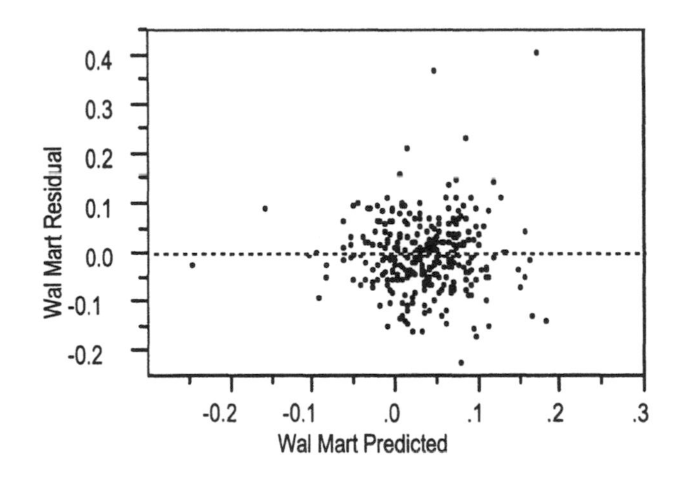

A question often asked of a stock in Finance is whether its beta is larger than one. Does the beta for Walmart (its regression slope) differ significantly from one? (Count the SE's, or use a confidence interval.)

The leverage plot from the multiple regression with one variable (the plot immediately below) is the same as the scatterplot produced by the *Fit Y by X* platform (shown at the bottom of the page). The latter scatterplot includes for comparison the confidence curves for the regression fit (not those for predicting an individual return).[1]

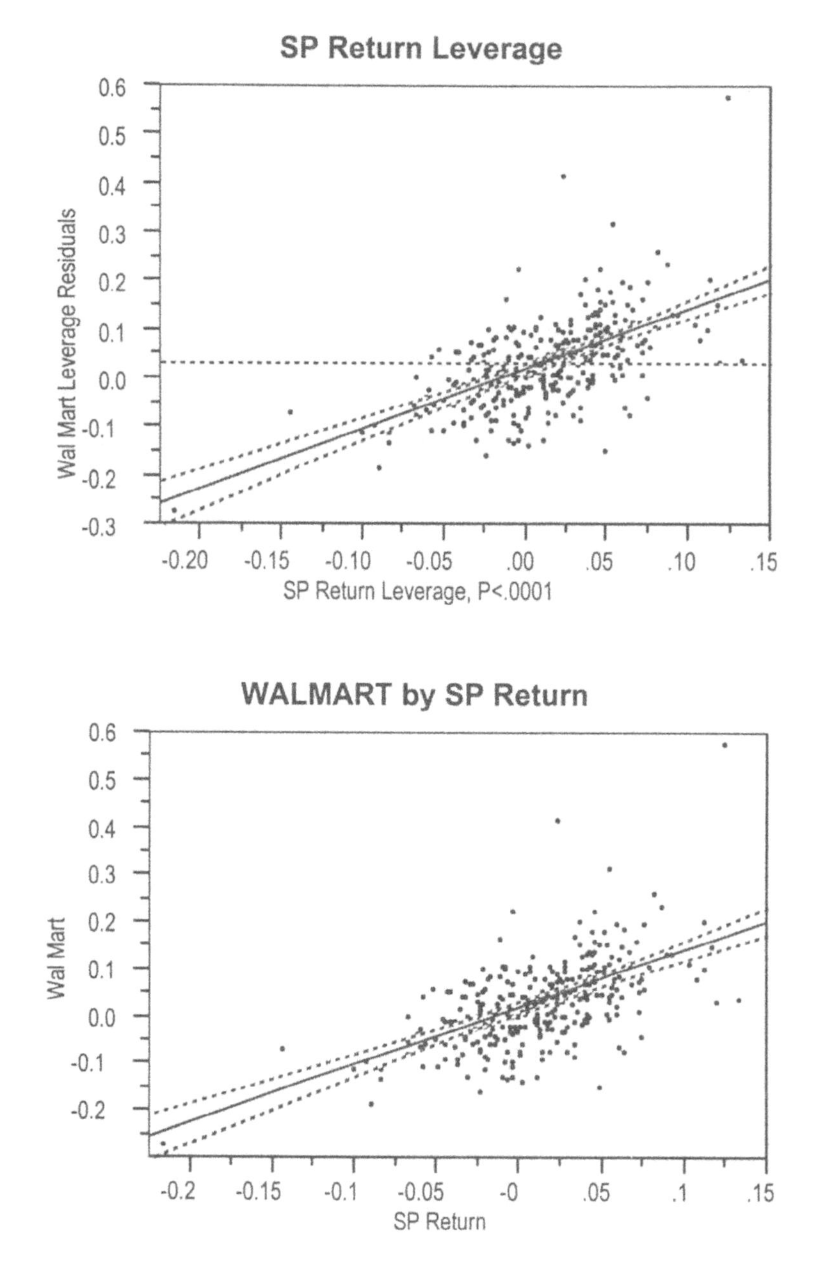

[1] To obtain these confidence curves, press the button that labels the linear fit. Then choose the "Confid Curves fit" option. Since these curves indicate confidence intervals for the fit, rather than individual observations, the intervals do not contain much of the data.

In contrast to this simple regression, the multiple regression using both market indices (S&P 500 and value-weighted) suggests at first glance that the relationship between Walmart's stock and the S&P 500 index is nonexistent (the slope for this coefficient is not significantly different from zero). Note the small, but significant, change in the overall R^2 of the model. (Look at the t-statistic for the single added variable.)

Response: WALMART

RSquare	0.361
Root Mean Square Error	0.073
Mean of Response	0.032
Observations	300

Analysis of Variance

Source	DF	Sum of Squares	Mean Square	F Ratio
Model	2	0.89601	0.44801	83.8123
Error	297	1.58757	0.00535	**Prob > F**
C. Total	299	2.48359		<.0001

Parameter Estimates

Term	Estimate	Std Error	t Ratio	Prob>|t|
Intercept	0.012	0.005	2.63	0.0091
SP Return	-0.871	0.604	-1.44	0.1502
VW Return	2.082	0.591	3.52	0.0005

The residual plot is very similar to that from the simple regression with just one predictor and is not shown here.

The leverage plots show the presence of several outliers and leveraged observations (use the point labeling to find the dates of these events). The substantial collapse of the points into the center of the horizontal axis in each case is the result of the high correlation between the two indices. Effectively, collinearity leads to less precise slope estimates because of a loss of uncorrelated variation along the x-axis.

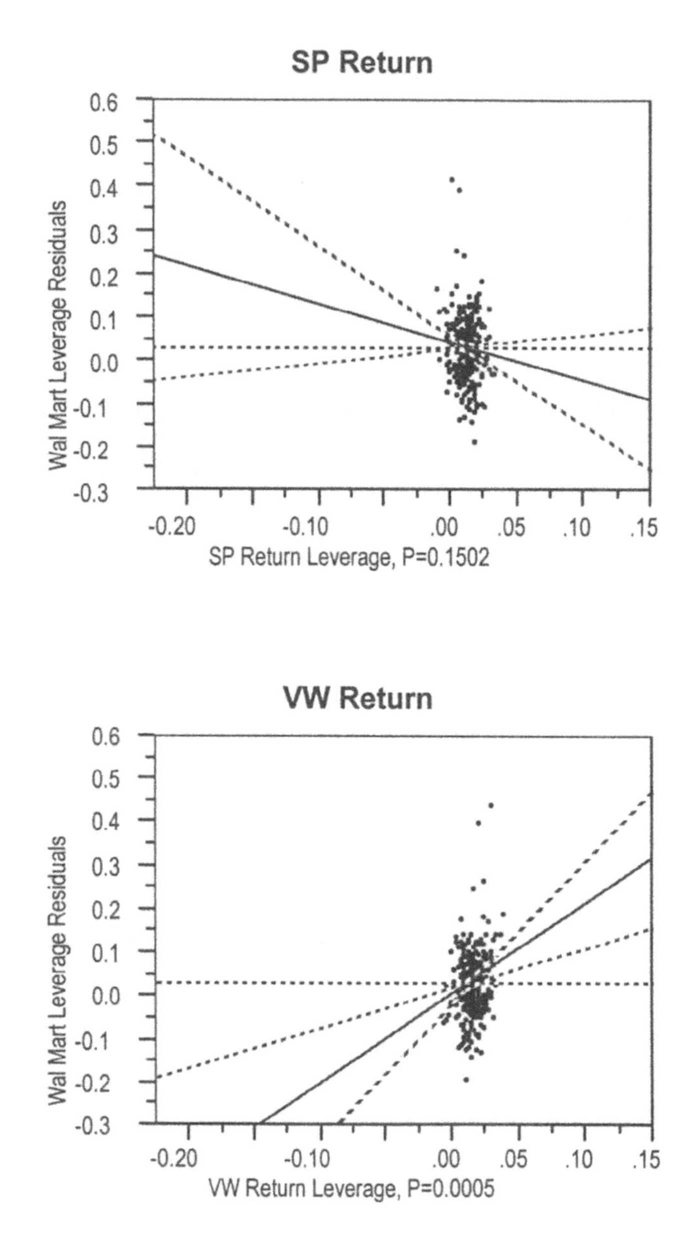

JMP supplies another diagnostic which is useful for measuring the degree of collinearity in a regression (all are not so obvious as in this example). The diagnostic is known as a *variance inflation factor* (VIF) and can be added to the table of the parameter estimates.

Parameter Estimates

Term	Estimate	Std Error	t Ratio	Prob>|t|	VIF
Intercept	0.012	0.005	2.63	0.0091	.
SP Return	-0.871	0.604	-1.44	0.1502	37.5
VW Return	2.082	0.591	3.52	0.0005	37.5

The VIFs in this example indicate that the variance of each slope estimate has been inflated multiplicatively by a factor of about 37.5 because of the severe collinearity. That implies that the standard error is larger by a factor of about 6.1 (= $\sqrt{37.5}$) due to collinearity

These results are consistent with financial theory. The value-weighted index is a more complete summary of the market, capturing the performance of numerous smaller companies omitted from the S&P index. Other standard residual plots also point to some outlying values but do not suggest the role of collinearity. As always with time series data, take a look at the residuals plotted over time.

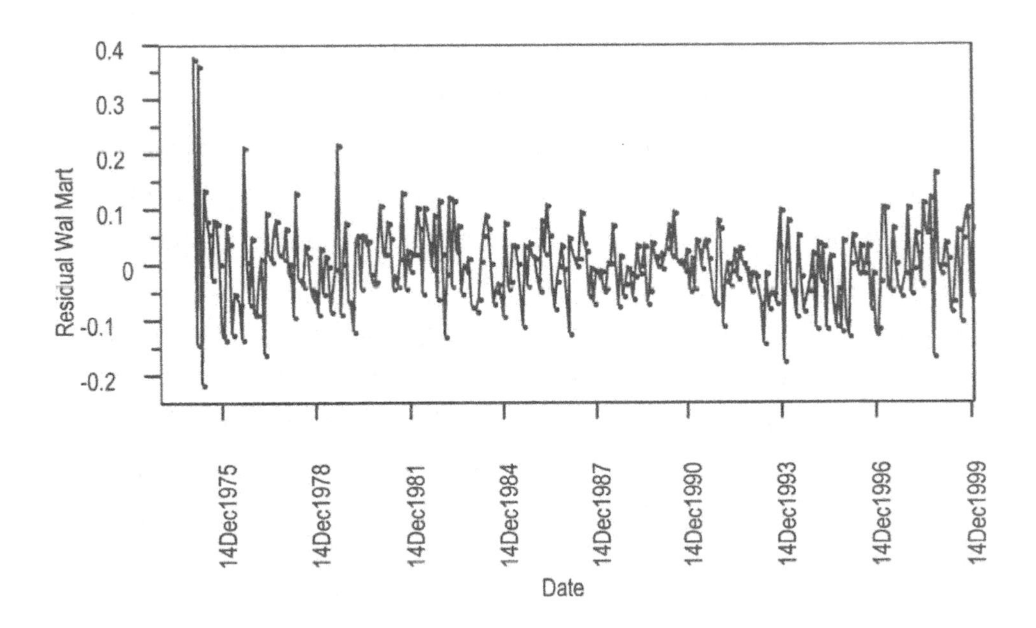

Finally, a check of the quantile plot shows that the residuals (but for the one large outlier visible above) appear close to normal.

How is the return on Walmart stock related to the return on various stock market indices, such as the S&P 500?

Multiple regression becomes very difficult to interpret when high collinearity appears among the predictors (the *X*'s). Using just the S&P 500, though, we can see that Walmart's stock has a high correlation with this index.

The regression of Walmart's stock price on either the S&P 500 or value-weighted index makes sense. Both regression models have similar slopes, 1.23 or 1.24. The slope associated with the regression of the price of an individual stock on a market index is known in finance as the beta of the stock.

An important question is whether the beta for Walmart, the coefficient of its return regressed on the return for the market, is different from one. Here is the relevant regression obtained by regressing returns for Walmart on just the value-weighted index alone:

Parameter Estimates

Term	Estimate	Std Error	t Ratio	Prob>|t|
Intercept	0.015	0.004	3.27	0.0012
VW Return	1.241	0.097	12.84	<.0001

Don't be confused by the large t-statistic for the value-weighted index. It measures the distance of the slope estimate from zero on a standard error scale. To judge whether the fitted beta is different from one, we need instead the number of standard errors between the estimate and one. From the output, we can see that the slope is about four standard errors from one, far above the threshold need for significance. The beta for Walmart is larger than one by a significant margin.

Summary. Methods for recognizing collinearity

(1) Test statistics:

The combination of a significant overall fit (big F-statistic) with small *t*'s is a common occurrence when the predictors are collinear.

(2) Graphical:

Narrow leverage plots indicate most of the variation in a factor is redundant.

(3) Variance inflation factors:

Summarize the degree of collinearity. What's a big VIF? Certainly, 10 indicates substantial collinearity, but the harm depends on the resulting standard error.

Summary. Methods for dealing with collinearity

(1) Suffer:

If prediction within the range of the data (interpolation) is the only goal, not the interpretation of coefficients, then leave it alone. Make sure, however, that the observations to be predicted are comparable to those used to construct the model. You will get very wide prediction intervals otherwise.

(2) Transform or combine:

Replacing *HP* by *HP/Weight* works well with the cars as a means to reduce collinearity. Knowing to do this with the car example presumes a lot of substantive knowledge.

(3) Omit one:

With the two stock indices, this makes a lot of sense. It is usually not so easy to simply remove one of the correlated factors. Often, for example, one objective of a regression analysis is to compare the effects of two factors by comparing their slopes in a multiple regression. In that case, you want to have both factors in the model.

Improving Parcel Handling

Parcel.jmp

A parcel delivery service would like to increase the number of packages that are sorted in each of its hub locations. Data recently sampled at 30 centers measure the number of sorts per hour. Three factors that the company can control and that influence sorting performance are the number of sorting lines, the number of sorting workers, and the number of truck drivers.

What should the company do to improve sorting performance?

In modeling problems of production, it is common to use a multiplicative model to represent the production function. Expressed in logs, the coefficients of such a regression model are interpreted as *elasticities*. For example, suppose that we assume that the multiplicative relationship between the response Y and two explanatory variables X_1 and X_2 is (here δ represents the random error)

$$Y = \alpha X_1^{\beta_1} X_2^{\beta_2} \delta$$

This equation is a standard Cobb-Douglas production function. If we take logarithms of both sides of the equation we obtain a more familiar, additive regression equation

$$\begin{aligned}\log Y &= \log(\alpha X_1^{\beta_1} X_2^{\beta_2} \delta) \\ &= \log\alpha + \log X_1^{\beta_1} + \log X_2^{\beta_2} + \log\delta \\ &= \beta_0 + \beta_1 \log X_1 + \beta_2 \log X_2 + \varepsilon\end{aligned}$$

In this form, the slopes of the multiple regression of log Y on log X_1 and log X_2 estimate the elasticities of the response with respect to changes in X_1 and X_2. For example, every 1% increase in the predictor X_1 on average produces a β_1% change in the response.[2]

[2] Since you take the log of *both* sides of this equation, you can use any type of log (e.g., base 10 or natural logs) without changing the interpretation of the slopes. For this case, we will use natural (base e) logs.

For this data set, we consider a multiplicative model using the three factors. Both the scatterplot matrix and correlation matrix indicate that the explanatory factors are inter-related, though not to the degree seen in the previous example with the two stock indices. If the variables are considered one at a time, the number of lines appears to have the highest correlation with the number of sorts, but the number of workers has the highest elasticity, as seen in the fitted model on the next page.

Correlations

Variable	Log Sorts/Hr	Log # Lines	Log # Sorters	Log # Truckers
Log Sorts/Hr	1.00	0.90	0.84	0.79
Log # Lines	0.90	1.00	0.80	0.65
Log # Sorters	0.84	0.80	1.00	0.60
Log # Truckers	0.79	0.65	0.60	1.00

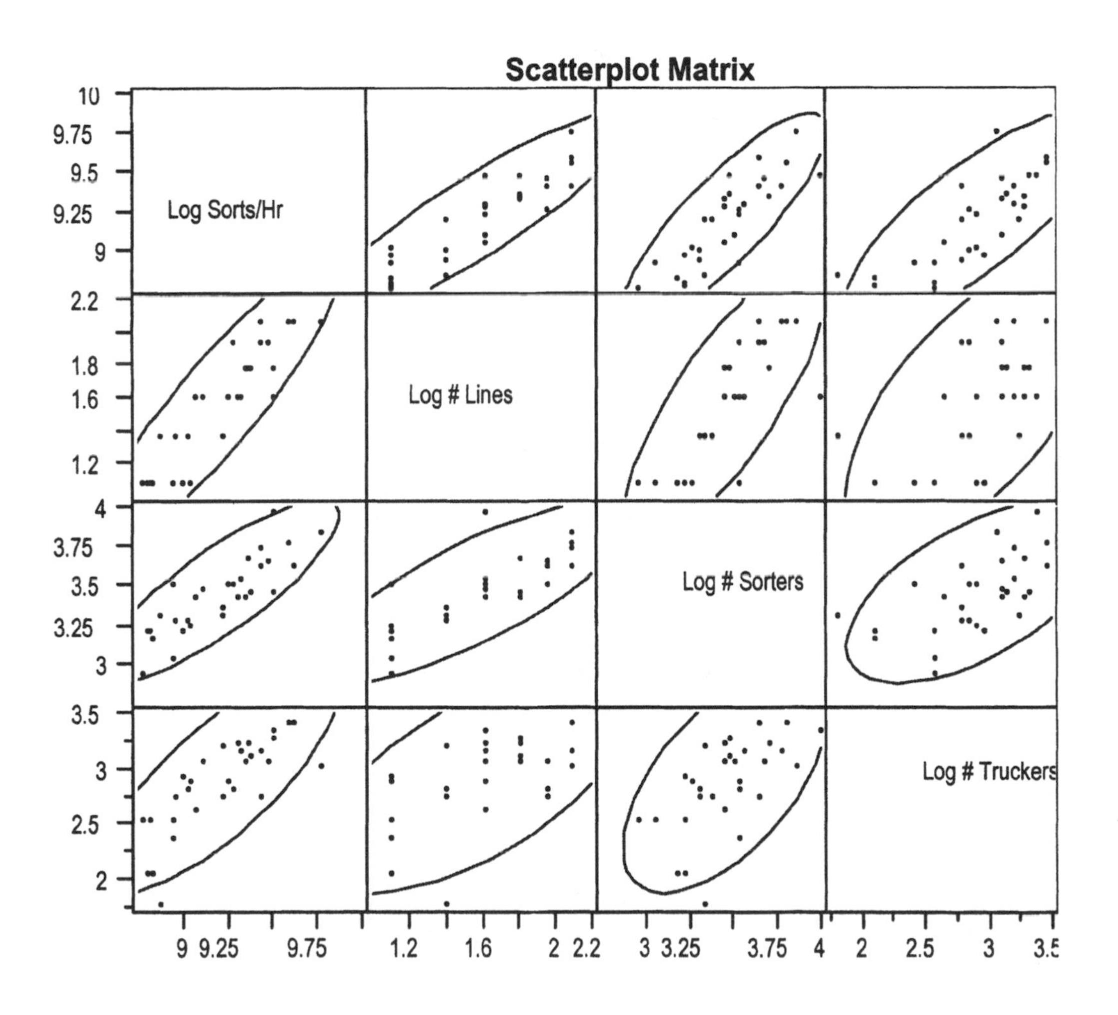

Here is the fit obtained by regressing the log of the number of sorts per hour on the log of the number of lines. The marginal elasticity of sorts per hour with respect to the number of lines is 0.7%. For every 1% increase in the number of lines, this model predicts a 0.7% increase in the number of sorts per hour.

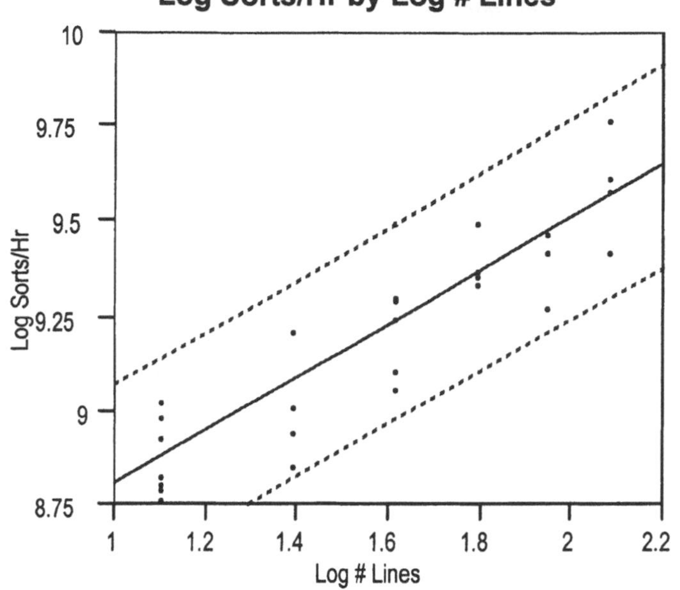

Summary of Linear Fit

RSquare	0.809
Root Mean Square Error	0.123
Mean of Response	9.197
Observations	30

Analysis of Variance

Source	DF	Sum of Squares	Mean Square	F Ratio
Model	1	1.783	1.783	118.3478
Error	28	0.422	0.015	**Prob>F**
C Total	29	2.204		<.0001

Parameter Estimates

Term	Estimate	Std Error	t Ratio	Prob>|t|
Intercept	8.10	0.10	78.65	<.0001
Log # Lines	0.70	0.06	10.88	<.0001

However, does it make sense to add more sorting lines without also adding the people to do the sorting or without considering the number of lines used? In either case, what about the need for trucks to carry away the sorted packages?

Multiple regression estimates the elasticities in the *joint* production function. These elasticities attempt to separate, for example, the effect of increasing the number of sorting lines from the concomitant differences in the number of workers and truckers that typically come with more sorting lines.

In the multiple regression, the partial elasticity of the number of sorts per hour with respect to the number of lines is about half what it was when considered alone (0.37 versus 0.70). Had management interpreted the elasticity of the original model as indicating that adding a line alone was going to produce the effect implied by the marginal elasticity, they would make a serious error. The effect of adding a line, without making changes in the numbers of works, is much smaller. The marginal elasticity from the bivariate model is so much larger since it includes all of the other things that typically happen when lines were added in the past, namely workers were added as well.

Response: Log Sorts/Hr

RSquare	0.911
Root Mean Square Error	0.087
Mean of Response	9.197
Observations	30

Term	Estimate	Std Error	t Ratio	Prob>|t|	VIF
Intercept	6.91	0.31	22.46	0.000	0.0
Log # Lines	0.37	0.08	4.51	0.000	3.2
Log # Sorters	0.32	0.12	2.73	0.011	2.9
Log # Truckers	0.22	0.05	4.18	0.000	1.8

Analysis of Variance

Source	DF	Sum of Squares	Mean Square	F Ratio
Model	3	2.0083	0.6694	88.75
Error	26	0.1961	0.0075	**Prob>F**
C Total	29	2.2044		<.0001

It appears clear in this example that the addition of the log of the number of workers and the log of the number of truckers improves the fit when the model is compared to that using only the log of the number of lines. In the multiple regression, the t-statistics for both added predictors are large, and R^2 increases from 0.809 to 0.911.

To be precise, we can test this conjecture. Since we have added *two* factors, we need to use the partial F-statistic to see if the addition of *both* factors has improved the fit. The partial F measures the size of the change in R^2 relative to the residual variation:

$$
\begin{aligned}
\text{Partial F} &= \frac{\text{Change in } R^2 \text{ per added variable}}{\text{Remaining variation per residual}} \\
&= \frac{(R^2_{complete} - R^2_{reduced})/(\#\text{ predictors added})}{(1 - R^2_{complete})/(\text{Residual d.f.})} \\
&= \frac{(0.911 - 0.809)/2}{(1 - 0.911)/26} \\
&= 14.9
\end{aligned}
$$

In the absence of an effect, the partial F is near 1. Though one should use a table or the JMP calculator to see if the observed F value is significant, a quick rule identifies significant values. If the F value is larger than 4, it is significant (at the usual 5% level).[3] In this example, the F value is thus clearly quite large and indicates that the added variables represent a significant improvement of the fit. After all, both have significant t-ratios in the multiple regression, so this should come as no surprise.

[3] This rule of thumb for significance of the F value does not find all significant results. In certain cases, an F value smaller than 4 can indicate a significant effect. Any that are larger than 4, however, are significant. You might miss some significant results, however, using this rule.

Residual plots show a single large outlying observation (marked with a + in the plot below). These plots do not, however, indicate how (if at all) this outlier affects the coefficient estimates.

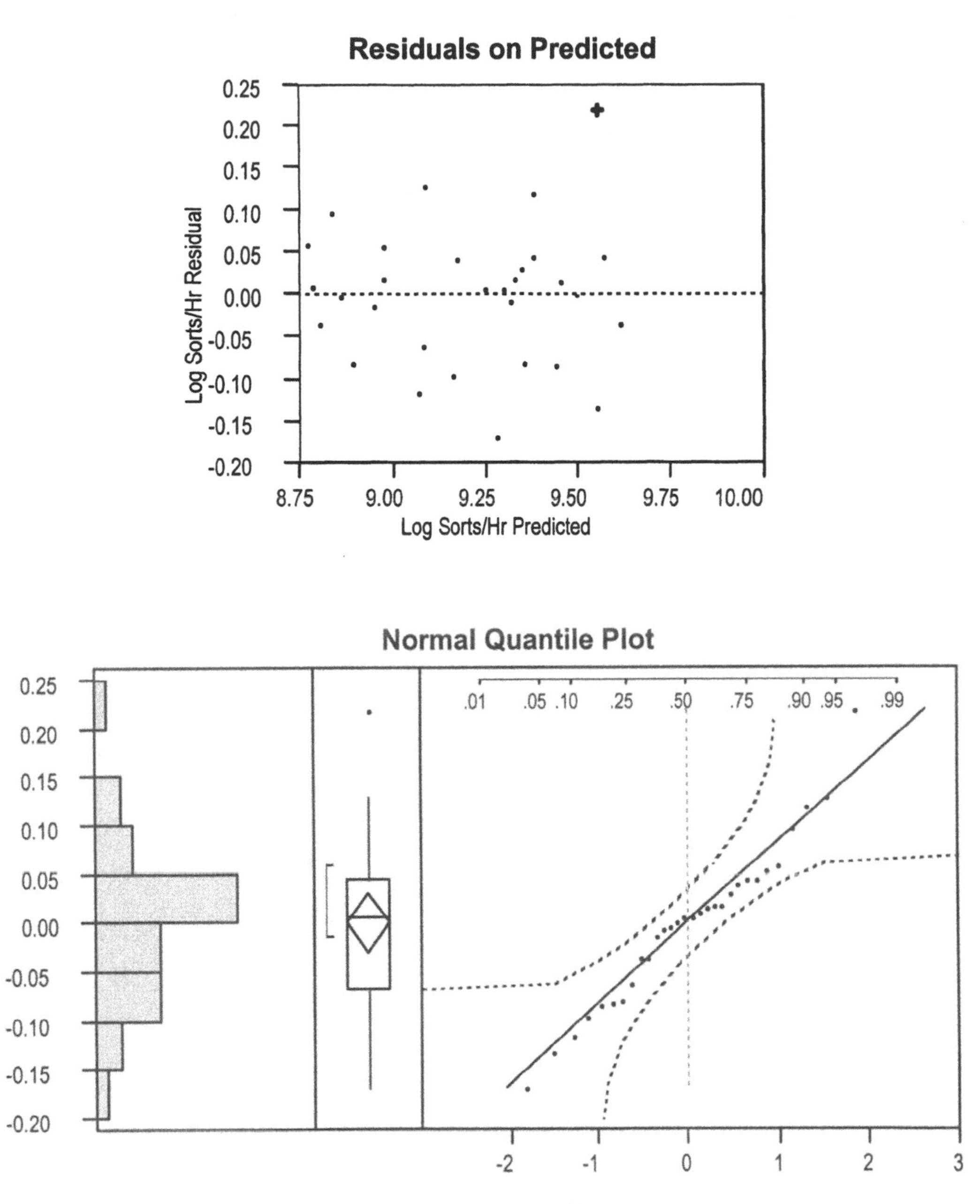

Leverage plots locate the large outlier and show that it is increasing the estimate of the elasticity for the number of sorters, but not by much. This outlier is also affecting the other estimates, though not so clearly. Another leveraged observation (marked with a x) is apparent, but its low residual implies that it is not influential. (It certainly makes the slope estimates for the elasticity of the number of lines and number of sorters more precise, however.)

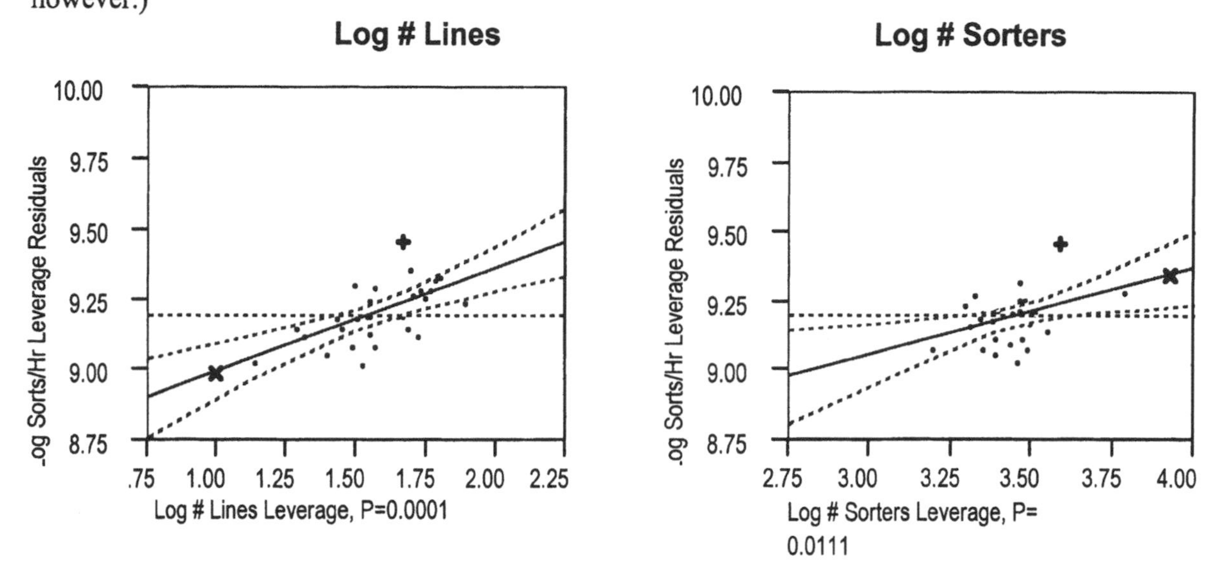

A different observation is highly leveraged, but not influential, for the log of the number of truckers.

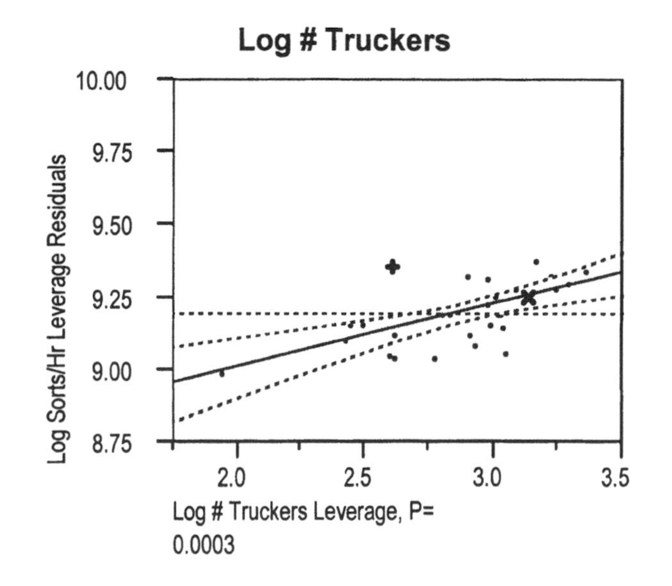

A parcel delivery service would like to increase the number of packages that are sorted in each of its hub locations. Data recently sampled at 30 centers measures the number of sorts per hour. Three factors that the company can control and that influence sorting performance are the number of sorting lines, the number of sorting workers, and the number of truck drivers.

What should the company do to improve sorting performance?

The estimated multiple regression indicates that increasing the number of sorting lines will have the highest impact upon sorting performance, but we cannot ignore the role of the number of workers and the number of truckers who assist these lines. In the sorting sites, there are about seven sorters per line and four truckers per line.

Class 6. Modeling Categorical Factors with Two Levels

Often we need to consider the effects of categorical information in a regression model. Inclusion of categorical variables allow us to model the differences between two (today) or more groups (Class 7). When the categorical variable has two levels, the group membership is easily coded in a special numerical variable, known as a dummy variable. Since these dummy variables are numerical representations of qualitative, not quantitative, information, the coefficients of dummy variables require special interpretation. The testing methods of Class 5 are valuable tools for comparisons among models using qualitative factors.

Topics

Dummy variables

Interactions and cross-product terms

Analysis of covariance

Example

Employee performance study (two groups)

Key application

Sports logos. Think back to the owners of logos for professional sports teams who were trying to determine what drives the sales of products carrying the logo. In the discussion for Class 4 , TV ratings and the number of games won were considered as possible predictors. Both are essentially continuous variables. It also seems likely that whether or not a team makes the playoffs influences sales, as would the introduction of a new design. But these variables are not continuous. Whether a team does or does not make the playoffs is answered with a "yes" or a "no," as is getting a new shirt design. Including such categorical variables in a regression, variables that take on attributes rather than numbers, is an important technique and the subject of this class.

Definitions

Analysis of covariance. Another name for a regression model that mixes continuous and categorical variables. Typically, one uses this term when the most interest lies in comparing averages of groups which are not directly comparable; the regression controls for other factors.

Categorical variable. A variable whose values determine group membership. Such variables are also known as "nominal" since they represent names. Indeed, to indicate to JMP that a variable is categorical, you pick the nominal item for that column in the spreadsheet.

Dummy variable. The calculation of the regression coefficients associated with a categorical variable is based on adding to the regression a collection of 0/1 or (in the case of JMP) –1/0/1 artificial variables known as dummy variables . Each dummy variable indicates whether an observation belongs to one of the groups associated with the categorical variable. JMP builds these automatically.

Nominal variable. A variable whose values denote names and lack a well-defined order. These are a special type of categorical variable.

Ordinal variable. A variable whose values denote ordered categories, but the differences between the categories are not necessarily comparable, as in class evaluation forms which have categories labeled "great," "OK," and "poor."

Concepts

Categorical variables. A categorical variable takes on attributes or categories rather than numbers. A "yes/no" response is a categorical variable as is "male/female." In JMP such variables should be in columns marked with an "N" indicating that they are *nominal*, a technical way of saying that we are dealing with names that have no ordering. Categorical variables are easily incorporated into a regression analysis. A categorical predictor with two levels is converted into a numerical *dummy variable* that represents group membership. We will not be too interested just yet in how this happens, but we will be extremely interested in interpreting the output from regressions that include categorical variables.

The simplest example in which to understand categorical variables occurs when you have a single categorical variable, a single continuous *X*-variable and the usual *Y*-

variable. Imagine the following spreadsheet with three columns: total merchandise sold in dollars, number of games won, and whether or not the team made the playoffs (either "yes" or "no"). Each row contains data from an NBA team. Take merchandise sold as the response variable. Running a regression of merchandise sold on number of games won tells you the effect of winning games. However, if you were to include the "making the playoffs" variable, what you are in effect doing is running two regressions, one regression for those teams that made the playoffs and one regression for those teams that didn't make the playoffs. A vital point to realize is that the two regressions are forced to produce parallel lines, lines that have the same slope. Each regression has its own intercept, but both share a common slope. The difference between intercepts is the vertical distance between the two parallel lines, and it tells you the effect of making the playoffs in terms of merchandise sold. Because these lines are parallel it implies that this difference does not depend on how many games the team has won. This is an important point to recognize; the impact of making the playoffs does not depend on the number of games won. That's all there is to including two-level categorical variables. You perform a separate regression for each category, but you force the slopes to be the same.

Interaction. What if you don't buy into the parallel lines assumption? That is, what if you believe that the impact of making the playoffs on merchandising depends on the number of games won? This also turns out to be simple to handle. You just include an interaction into the model (interacting the categorical variable "making the playoffs" with the continuous variable "number of games won"). The interaction allows each regression line its own slope as well as its own intercept. Parallel-lines regression allows different intercepts but forces a common slope. Interaction is an important conceptual (and unfortunately difficult) part of regression. It reappears in Class 10.

Heuristics

Start by fitting a model to each group.

Sometimes it is easier to start by fitting a separate regression to each of the different groups under study. JMP makes this very easy in the *Fit Y by X* view; just choose the *Grouping variable* option from the *Fitting* button at the bottom of the plot. If the fits are similar, then use the categorical variable in a multiple regression to compare the fits (i.e., get the associated SE's).

Understanding interaction. Here is a generic definition that may help to translate the concept of interaction into words. Interaction is a three-variable concept. One of these is the response variable (Y) and the other two are predictors (X). Assume that one of the X variables, call it X_1, is categorical and the other X variable X_2 is continuous. To believe that there is interaction between X_1 and X_2 is to believe the following statement: the impact of X_2 on Y depends on the level of X_1.

To make this a little less abstract, we can put in some names for these variables. Assume that we are looking at a surgical technique for removal of cancer. Call the response variable Y survival time, the X_1 variable sex (male/female), and the X_2 variable age. Interaction in this situation translates to the following clear statement:

"The impact of age (X_2) on survival time (Y) depends on your sex (X_1)"

It is a little surprising that a recent review of three leading surgical journals did not find a single regression model that included an interaction. Perhaps none of the interactions was significant; perhaps no one thought to check.

It is unfortunately the case that interaction is not a well understood idea though the above statement hopefully gives you the feeling that it is potentially very important. From the management perspective, identifying interaction is similar to identifying synergy, the whole being more than the sum of the parts.

Potential Confusers

Everything! Some people really start to lose it here. Try to separate the concepts of categorical variables, parallel regressions and interactions, from the details in JMP. The output can be pretty confusing. Try to understand the pictures first, then worry about tying the JMP output to the picture.

Employee Performance Study

Manager.jmp

A firm either promotes managers in-house or recruits managers from outside firms. It has been claimed that managers hired outside the firm perform better than those promoted inside the firm, at least in terms of internal employee evaluations.

Two candidates for promotion have comparable backgrounds, but one comes from within the firm and the other comes from outside. Based on this sample of 150 managers, which should you promote?

The results of a two-sample comparison of in-house promotions (*Origin* = "Internal") to those brought in from outside the firm (*Origin* = "External") support the conjecture that those recruited outside the firm perform better.

Mngr Rating by Origin

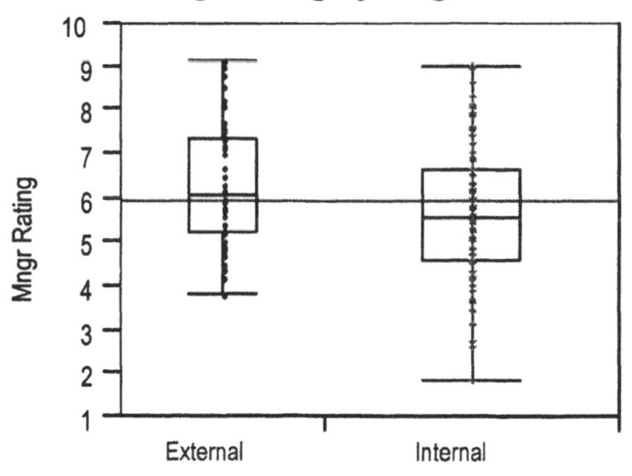

Origin

Difference	t-Test	DF	Prob>|t|	
Estimate	0.72	2.984	148	0.0033
Std Error	0.24			
Lower 95%	0.24			
Upper 95%	1.19		Assuming equal variances	

Level	Number	Mean	Std Error (pooled)
External	62	6.32	0.18
Internal	88	5.60	0.15

Surprising though it may seem, regression analysis with a categorical factor gives the same results as the initial *t*-test! To force JMP to do the regression (rather than a two-sample *t*-test), use the *Fit Model* platform with *Origin* as the single predictor.

Response: Mngr Rating

Term	Estimate	Std Error	t Ratio	Prob>|t|
Intercept	5.96	0.12	49.67	<.0001
Origin[External]	0.36	0.12	2.98	0.0033

The leverage plot is also of interest with a categorical predictor. The vertical scale is that from the prior comparison boxplot, and the groups are located at the so-called "least squares means," which in this case are the original group means (compare to the *t*-test output).

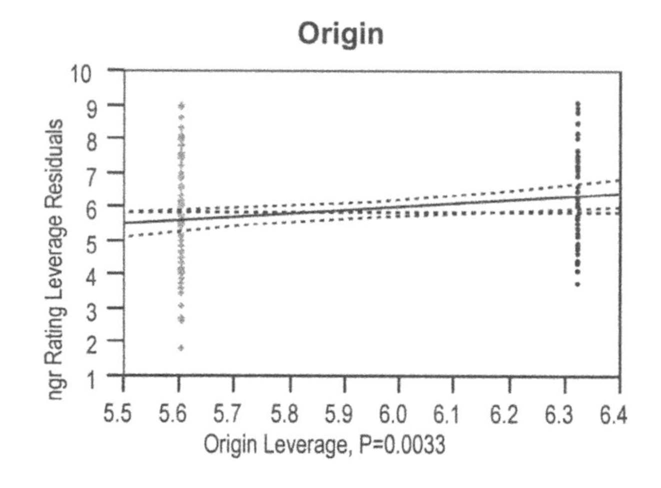

Least Squares Means

Level	Least Sq Mean	Std Error	Mean
External	6.321	0.184	6.321
Internal	5.605	0.154	5.605

Both analyses lead to the same conclusion. The t-test indicates that the external managers are doing significantly better:

Difference	t-Test	DF	Prob>|t|	
Estimate	0.72	2.984	148	0.0033

Level	Number	Mean	Std Error (pooled)
External	62	6.32	0.18
Internal	88	5.60	0.15

Equivalently, via regression (from the *Fit Model* output), we obtain the same t-statistic for the size of the difference in the averages. In this case, though, the estimated regression coefficient for *Origin* is one-half of the difference 0.72 between the two means.

Term	Estimate	Std Error	t Ratio	Prob>|t|
Intercept	5.96	0.12	49.67	<.0001
Origin[External]	0.36	0.12	2.98	0.0033

To understand the interpretation of the coefficient estimate attached to the categorical factor, write out the model for each group represented by the categorical factor. In this case, only one of the two groups appears (external, since it's first when the labels are sorted alphabetically).

External Fit = 5.96 + 0.36 = 6.32 = average for "External"

What about internal managers? To get the fit for these most easily, we have JMP-IN produce an "expanded" set of estimates. (Use the pop-up at the top of the output, and follow the estimates hierarchical menu.) With this option chosen, an additional table appears in the output as follows.

Expanded Estimates

Nominal factors expanded to all levels

Term	Estimate	Std Error	t Ratio	Prob>|t|
Intercept	5.96	0.12	49.67	<.0001
Origin[External]	0.36	0.12	2.98	0.0033
Origin[Internal]	-0.36	0.12	-2.98	0.0033

From this expanded table, we see that the fit for the internal managers is

Internal Fit = 5.96 – 0.36 = 5.60 = average for "Internal"

This fit simply subtracts 0.36 from the constant rather than adds it. Since the two coefficients simply have opposite signs, we don't really need the expanded estimates. However, in more complex models they are convenient.

Finally, we note that the coefficient of the categorical factor for both groups shows how the fitted intercept for each group differs from the overall intercept. The notation Origin[External] indicates that we are judging the effect of *Origin* being zero relative to the average of both groups, not to the other group directly.

Other factors appear to influence the rating score as well. For example, the rating is correlated with the salary of the employee. The coded scatterplot (use the *Color marker by Col...* command from the *Rows* menu to use *Origin* as the color and marker variable) suggests that the internal managers are paid less (internal managers are coded with the green +'s). Evidently, external managers were recruited at higher salaries.

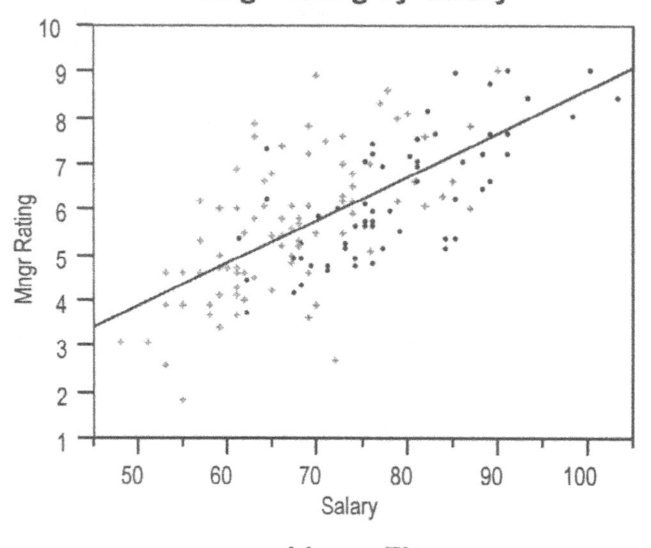

Linear Fit

RSquare	0.467
Root Mean Square Error	1.088
Mean of Response	5.901
Observations	150

Term	Estimate	Std Error	t Ratio	Prob>|t|
Intercept	-0.898	0.603	-1.49	0.139
Salary	0.095	0.008	11.40	0.000

To confirm the impression from the coded scatterplot, here is the comparison of salaries of the two groups. Indeed, the externally recruited managers have significantly higher salaries.

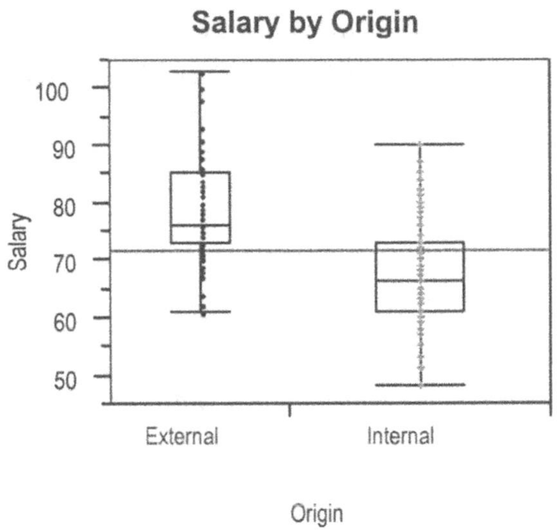

Salary by Origin

Salary

External Internal

Origin

t-Test

Difference	t-Test	DF	Prob>|t|	
Estimate	11.457	7.581	148	<.0001
Std Error	1.511			
Lower 95%	8.470			
Upper 95%	14.444		Assuming equal variances	

Level	Number	Mean	Std Error
No	62	78.35	1.16
Yes	88	66.90	0.97

Thus, it appears that our initial assessment of performance has mixed two factors. The performance evaluations are related to the salary of the manager. Evidently, those more well paid are in higher positions and receive higher evaluations. Since the externally recruited managers occupy higher positions (presumably), we do not have a "fair" comparison.

The easiest way to adjust statistically for the differences in salary paid to the internal and external managers is to fit separate lines in the plot of *Rating* on *Salary*. (Use the grouping option at the bottom of the options offered by the fitting button in the *Fit Y by X* view.) The top fit in this plot is for the internal managers. The plot suggests that at comparable salaries (as a proxy for position in the company), the internal managers are doing better than those externally recruited. Summaries of the two fitted lines appear on the next page.

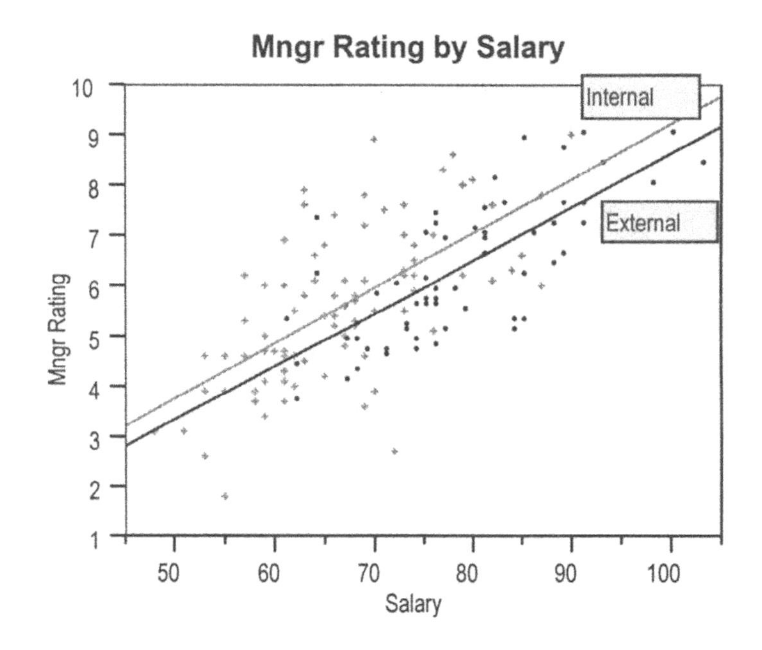

This plot suggests two simple questions that are not so easy to answer. First, are the slopes of the two lines shown here the same, or are the differences statistically significant? That is, are the two lines parallel? Second, assuming that they are parallel, are the intercepts different?

Linear Fit, External Managers

Mngr Rating = -1.9369 + 0.10539 Salary

RSquare	0.542
Root Mean Square Error	0.915
Mean of Response	6.321
Observations	62

Term	Estimate	Std Error	t Ratio	Prob>|t|
Intercept	-1.937	0.986	-1.96	0.0542
Salary	0.105	0.012	8.43	<.0001

Linear Fit, Internal Managers

Mngr Rating = -1.6935 + 0.10909 Salary

RSquare	0.412
Root Mean Square Error	1.171
Mean of Response	5.605
Observations	88

Term	Estimate	Std Error	t Ratio	Prob>|t|
Intercept	-1.694	0.949	-1.78	0.0779
Salary	0.109	0.014	7.76	<.0001

The difference between the two lines is roughly constant (the slopes are about the same). The difference between the intercepts is $-1.937 - (-1.694) = -0.243$. Is this a significant difference?

To determine if the difference in intercepts is significant (assuming equal slopes), we use an *analysis of covariance*, a multiple regression analysis that combines a categorical variable with other covariates. In this case, the coefficient of the categorical variable *Origin* indicates that the in-house managers are performing better if we adjust for salary. Again, keep the JMP coding convention in mind. The notation Origin[External] indicates that we are judging the effect of *Origin* being zero relative to an overall model, not to the other group directly. Since the coefficient associated with Origin[External] is the distance from the average, and there are only two groups, the distance for the other group has the opposite sign.

Response: Mngr Rating

RSquare	0.488
Root Mean Square Error	1.070
Mean of Response	5.901
Observations	150

The standard output shows the minimal number of coefficients with just one group.

Term	Estimate	Std Error	t Ratio	Prob>|t|
Intercept	-1.843	0.706	-2.61	0.0100
Origin[External]	-0.257	0.105	-2.46	0.0149
Salary	0.107	0.010	11.14	<.0001

The expanded output shows both groups, with the redundant coefficient for the internal managers having the opposite sign.

Expanded Estimates

Term	Estimate	Std Error	t Ratio	Prob>|t|
Intercept	-1.843	0.706	-2.61	0.0100
Origin[External]	-0.257	0.105	-2.46	0.0149
Origin[Internal]	0.257	0.105	2.46	0.0149
Salary	0.107	0.010	11.14	<.0001

Using either of the two forms for the output, if we assume a common slope for *Salary*, the difference in intercepts is significant. The figure at the top of the next page illustrates how both terms for origin determine the overall difference.

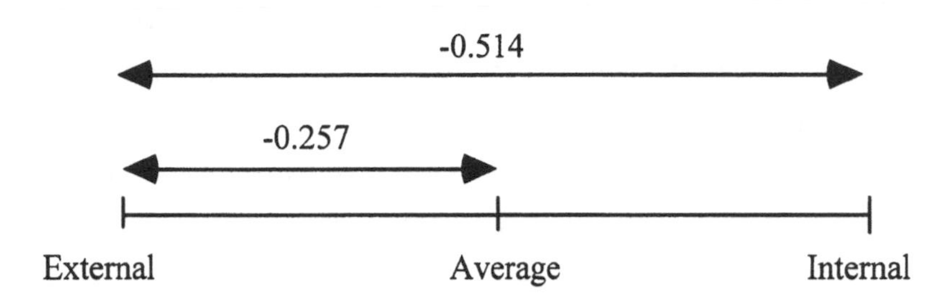

As before with just the single categorical predictor, it is helpful to write out the fitted model for each group. When the model is augmented by adding *Salary* to the equation, the fitted model has a new interpretation. The effect of adding just *Salary* and not its interaction with the categorical variable *Origin* is to fit parallel lines to the two groups. Here is the repeated summary from the *Fit Model* command with *Origin* and *Salary*:

Term	Estimate	Std Error	t Ratio	Prob>|t|
Intercept	-1.843	0.706	-2.61	0.0100
Origin[External]	-0.257	0.105	-2.46	0.0149
Origin[Internal]	0.257	0.105	2.46	0.0149
Salary	0.107	0.010	11.14	<.0001

The fits for thc two groups are

External Fit = (-1.843 – 0.257) + 0.107 *Salary* = –2.1 + 0.107 *Salary*

Internal Fit = (-1.843 + 0.257) + 0.107 *Salary* = –1.586 + 0.107 *Salary*

The fitted lines are parallel because we have made an important assumption that *forces* them to be parallel. The coefficient of the categorical factor is *one-half of the difference* between the intercepts of the two fits. As in the first example (where the regression contained only the variable *Origin*), the coefficient of the categorical factor represents the difference of the intercept of each group from the overall intercept. From this result, the difference between the fitted intercepts is significant. Only now the model adjusts for the *Salary* effect, and the internal managers are doing significantly better (higher) than the external managers (t=-2.46). The adjustment for *Salary* reverses the impression conveyed by the initial t-test.

Leverage plots are still useful, though they are a bit harder to understand when other predictors are in the model. Before, the leverage plot for *Origin* showed two discrete categories; the plot showed two columns. Now, with the other factor added, the categories are less well defined (albeit still rather distinct when viewed in color).

Why? As we have seen, *Salary* and *Origin* are related. Now the values along the horizontal axis of the leverage plot are adjusted for the impact of *Salary*.

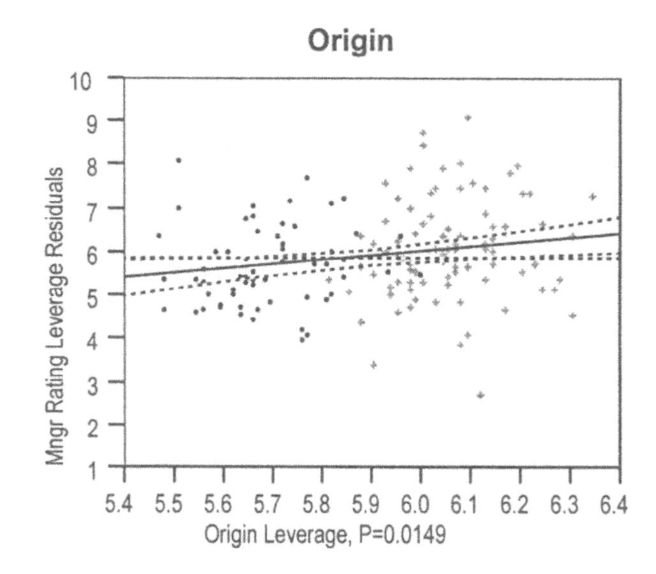

Note the inversion in the ordering of the means. Initially, the mean for in-house managers was smaller (5.60 versus 6.32). After adjusting for the salary effect as given by the least squares means, the ordering is reversed (6.11 versus 5.60)

Origin	Least Sq Mean	Std Error	Mean
External	5.60	0.15	6.32
Internal	6.11	0.12	5.60

The leverage plot for *Origin* provides further confirmation that the difference in intercepts is significant, though small relative to the variation.

The residual plot and the normal probability plot of the residuals suggest that the assumptions underlying this fitted model are reasonable.

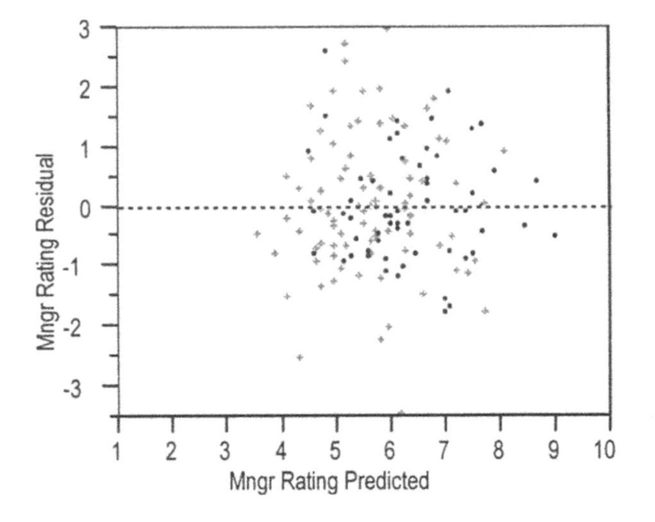

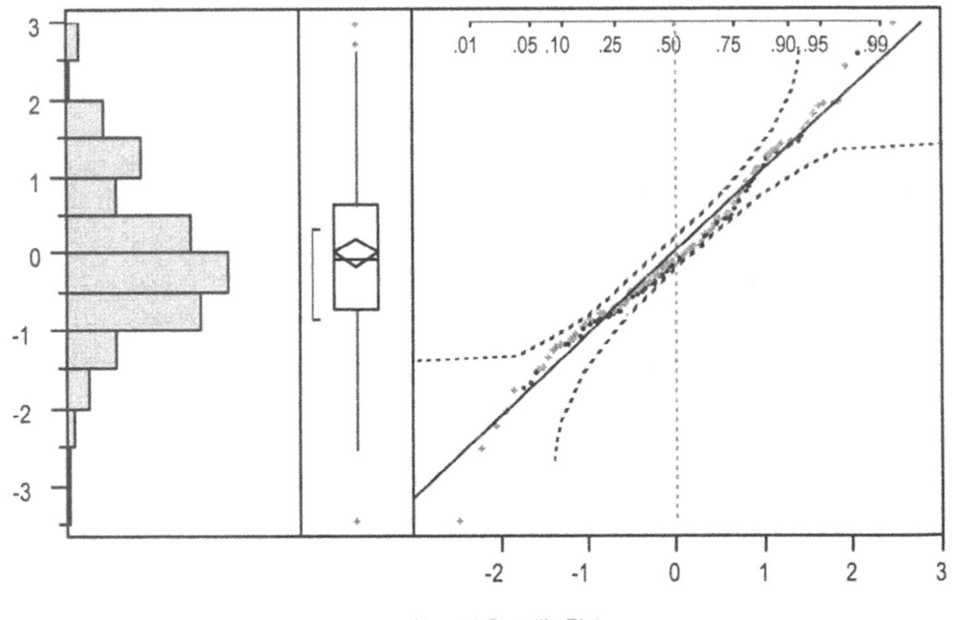

But what of the important assumption regarding parallel fits? Though clearly similar, we should check this assumption. Lines fit to separate groups are not always going to be so similar, and we need a method to check this crucial assumption of the analysis. Doing so requires that we add an *interaction* to the analysis (also known as a *cross-product* term). An interaction allows the slopes to differ in the multiple regression fit.

An interaction is added to the fitted model by the following sequence of steps in the *Fit Model* dialog, assuming that the dialog from the prior fit is still available (with *Mngr Rating* as the response and *Origin* and *Salary* as predictors). The steps are

(1) Add *Salary* to the set of predictors. *Salary* shows up as a predictor twice;

(2) Select the *Salary* term just added to the collection of predictors (click on it);

(3) Select the categorical factor *Origin* from the list of variables in the upper left corner of the dialog box;

(4) With both selected, the "Cross" button becomes highlighted. Click on this button to add the cross-product or interaction term to the model; and

(5) Run the model as usual.

Here is the summary of the fit, obtained with the expanded estimates. The added variable is not significant, implying that the slopes are indeed roughly parallel. Notice that, as we saw with polynomials previously, JMP centers the continuous variable *Salary* in the interaction term (i.e., JMP has subtracted the mean 71.63 prior to forming the interaction term). This centering reduces collinearity in the model (you can check that the VIF terms are small). Consequently, the slope for *Origin* is similar to that in the model without an interaction (0.257). The centering does, however, complicate the interpretation.

Response: Mngr Rating

RSquare	0.489
Root Mean Square Error	1.073
Mean of Response	5.901
Observations	150

Term	Estimate	Std Error	t Ratio	Prob>|t|
Intercept	-1.815	0.724	-2.51	0.0132
Origin[External]	-0.254	0.106	-2.39	0.0179
Origin[Internal]	0.254	0.106	2.39	0.0179
Salary	0.107	0.010	10.99	<.0001
(Salary-71.63)*Origin[External]	-0.00185	0.010	-0.19	0.8499
(Salary-71.63)*Origin[Internal]	0.00185	0.010	0.19	0.8499

Once again, to interpret the coefficients in this model, write out the fits for the two groups.

External Fit = (-1.815-0.254+0.133) + (0.107 - 0.002) Salary = –1.936 + 0.105 Salary

Internal Fit = (-1.815+0.254-0.133) + (0.107 + 0.002) Salary = –1.694 + 0.109 Salary

These are the same two fits that we obtained by fitting regression lines to the two groups separately at the start of the analysis. The addition of the interaction allows the slopes to differ rather than be forced to be equal (parallel fits). The small size of the interaction ($t = -0.19$) indicates that the slopes are indeed essentially the same (as noted in the graph of the two fits). Since the fits are evidently parallel, we should remove the interaction and focus our interpretation upon the model with just *Salary* and *Origin.* Since JMP has centered the interaction term, reducing the collinearity in the model, this step is less important than when fitting such a model that has not centered the continuous term in the interaction.

The residual plots from this last regression look fine with consistent variance in the two groups and a reasonable approximation to normality. When analyzing grouped data, we need to confirm that the variability is consistent across the groups, in keeping with the assumption of equal error variance for all observations. In this example, the two sets of residuals appear have similar variation. Remember, with groups of differing size, compare the heights of the boxes, not the range of the data. Once we have confirmed that the groups have similar variances, we can go on to check for normality via a quantile plot of all of the model residuals. A single normal quantile plot would be inappropriate if the variances of the two groups differed; we might interpret a deviation from the diagonal as indicating a lack of normality rather than more simply being a difference in variation.

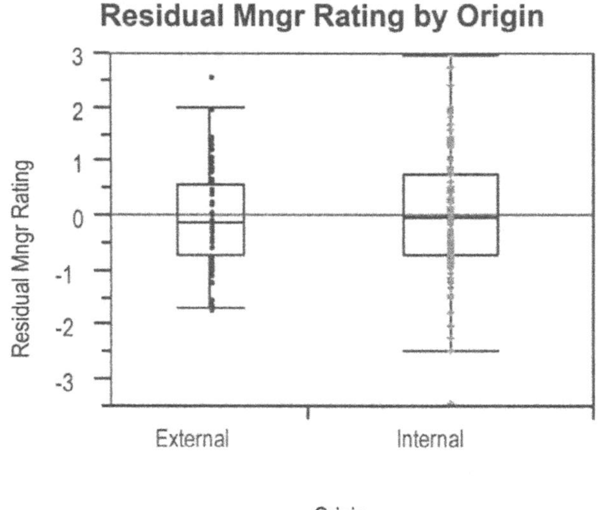

Means and Std Deviations

Level	Number	Mean	Std Dev	Std Err Mean
External	62	-0.000	0.908	0.115
Internal	88	-0.000	1.164	0.124

A firm either promotes managers in-house or recruits managers from outside firms. It has been claimed that managers hired outside of the firm perform better than those promoted inside the firm, at least in terms of internal employee evaluations.

Do the available data on a sample of 150 managers support this claim?

We find that the fits to the two groups are indeed parallel and conclude from the prior fit (with *Salary* and *Origin*) that internal managers are doing better once we adjust for differences in position within the firm. That is, if we adjust for differences in salary, the in-house managers are doing better (t=-2.5, p=0.01) than their colleagues who are recruited from the outside at higher salaries.

On the other hand, why control for only differences in salary? The following regression shows what occurs if years of experience is used as an additional control. The internal and external managers are doing about the same!

Response: Mngr Rating

RSquare	0.560
Root Mean Square Error	0.995
Mean of Response	5.901
Observations	150

Expanded Estimates

Term	Estimate	Std Error	t Ratio	Prob>|t|
Intercept	-2.948	0.695	-4.24	<.0001
Origin[External]	-0.014	0.109	-0.13	0.8979
Origin[Internal]	0.014	0.109	0.13	0.8979
Salary	0.110	0.009	12.22	<.0001
Years Exp	0.120	0.025	4.88	<.0001

Something extra to think about

Why did that last result occur? Collinearity. The variable added to the equation, *Years Exp*, is related to the other factors in the model. Oddly, *Years Exp* is negatively related to *Salary* when the two groups are not distinguished.

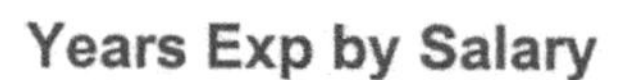

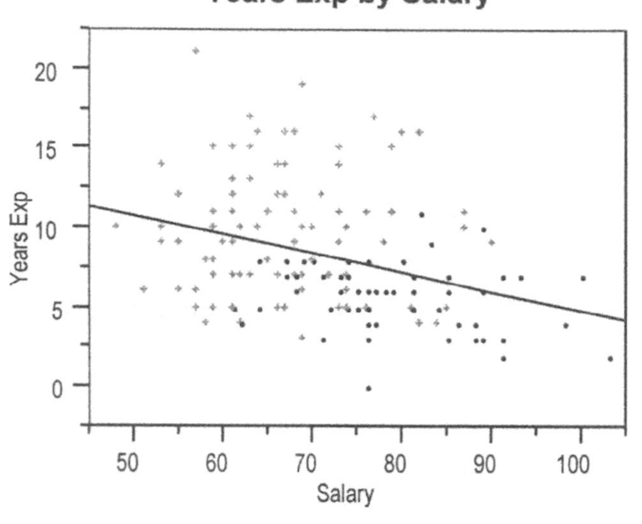

If w stinguish the two groups and fit separate lines, one for each, the correlation goes away! Neither slope in the next figure is significant.

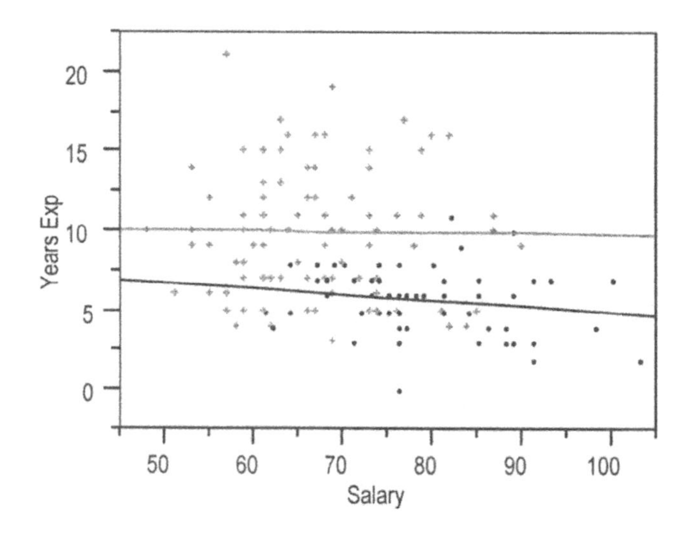

Class 7. Modeling Categorical Factors with Two or More Levels

This class continues our study of categorical factors. We begin with a review example that covers the topics from Class 6. Our second example uses a model in which the categorical factor has more than two levels.

Examples

1. Wage discrimination, revisited
2. Timing production runs (three levels)

Key application

Measuring the impact of different managers on the efficiency of a production process. A plant could have many managers, so it's not obvious how we can include "manager" as a factor in the regression modeling paradigm. All we have seen in Class 6 are two-group categorical variables ("yes/no," "male/female"). To handle categorical factors with two groups, we simply treat them as categorical variables and add them to the regression. Similarly, if the categorical variable has more levels, the approach is just the same. The differences appear in the complexity of the output; now we have more regression coefficients to interpret.

Concepts

A categorical variable with more than two levels. All of the categorical variables we have seen so far have only two levels. It is not at all uncommon, however, to face categorical variables that take on more than two levels, for example, a categorization of firms into size based on small/medium/large might be useful in an analysis of their risk taking behavior. Another example of a multilevel categorical variable would be the supervisor on duty in a plant, as noted already. In the regression project there is such a variable with levels $manager_1$, $manager_2$, $manager_3$, $manager_4$, and $manager_5$, a five-level categorical variable. The advantage of including such a variable in a regression is that you can gauge the impact of each manager on the production process, clearly a useful thing to be able to do.

When dealing with a multilevel categorical variable in JMP, just as with two-level variables, the software translates the categories to numbers using dummy variables.

What happens is that JMP makes a series of new columns in the spreadsheet. These are the dummy variables; you can't see them. The rule is that if the categorical variable has k levels, then JMP has to make $k-1$ new columns. When these k-1 additional terms are added to the regression, you will see $k-1$ extra regression coefficients in the output.

Just as with two-level categorical variables, including a multilevel categorical variable in the regression allows you to fit separate regressions to each level of the categorical, so in the case of one coded "low/medium/high," you will get one regression for those firms that fall in the "low" category, one regression for those that fall in the "medium" category, and one regression for firms falling in the "high" category. These regressions will be forced to be *parallel* unless you allow the slopes of the regressions to be different by adding an interaction with the appropriate continuous predictor(s). Including a categorical variable and an interaction thus allows you to compare the slopes and intercepts for different categories.

Of course, whenever you compare something the question is "Comparison to what?" Although there are many ways to do the comparison, JMP takes the following approach. It compares each level of the categorical variable to the *average* of all the levels, average being the operative word. This means that the regression coefficient you see for a specific level of a categorical variable tells you how different this level of the categorical variable is from the *average* of all the levels. If it's an intercept coefficient, it tells you how different the intercept of a particular level, say medium, is from the average intercept of "low/medium/high" taken together. The same applies to regression slopes that you get via the interaction. The coefficient of the interaction tells you how the slope for a particular level of the categorical variable differs from the average slope for all the levels taken together.

Potential Confusers

The JMP output. When a categorical factor is used in regression, the default JMP output can be pretty hard to interpret because one of the groups will be left out. The easiest thing to do is to get JMP to compute what it calls the "Expanded Estimates". Then you will see a regression coefficient for each of the categories represented by that variable.

Why not show the expanded estimates as the default? It's a tradition in statistics (going back to when one did all of the underlying manipulations "by hand") to leave out one of the groups when building regression models that involve categorical predictors. In fact, there was a time when the standard software would not work if you tried to fit a model that included all of the groups. The reason for this behavior is that the expanded coefficients that we shall use are redundant. They have to sum to zero. Thus, if you were to remove any one of them, you could figure out its coefficient by adding up all of the others and changing the sign. Fortunately, we do not have to do this any more, but it still has a legacy.

Wage Discrimination

Salary.jmp

A business magazine has done a large survey that appears to indicate that midlevel female managers are being paid less well than their male counterparts.

Is this claim valid?

This data set is introduced in Class 9 of *Basic Business Statistics*. For review, data were gathered on a sample of 220 randomly selected managers from firms. The data file includes row markers that define colors and markers that appear in plots (red boxes for the women and blue +'s for the men).

Salary	Sex	Position	YearsExper	Sex Codes	Mid-Pos?
148	male	7	16.7	1	0
165	male	7	6.7	1	0
145	male	5	14.8	1	1
139	female	7	13.9	0	0
...					
147	male	5	8.8	1	1
156	male	7	15.1	1	0
132	male	4	4.7	1	1
161	male	7	16.5	1	0

The variables represented by these columns are defined as follows:

Salary	Base annual compensation in thousands of $US.
Sex	Pretty obvious.
Position	An index for the rank of the employee in the firm; roughly corresponds to the number of employees supervised by the individual, size of budget controlled, etc.
Years Exp	Number of years of relevant experience for current position.
Sex Code	A dummy variable, with 0 for females and 1 for males (Formula).
Mid-Pos?	A dummy variable, coded 1 for positions 4 to 6, zero otherwise (Formula).

The two-sample t-test comparison indicates that men are paid more than \$3,600 more on average than the women in this sample. The two-sample comparison via the t-test or confidence interval indicates that this difference is significant ($p \approx 0.04$).

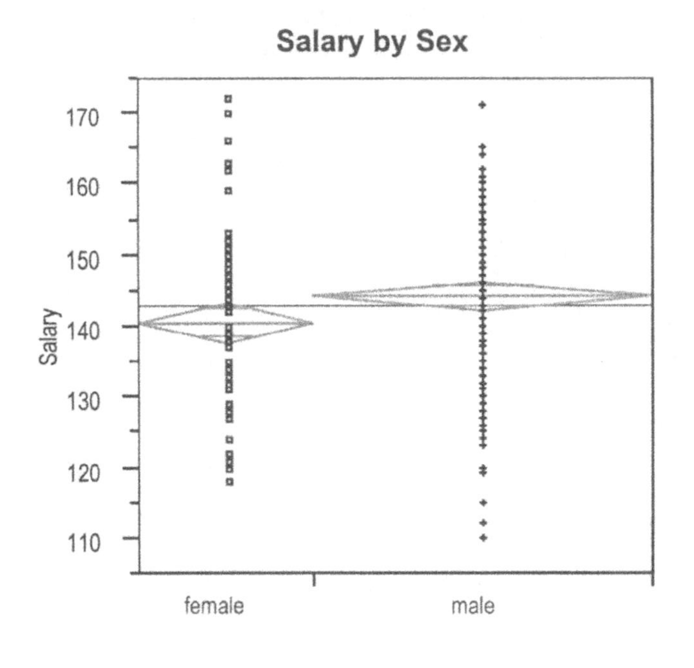

t-Test

	Difference	t-Test	DF	Prob>|t|
Estimate	-3.64	-2.061	218	0.0405
Std Error	1.77			
Lower 95%	-7.13			
Upper 95%	-0.16		Assuming equal variances	

Level	Number	Mean	Std Error
female	75	140.467	1.4351
male	145	144.110	1.0321

A more thorough analysis recognizes that there are other differences between these men and women aside from their sex alone. For example, the men have more years of experience than the women and tend to occupy higher positions. Such differences between the two groups would be irrelevant were it not for the fact that these other factors are related to salary (as well as to each other).

The plots of salary by years of experience and position on years of experience show that indeed both salary and position are related to experience.

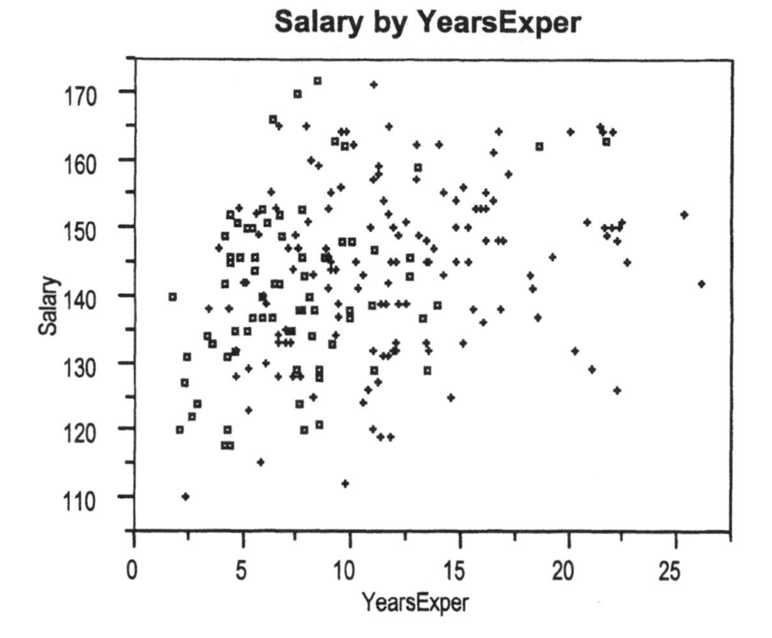

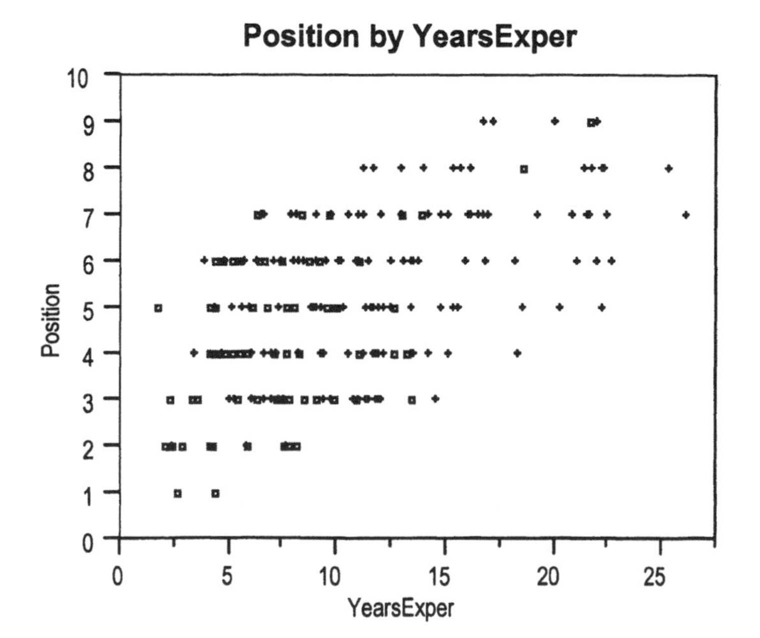

It appears that the two samples are in some ways not comparable. For example, men tend to occupy positions that pay more, and they have more experience. In *Basic Business Statistics*, we handled this lack of comparability by focusing on a smaller, more homogeneous subset of the data (the 116 employees for which *Position* lies in the range 4 to 6).

With the analysis restricted to this subset, we get a different impression of the differences in compensation. In this subset, men are paid more than $2,000 *less* than the women! The difference is not significant, but quite different from that in the overall analysis.

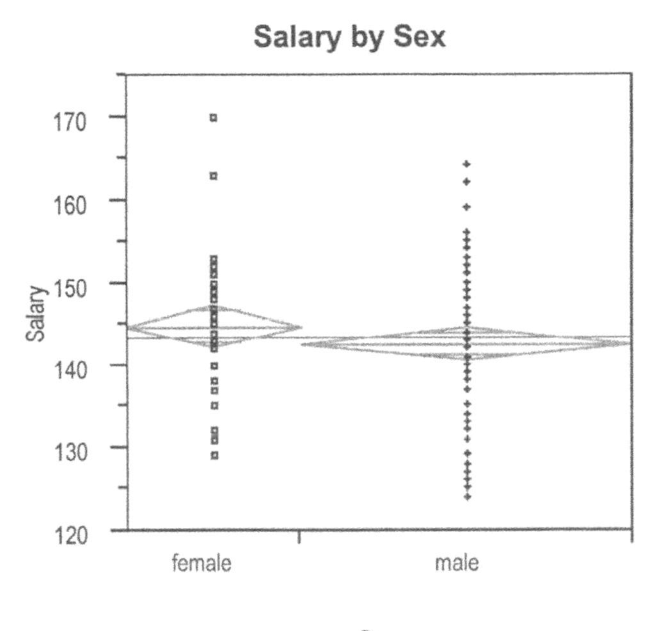

t-Test

	Difference	t-Test	DF	Prob>|t|
Estimate	2.15	1.289	114	0.2000
Std Error	1.67			
Lower 95%	-1.16			
Upper 95%	5.46	Assuming equal variances		

Means and Std Deviations

Level	Number	Mean	Std Dev	Std Err Mean
female	40	144.600	8.23	1.30
male	76	142.447	8.71	1.00

Rather than subset the data, we can use regression with categorical predictors to examine this matter. Here is the summary of a regression of *Salary* on *Sex, Position* and *Years Exper*. The coefficient of *Sex* indicates that, when adjusted in this manner for *Position* and *YearsExper*, women in this sample are making about $2,200 more than their male counterparts (twice the coefficient of the *Sex* variable).

Response: Salary

RSquare	0.712
Root Mean Square Error	6.768
Mean of Response	142.868
Observations	220

Expanded Estimates

Term	Estimate	Std Error	t Ratio	Prob>|t|
Intercept	114.16	1.44	79.17	<.0001
Sex[female]	1.10	0.54	2.04	0.0428
Sex[male]	-1.10	0.54	-2.04	0.0428
Position	6.71	0.31	21.46	<.0001
YearsExper	-0.47	0.11	-4.17	<.0001

This value ($2,200) is similar to what we found by focusing on the subset. Since we have used all of the data, however, the difference found here *is significant*. It's not significant when restricted to the subset.

Of course, this analysis presumes that we can ignore possible interactions between *Sex* and the other factors, *Position* and *YearsExper*. We'll come to that after an excursion to the leverage plot for *Sex*.

The leverage plot for *Sex* shows the "blurring" due to the interaction of this indicator with the other predictors *Position* and *YearsExper* in this model. Note the reversed means. Unadjusted, the men's salaries are higher. After the adjustment for *Position* and *YearsExper*, on average the men make less. The least squares means are "corrected" for differences between these men and women in average years of experience and position.

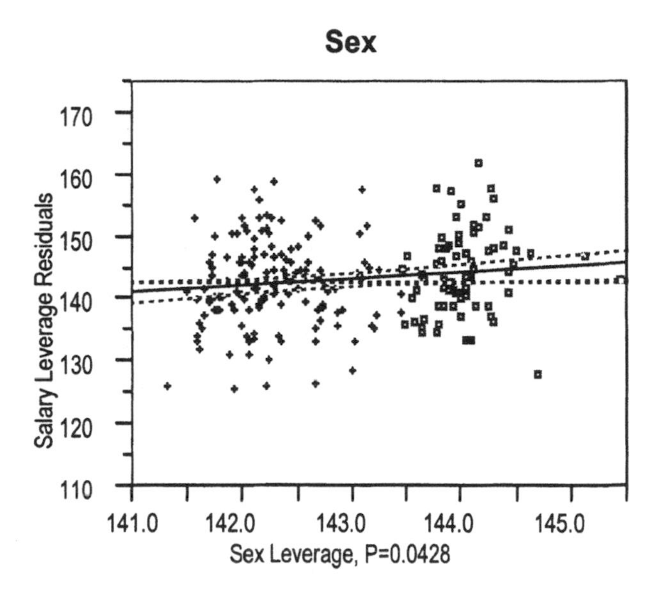

Least Squares Means

Level	Least Sq Mean	Std Error	Mean
female	144.32	0.85	140.467
male	142.12	0.59	144.110

As we noted, this analysis assumes that the slopes are parallel for both *Position* and *YearsExper*. By fitting a model with interactions, we can see that this assumption is reasonable; neither of the interaction terms is significant.

Response: Salary

RSquare	0.715
Root Mean Square Error	6.767
Mean of Response	142.9
Observations	220

Expanded Estimates

Term	Estimate	Std Error	t Ratio	Prob>|t|
Intercept	113.84	1.49	76.33	<.0001
Sex[female]	1.44	0.60	2.38	0.0181
Sex[male]	-1.44	0.60	-2.38	0.0181
Position	6.63	0.33	20.26	<.0001
YearsExper	-0.36	0.14	-2.61	0.0098
Sex[female]*(Position-5.06818)	-0.21	0.33	-0.65	0.5142
Sex[male]*Position	0.21	0.33	0.65	0.5142
Sex[female]*(YearsExper-10.4786)	0.20	0.14	1.43	0.1550
Sex[male]*YearsExper	-0.20	0.14	-1.43	0.1550

Although there is not much interaction between *Sex* and the other covariates, why should we ignore the possibility of an interaction between *Position* and *YearsExper*? Indeed, this interaction is very significant. Residual analysis (not shown here) suggests that the model with this interaction added is okay.

Response: Salary

RSquare	0.722
Root Mean Square Error	6.667

Term	Estimate	Std Error	t Ratio	Prob>|t|
Intercept	113.80	1.426	79.79	<.0001
Sex[female]	1.41	0.544	2.59	0.0104
Sex[male]	-1.41	0.544	-2.59	0.0104
Position	6.68	0.308	21.69	<.0001
YearsExper	-0.35	0.121	-2.88	0.0044
(Position-5.06818)*(YearsExper-10.4786)	-0.13	0.049	-2.76	0.0064

Once again, the simplest way to interpret this interaction term between *Position* and *YearsExper* is to write out the fitted equation under different conditions. Let's focus on the effect of years of experience for varying positions. Rearranging the regression equation makes it simpler to interpret. For women,

$$\begin{aligned}\text{Fit} &= (106.6 + 1.4) + 8.1\ \textit{Position} + .34\ \textit{Years} - 0.13(\textit{Years} * \textit{Position}) \\ &= (108 + 8.1\ \textit{Position}) + (0.34 - 0.13\ \textit{Position})\ \textit{Years}\end{aligned}$$

The higher the position, the greater the intercept, but the smaller the slope for years of experience. It seems from this fit that those who stay too long in higher positions (Position > 2) have less pay than those who perhaps are climbing quickly through the ranks.

Here are two examples. In low grades, it's useful to have more experience. All of the fits shown below are for women.

Position 1, Years Experience 1

$$\begin{aligned}\text{Fit} &= 108 + 8.1\ (1) + (0.34 - 0.13 \times 1)\ 1 \\ &= 116.3\end{aligned}$$

Position 1, Years Experience 10

$$\begin{aligned}\text{Fit} &= 108 + 8.1\ (1) + (0.34 - 0.13 \times 1)\ 10 \\ &= 118.2\end{aligned}$$

At higher levels, it is not.

Position 5, Years Experience 1

$$\begin{aligned}\text{Fit} &= 108 + 8.1\ (5) + (0.34 - 0.13 \times 5)\ 1 \\ &= 148.2\end{aligned}$$

Position 5, Years Experience 10

$$\begin{aligned}\text{Fit} &= 108 + 8.1\ (5) + (0.34 - 0.13 \times 5)\ 10 \\ &= 145.4\end{aligned}$$

A business magazine has done a large survey which appears to indicate that midlevel female managers are being paid less well than their male counterparts.

Is this claim valid?

An initial analysis using all data indicates that women in this sample are paid less, significantly so. Restricting attention to those observations that occupy mid-level positions suggests that women are paid a bit more than their male counterparts, but the difference is not significant. By using regression to adjust for the other factors, the reversal does appear significant, but only marginally so.

Timing Production Runs

ProdTime.jmp

> An analysis has shown that the time required in minutes to complete a production run increases with the number of items produced. Data were collected for the time required to process 20 randomly selected orders as supervised by three managers.
>
> How do the managers compare?

Suppose that we begin this analysis by ignoring the stated effect of run size and look first only at the length of time required by the three managers. Comparison of the boxplots shown below suggests that Manager *c* is doing very well with the shortest run times, with Manager *b* second and Manager *a* third. But is this the right way to look at the data in this problem?

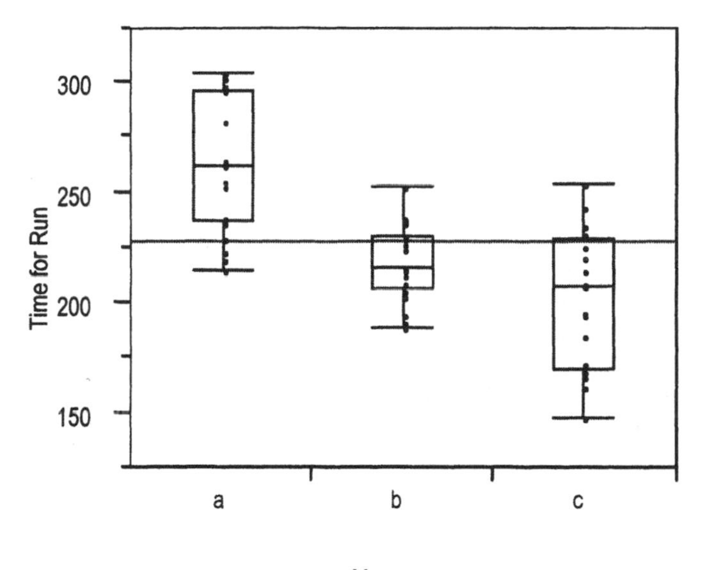

Such a marginal comparison can be very misleading. In this example, we have been informed that, not surprisingly, large production runs with more items take longer than small runs of few items. How can we be sure that Manager *c* has not simply been supervising very small production runs? Of course, we could have avoided this issue altogether by having done a randomized experiment, but it can be hard to implement such idealized data collection procedures in a real-world production environment.

We need to take the size of the supervised product runs into account when comparing the managers. To begin to get a handle on this, we first check to see whether indeed there is a relationship between run size and the time for the run. The following plot and bivariate regression fit make it clear that the two are indeed related. The overall fit is indeed positive in slope, with an average time per unit (the slope) of 0.22 per unit.

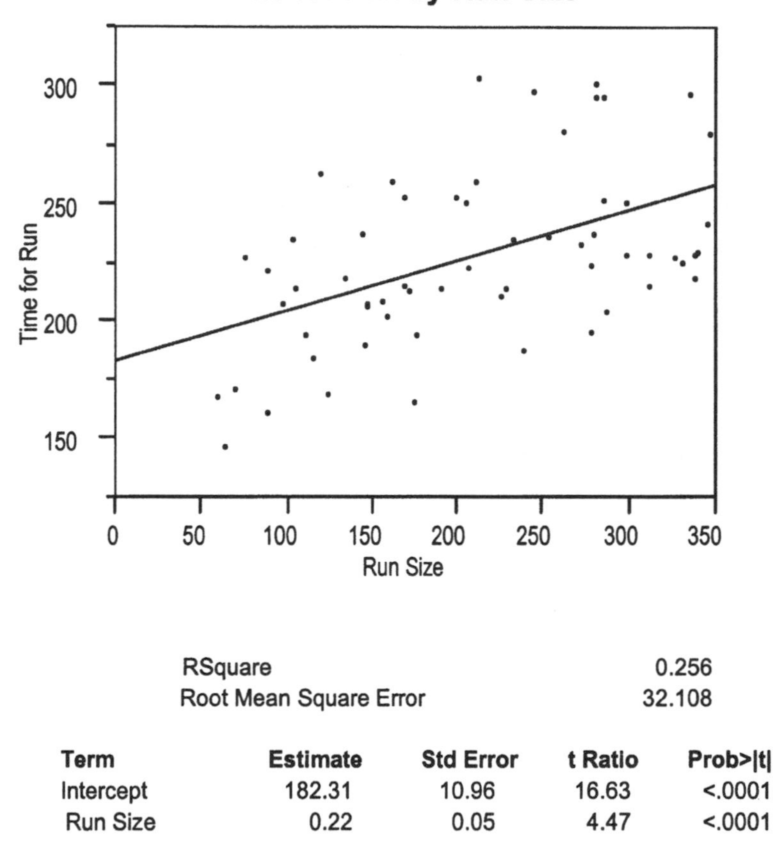

RSquare	0.256
Root Mean Square Error	32.108

Term	Estimate	Std Error	t Ratio	Prob>|t|
Intercept	182.31	10.96	16.63	<.0001
Run Size	0.22	0.05	4.47	<.0001

What does this analysis say about the mangers? Not much, at least not until we color code the points to distinguish the data for the three managers. The initial plot is vague until coded using the *Manager* indicator variable.[1] Once the points are color coded in this manner, we can easily distinguish the differences among the three managers, as seen on the next page where we fit separate linear models for each manager.

[1] To get the color coding added, go to the "Rows" menu and pick the "Color or Mark by column" item. Then choose the Manager variable to color and mark the separate observations.

In contrast to one overall regression, might the fit differ if restricted to the 20 runs available for each manager? When a regression is fit separately for each manager as shown next, it is clear that the fits are different. Indeed, it does appear to confirm our initial impression that Manager *a* is slower than the other two. The comparison between Manager *b* and Manager *c* is less clear.

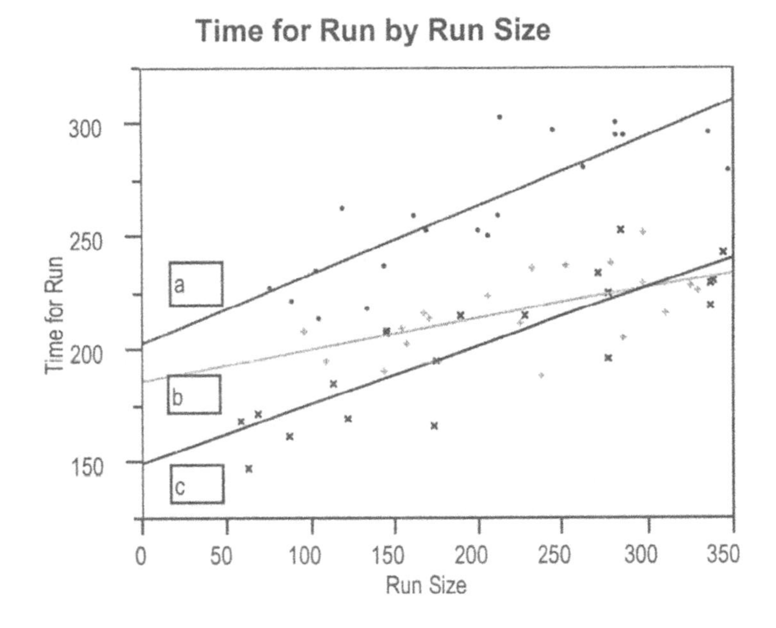

Are the differences among these three fitted lines large enough to be called significant, or might they be the result of random variation? After all, we haven't much data for each manager. Summaries of the three fitted regressions appear on the next page.

What is the interpretation of the slope and intercept in each equation?

Linear Fit Manager=a (Red)

RSquare	0.710
RSquare Adj	0.694
Root Mean Square Error	16.613
Mean of Response	263.050
Observations	20

Term	Estimate	Std Error	t Ratio	Prob>|t|
Intercept	202.53	9.83	20.59	<.0001
Run Size	0.31	0.05	6.65	<.0001

Linear Fit Manager=b (Green)

RSquare	0.360
RSquare Adj	0.325
Root Mean Square Error	13.979
Mean of Response	217.850
Observations	20

Term	Estimate	Std Error	t Ratio	Prob>|t|
Intercept	186.49	10.33	18.05	<.0001
Run Size	0.14	0.04	3.18	0.0051

Linear Fit Manager=c (Blue)

RSquare	0.730
RSquare Adj	0.715
Root Mean Square Error	16.252
Mean of Response	202.050
Observations	20

Term	Estimate	Std Error	t Ratio	Prob>|t|
Intercept	149.75	8.33	17.98	<.0001
Run Size	0.26	0.04	6.98	<.0001

To see if these evident differences are significant, we embed these three models into one multiple regression.

The multiple regression tool *Fit Model* lets us add the categorical variable *Manager* to the regression. Here are the results of the model with *Manager* and *Run Size*, again showing the expanded parameter estimates that show slopes for all 3 managers. Since we have more than two levels of this categorical factor, the "Effect Test" component of the output becomes relevant. It holds the partial F-tests. We ignored these previously since these are equivalent to the usual t-statistic if the covariate is continuous or is a categorical predictor with two levels.

Response: Time for Run

RSquare	0.813
Root Mean Square Error	16.38
Mean of Response	227.65
Observations	60

Expanded Estimates

Term	Estimate	Std Error	t Ratio	Prob>|t|
Intercept	176.71	5.66	31.23	<.0001
Run Size	0.24	0.03	9.71	<.0001
Manager[a]	38.41	3.01	12.78	<.0001
Manager[b]	-14.65	3.03	-4.83	<.0001
Manager[c]	-23.76	3.00	-7.93	<.0001

Effect Test

Source	Nparm	DF	Sum of Squares	F Ratio	Prob>F
Run Size	1	1	25260.250	94.2	<.0001
Manager	2	2	44773.996	83.5	<.0001

The three coefficients labeled as Manager[a], Manager[b], and Manager[c] represent the differences in the intercepts for managers *a*, *b*, and *c* from the average intercept of *three parallel* regression lines. Notice that the effect for manager *c* is the negative of the sum of those for Managers *a* and *b* (so that the sum of these three terms is zero). That is,

$$\text{Effect for Manager } c = -(38.41 - 14.65) = -23.76$$

By default, JMP-IN does not show all three since the last one is redundant, but you can see all three using the expanded estimates option as displayed here.

The effect test for *Manager* indicates that the differences among these three intercepts are significant. It does this by measuring the improvement in fit obtained by adding the *Manager* terms to the regression simultaneously. One way to see how it's calculated is to compare the R^2 without using *Manager* to that with it. If we remove *Manager* from the fit, the R^2 drops to 0.256. Adding the two terms representing the two additional intercepts improves this to 0.813. The effect test determines whether this increase is significant by looking at the improvement in R^2 per added term. Here's the expression: (another example of the partial *F*-test appears in Class 5).

$$\begin{aligned}\text{Partial F} &= \frac{\text{Change in } R^2 \text{ per added variable}}{\text{Remaining variation per residual}} \\ &= \frac{(R^2_{complete} - R^2_{reduced})/(\#\ \text{variables added})}{(1 - R^2_{complete})/(\text{Error degrees of freedom})} \\ &= \frac{(0.813 - 0.256)/2}{(1 - 0.813)/56} \\ &= 83.5\end{aligned}$$

What about the clear differences in slopes?

We need to add interactions to capture the differences in the slopes associated with the different managers.

Response: Time for Run

RSquare	0.835
Root Mean Square Error	15.658
Mean of Response	227.65
Observations (or Sum Wgts)	60

Expanded Estimates

Term	Estimate	Std Error	t Ratio	Prob>|t|
Intercept	179.59	5.62	31.96	<.0001
Run Size	0.23	0.02	9.49	<.0001
Manager[a]	38.19	2.90	13.17	<.0001
Manager[b]	-13.54	2.94	-4.61	<.0001
Manager[c]	-24.65	2.89	-8.54	<.0001
Manager[a]*(Run Size-209.317)	0.07	0.04	2.07	0.0437
Manager[b]*(Run Size-209.317)	-0.10	0.04	-2.63	0.0112
Manager[c]*(Run Size-209.317)	0.02	0.03	0.77	0.4444

Effect Test

Source	Nparm	DF	Sum of Squares	F Ratio	Prob>F
Run Size	1	1	22070.614	90.0	<.0001
Manager	2	2	4832.335	9.9	0.0002
Manager*Run Size	2	2	1778.661	3.6	0.0333

In this output, the coefficients of the terms labeled Manager[a]*(Run Size-209) and so forth indicate the differences in slopes from the average manager slope. As before, the term for Manager *c* is the negative of the sum of the values for Managers *a* and *b*, and JMP has centered *Run Size* in the interactions (subtracting the mean 209.317). The differences are significant, as seen from the effect test for the interaction. Most of the significance of this test derives from the difference of Manager *b* from the others.

The residual checks indicate that the model satisfies the usual assumptions.

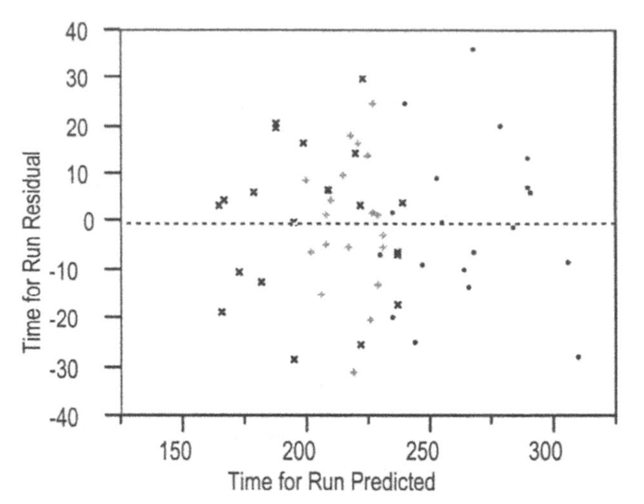

Since the model has categorical factors, we also need to check that the residual variance is comparable across the groups. Here, the variances seem quite similar.

Residual Time for Run by Manager

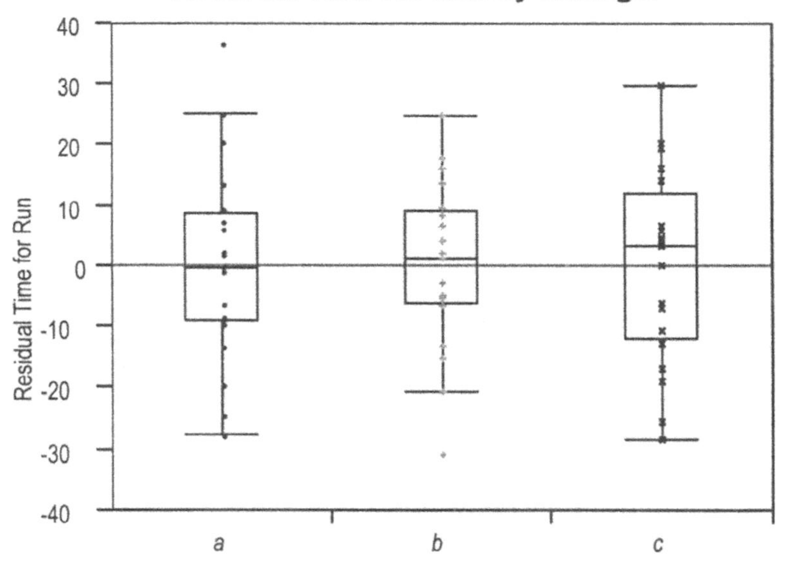

Manager

Manager	#	Mean	Std Dev	Std Err Mean
a	20	0.00	16.2	3.6
b	20	0.00	13.6	3.0
c	20	0.00	15.8	3.5

Finally, since the residuals appear to have consistent levels of variation, we can combine them and check for normality. This plot suggests that the model errors are indeed close to normally distributed.

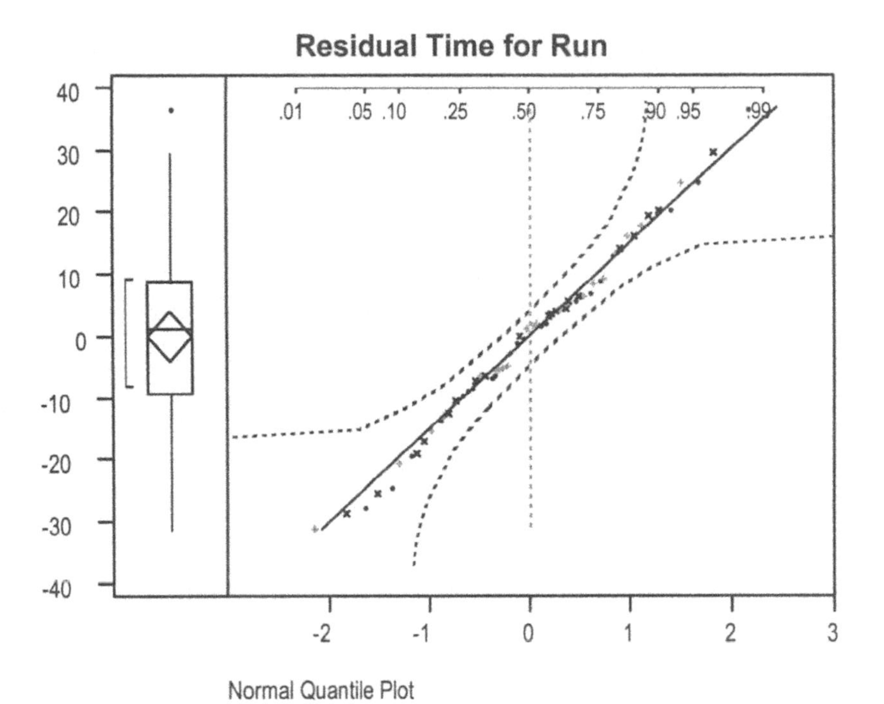

An analysis has shown that the time required in minutes to complete a production run increases with the number of items produced. Data were collected for the time required to process 20 randomly selected orders as supervised by three managers.

How do the managers compare?

The runs supervised by Manager *a* appear abnormally time consuming. Manager *b* has high initial fixed setup costs, but the time per unit for this manager is the best of the three. Manager *c* has the lowest fixed costs with a typical per unit production time.

The adjustments obtained via regression in this example only control for possible differences in size among the production runs. Other differences might also be relevant. For example, it might be the case that some of the production runs were inherently more difficult to produce. Unless these production runs were randomly assigned to the different managers, we should control for this effect as well. It could be the case that Manager *a* supervises the most difficult and tedious production runs.

Class 8. Summary Regression Case

The following analysis revisits the Forbes executive compensation data that was studied in Classes 2 and 3 of *Basic Business Statistics*. Rather than simply characterize the variation in total compensation, this regression analysis builds a model that attempts to "explain" the differences in compensation. We have come a long way since that introductory material.

Examples

1. Executive compensation
2. Prediction from stepwise regression

Key application.

Building and validating a complex regression model. You will often be in the position of either building or interpreting a regression model that has been constructed from a large database consisting of many variables. Having gone through the process of building such a model, you will have a better sense of what sorts of subjective choices are made along the way. As we have seen in past classes, collinearity complicates our ability to interpret the fitted model since the coefficients change as variables are added or removed from the fit.

Definitions

Cross-validation. One of a collection to methods for attempting to validate a regression model. In cross-validation, a subsample is set aside prior to starting the data analysis. This subsample is then used to test the fitted model, producing an independent measure of its predictive ability, for example. Cross-validation is hard to do without having lots of data because it is hard to be willing to set aside enough data to permit an accurate assessment of the fit of the model. For many problems, useful cross-validation requires just as much data to test the model as was used to build it.

Stepwise regression. An automated, "greedy" tool that adds variables to a regression, one at a time, to make the R^2 statistic grow as rapidly as possible. It can also be configured to remove variables from a model, starting from something complex and reducing it to a simpler, more parsimonious model.

Response surface. For us, this term refers to regression models that include all of the possible interactions and quadratic terms for each predictor in the included set. Assuming that all of the factors are continuous (no categorical terms), a collection of 10 predictors implies — when using the response surface approach — that you really are considering the original 10 factors *plus* the 10 quadratic (X^2) terms *plus* the $\binom{10}{2}$ = 45 interaction terms. That's a total of 65 possible predictors from the set that started with 10. Fitting a collection of so many predictors requires an automated routine like stepwise regression.

Concepts

Automated model fitting. Automated regression modeling (or "data mining") has been around for quite some time. It is fairly easy to program an algorithm that adds variables to a given regression so as to obtain the maximum improvement in R^2. Such greedy algorithms can miss things in the presence of collinearity. For example, consider what happens with categorical predictors associated with large interactions. Using just the categorical factor or the continuous predictor alone might not be very useful, as in the case where the fits for the separate groups "criss-cross" with comparable intercepts. An algorithm that adds only one variable at a time and ignores the potential for interaction would miss this synergistic combination.

Overfitting. The problem with automated fitting methods is that they may work too well. It is possible to get a high R^2 using stepwise regression that is not meaningful. Used carelessly, the algorithm can produce good fits that make you think you can do prediction well, but in fact don't work well at all when applied to data outside the set used to build the model.

Validation. Since methods like stepwise regression are capable of finding structure where there is none, one needs to validate the results of such optimization routines before taking them too seriously. One approach is to hold back some of the data being studied, saving it as a validation sample. The idea is to fit whatever model you want to the rest of the data, then check the fit of that model to the hold-back sample. Typically, the model will not predict observations in the hold-back sample as well as it fits the data used in its creation. The RMSE, or residual SD, of a model built by stepwise regression is likely to be quite a bit smaller than the RMSE you obtain when predicting new observations. Unless enough data is reserved for such validation, however, random variation in the hold-back sample makes it hard to judge the models. The RMSE based on a hold-back sample of just a few observations is far too variable to be a reliable measure of model fit.

An alternative approach is to exploit the results we have derived previously with the Bonferroni inequality. These imply that if you are doing a regression with 10 predictors, for example, and want to use 95% procedures, then you need to compare the usual stepwise p-value to 0.05/10 rather than 0.05. This stricter rule for assessing significance makes it harder for a predictor to join the model, and thus avoids overfitting. We will discuss the use of the Bonferroni method further in Class 9.

Executive Compensation

Forbes94.jmp

Highly publicized salaries for CEO's in the US have generated sustained interest in the factors related to total compensation. Does it matter how long the executive has been at the firm? Do CEO's pay themselves less if they have a large stake in the stock of the company? And, does having an MBA increase executive salaries?

The data for this analysis appear in the May 1994 issue of *Forbes* magazine. Each year, *Forbes* magazine publishes data giving the compensation package of 800 top CEO's in the US. The response variable of interest is total compensation, defined as the sum of salary plus any bonuses, including stock options.

Our analysis begins with a graphical exploration of the data. The intent of this preview is to introduce the data, locate some interesting observations, and examine the distinction among several industry segments, particularly the relatively large retail and financial segments. The remainder of the analysis focuses upon the 168 companies in the financial segment.

We are going to focus on the effect of having an MBA degree on the compensation of these executives. The plots in the following figures are coded, using the *MBA?* column to define the markers (+'s for those with an MBA). It will become pretty clear that we are not sure which variables to include or how they might impact the analysis. In this case, stepwise regression becomes a useful tool to help explore the collection of potential predictors. Stepwise regression also makes it feasible to check for the many possible interactions and nonlinearities that would otherwise be too cumbersome to explore, one at a time.

We begin by considering the relationship between total executive compensation and annual sales of the company. Annual sales serves as a proxy for the size and complexity of the organization. Several outlying values dominate the initial plot; these companies would essentially dominate a regression analysis of the raw compensation data. (Yes, the top compensation package in 1994 paid $200 million.)

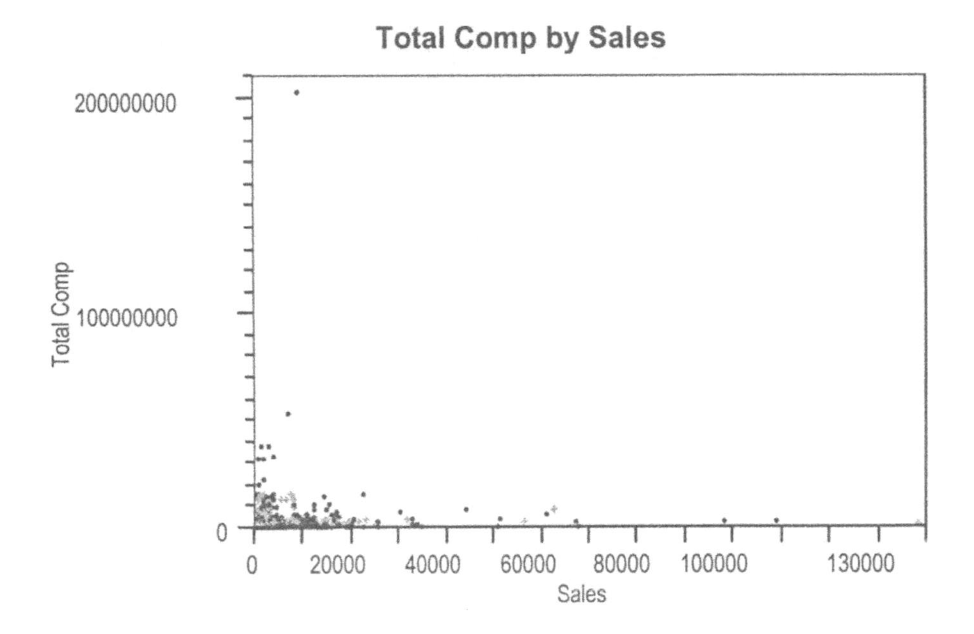

Point labeling with the name of either the executive or the company is rather interesting in these figures.

Before going on, sketch in the line that you believe would best summarize the relationship shown in this plot.

By comparison, the $\log_{10}$ transformation[1] pulls these outliers toward the rest of the data and reveals more of the structure in the data. Some interesting outliers exist around the fringes of the plot, but the structure is much more revealing. Notice that sales data for 10 companies is missing (so we have a total of 790 points in the figure). Notice that the relationship between log compensation and log sales is very positive, something that is hard to see in the plot on the previous page.

Since both of the variables are expressed on log scales, we can interpret the regression coefficient as an *elasticity*. The slope 0.27 implies that the typical compensation package increases 0.27% for each 1% increase in company sales.

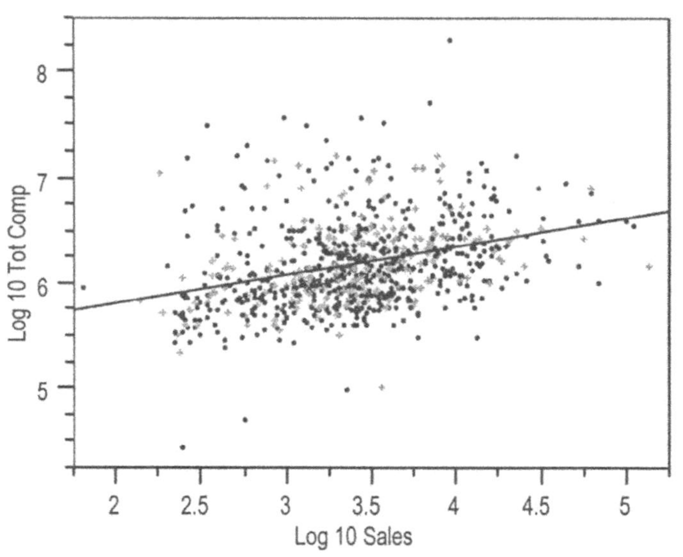

Linear Fit

RSquare	0.108
Root Mean Square Error	0.389
Mean of Response	6.178
Observations	790

Term	Estimate	Std Error	t Ratio	Prob>|t|
Intercept	5.28	0.093	56.99	<.0001
Log 10 Sales	***0.27***	0.027	9.78	<.0001

[1] Why use the base 10 log now? Though natural logs are convenient for certain manipulations, common logs provide a very recognizable scale, counting the number of digits.

Looking at all 790 companies, how do the MBA's do? It seems from this figure that they do about as well as the others. The two fitted lines in the figure are almost identical. Relative to the variation in the plot and the coefficient standard errors, the fitted models are virtually the same.

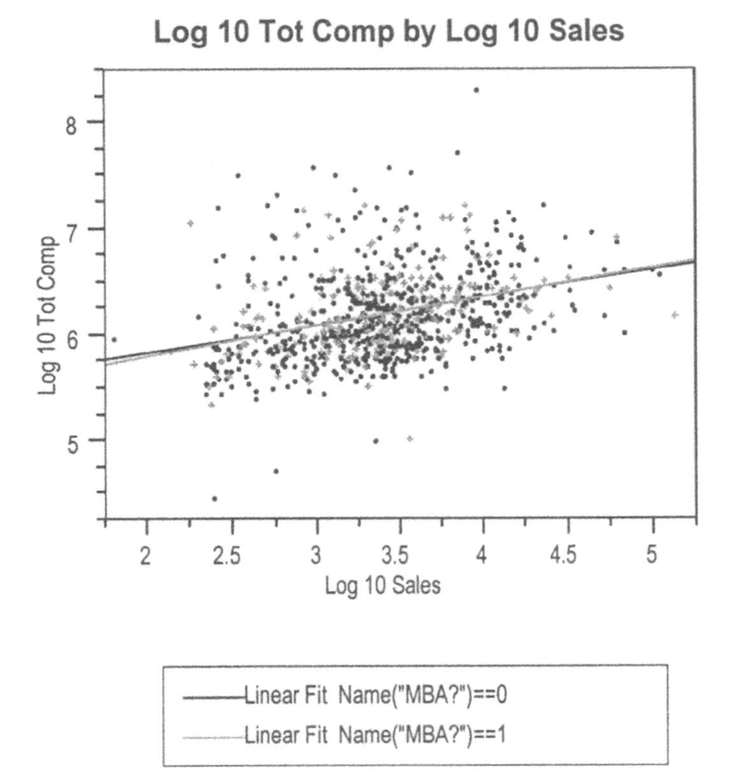

Linear Fit MBA?=0 (no MBA)

Term	Estimate	Std Error	t Ratio	Prob>|t|
Intercept	5.30	0.11	47.20	<.0001
Log 10 Sales	0.26	0.03	7.97	<.0001

Linear Fit MBA?=1 (has MBA)

Term	Estimate	Std Error	t Ratio	Prob>|t|
Intercept	5.24	0.16	32.66	<.0001
Log 10 Sales	0.28	0.05	5.86	<.0001

The nature of executive compensation varies by industry. Some industries pay more handsomely than others. The manner in which the compensation is determined also varies. To bring a bit of homogeneity to our analysis, now we are going to focus on the 168 companies in the financial segment.

In the figure shown below, the brush tool was used to select the points associated with the financial segment (the largest segment in this collection). Use the shift key to extend the selection to all of the points in this column. (If you expand the horizontal axis you can keep the group labels from overprinting, but then it won't fit across this page.)

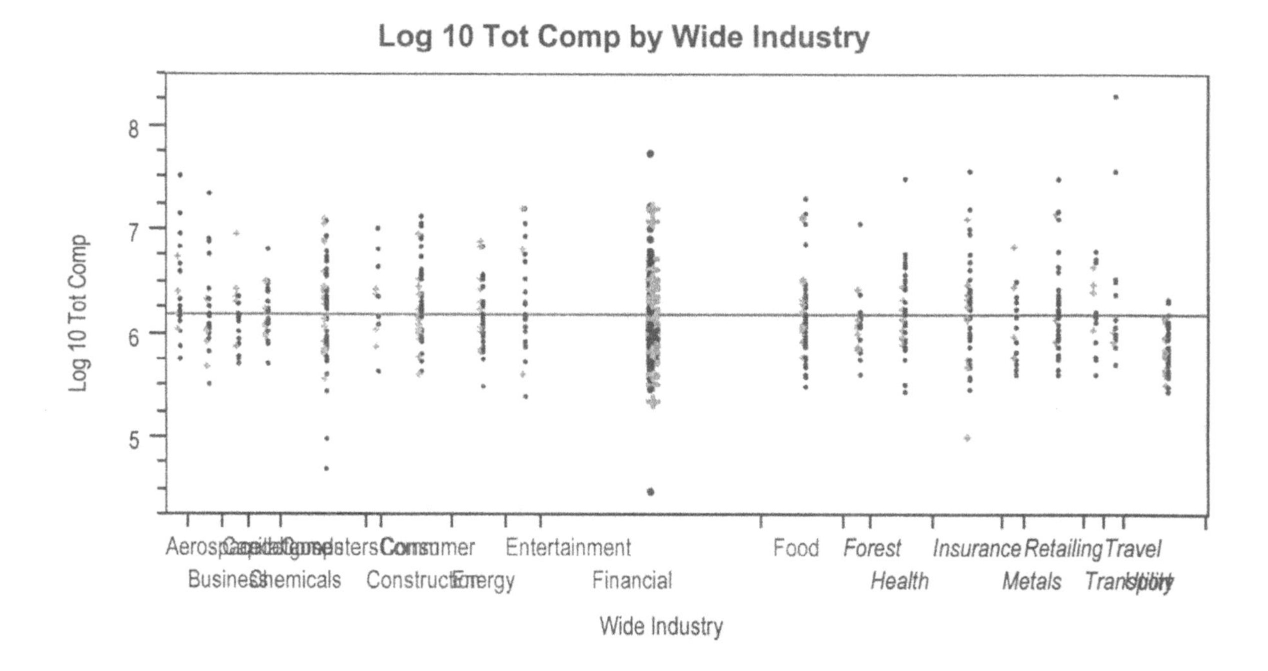

To exclude companies from other segments, pick the *Row Selection* item from the *Rows* menu, then choose the *Invert selection* command from next menu. It is easier to select the rest this way rather than try to brush all but the financial companies. Now use the *Exclude/Unexclude* as well as the *Hide/Unhide* commands from the *Rows* menu to exclude the others from the rest of the analysis and have them not appear in the plots.

With the analysis restricted to the financial segment, we obtain the following relationship between log total compensation and log sales. The elasticity is quite a bit higher in the financial segment than overall. The elasticity for these 168 companies is 0.53% for each percentage gain in sales, compared to 0.27% overall.

A distinct outlier appears at the lower edge of the plot. What company is this? The point codes make it easy to see that this executive does not have an MBA.

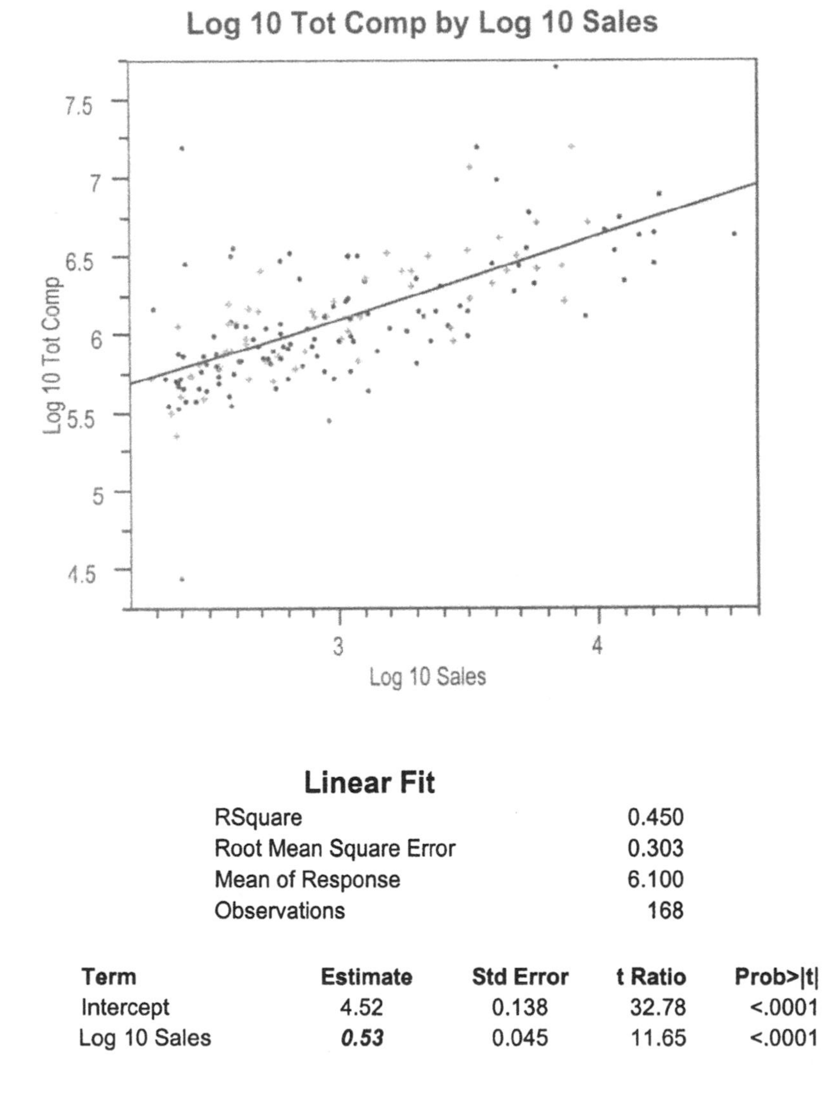

Linear Fit

RSquare	0.450
Root Mean Square Error	0.303
Mean of Response	6.100
Observations	168

Term	Estimate	Std Error	t Ratio	Prob>|t|
Intercept	4.52	0.138	32.78	<.0001
Log 10 Sales	***0.53***	0.045	11.65	<.0001

So how do the MBAs in the financial segment do? The differences are again slight, but a bit more apparent than overall. Compensation for the 56 MBA's rises more rapidly with sales than does that of the 112 financial executives who lack this degree. Also, the SD about the fit for the MBAs is about 70% of that for the others. The root mean squared error for the fit to the MBAs is .235 versus .332 for the fit to the other executives.

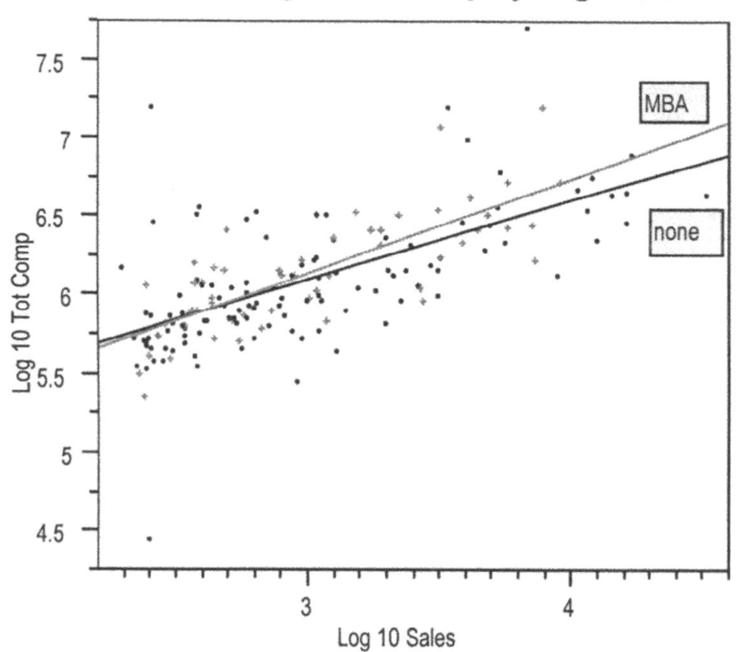

Linear Fit MBA?=0 (none)

RSquare	0.396
Root Mean Square Error	***0.332***
Mean of Response	6.083
Observations	***112***

Term	Estimate	Std Error	t Ratio	Prob>|t|
Intercept	4.591	0.178	25.74	<.0001
Log 10 Sales	***0.500***	0.059	8.49	<.0001

Linear Fit MBA?=1 (have MBA)

RSquare	0.605
Root Mean Square Error	***0.235***
Mean of Response	6.136
Observations	***56***

Term	Estimate	Std Error	t Ratio	Prob>|t|
Intercept	4.336	0.201	21.63	<.0001
Log 10 Sales	***0.598***	0.066	9.09	<.0001

A multiple regression adding the dummy variable *MBA?* and the interaction (*MBA? * Log 10 Sales*) indicates that these two variables do not improve upon the fit obtained with $\log_{10}$ *Sales* alone. The partial F-statistic for the addition of these two factors is

$$\begin{aligned}\text{Partial F} &= \frac{\text{Change in } R^2 \text{ per added predictor}}{\text{Variation remaining per residual}} \\ &= \frac{(0.455-0.450)/2}{(1-0.455)/164} \\ &= 0.75\end{aligned}$$

which is quite small and not significant.

Response: Log 10 Tot Comp

RSquare	0.455
Root Mean Square Error	0.303
Mean of Response	6.100
Observations	168

Expanded Estimates

Term	Estimate	Std Error	t Ratio	Prob>|t|
Intercept	4.464	0.153	29.15	<.0001
Log 10 Sales	0.549	0.050	10.91	<.0001
MBA?[0]	-0.019	0.025	-0.76	0.4481
MBA?[1]	0.019	0.025	0.76	0.4481
(Log 10 Sales-2.99)*MBA?[0]	-0.049	0.050	-0.97	0.3323
(Log 10 Sales-2.99)*MBA?[1]	0.049	0.050	0.97	0.3323

Analysis of Variance

Source	DF	Sum of Squares	Mean Square	F Ratio
Model	3	12.607	4.20224	45.6459
Error	164	15.098	0.09206	Prob>F
C Total	167	27.705		<.0001

Notice that you can recover the original separate fits from this one multiple regression. For example, the slope for the MBAs (coded as MBA = 1) is

MBA slope = Overall slope + MBA effect = 0.549 + 0.049 = 0.598.

Before continuing, residual analysis for this model reveals a distinct negative outlier. This is the same point that we noticed in the original scatterplot of these data. Since this observation is also somewhat leveraged, its presence accentuates the overall slope for $\log_{10}$ of sales by "pulling" the line lower on the left, as seen in the leverage plot. (What is unusual about this company, Citizens Bancorp?)

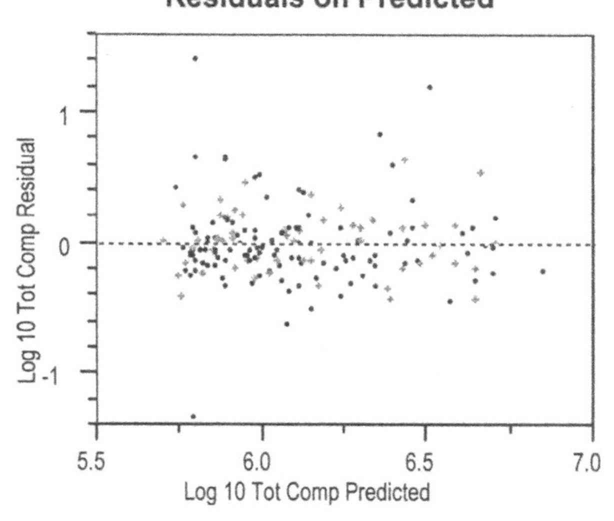

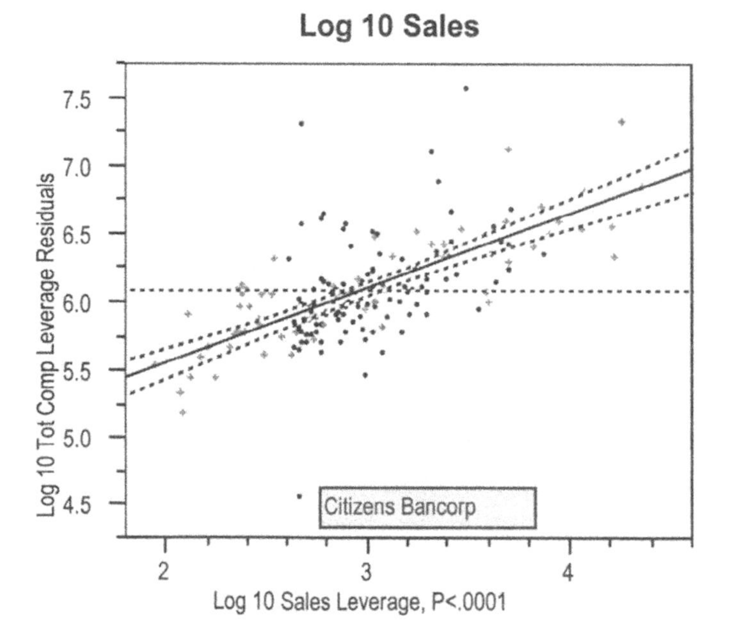

How much does this one outlier affect the analysis? With 168 observations, the slope and intercept do not change by much. However, the changes are rather large considering this one observation is only 0.6% of the data.

Response: Log 10 Tot Comp (with Citizens Bancorp)

RSquare	0.455
Root Mean Square Error	0.303
Observations	168

Expanded Estimates

Term	Estimate	Std Error	t Ratio	Prob>|t|
Intercept	4.464	0.153	29.15	<.0001
Log 10 Sales	0.549	0.050	10.91	<.0001
MBA?[0]	-0.019	0.025	-0.76	0.4481
MBA?[1]	0.019	0.025	0.76	0.4481
(Log 10 Sales-2.99)*MBA?[0]	-0.049	0.050	-0.97	0.3323
(Log 10 Sales-2.99)*MBA?[1]	0.049	0.050	0.97	0.3323

Excluding Citizens Bancorp

RSquare	0.468
Root Mean Square Error	0.286
Observations	167

Expanded Estimates

Term	Estimate	Std Error	t Ratio	Prob>|t|
Intercept	4.508	0.145	31.19	<.0001
Log 10 Sales	0.536	0.047	11.30	<.0001
MBA?[0]	-0.013	0.023	-0.56	0.5740
MBA?[1]	0.013	0.023	0.56	0.5740
(Log 10 Sales-2.99)*MBA?[0]	-0.062	0.047	-1.30	0.1959
(Log 10 Sales-2.99)*MBA?[1]	0.062	0.047	1.30	0.1959

Other outliers remain, but the leverage plot for the log of sales shows that they are not so highly leveraged.

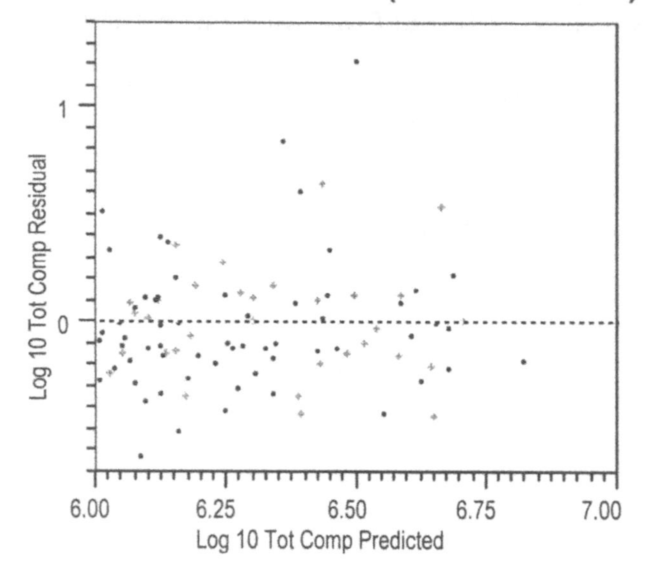

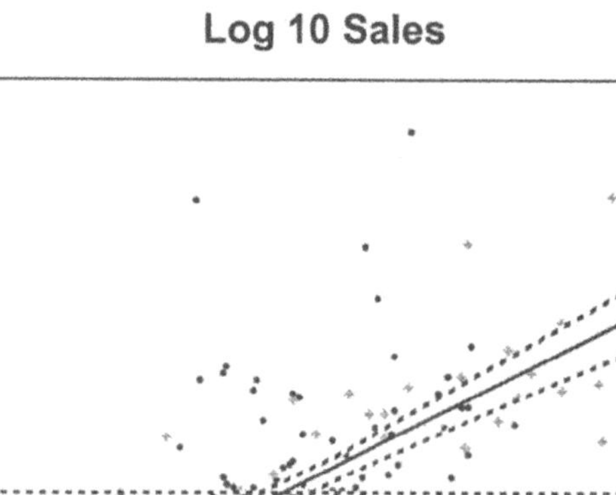

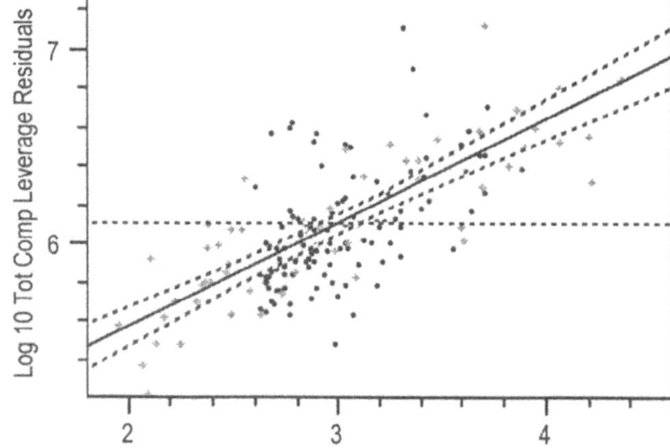

To get a more precise estimate of the effect of having an MBA degree, we need a better model that can explain some of the substantial remaining residual variation. What other factors might explain more of the residual variation? Common sense suggests other relevant factors, such as those involving the age/experience of the CEO and number of years at the firm. Also, we would expect the CEOs' salaries to be affected by the long term growth of the company as measured by the five-year rate of return on its stock. A scatterplot matrix is useful to examine other potential factors. After a bit of trial and error, we found a model that includes, in addition to the $\log_{10}$ of sales, factors related to

company stock (% stock owned by CEO, stock return over the last 5 years)

and

age of the CEO (age and years at firm).

Other factors considered, profits and years as CEO do not improve this fit. Each of the coefficients is significant, though the addition of these three variables explains only about 6% more variation. Missing values reduce the sample size to 163 from 167 (excluding Citizens Bancorp).

Response: Log 10 Tot Comp

RSquare	0.527
Root Mean Square Error	0.270
Mean of Response	6.113
Observations	163

Term	**Estimate**	**Std Error**	**t Ratio**	**Prob>|t|**
Intercept	3.864	0.222	17.37	0.00
Log 10 Sales	0.501	0.042	11.96	0.00
Return over 5 yrs	0.005	0.001	3.89	0.00
Age	0.014	0.004	3.59	0.00
Years Firm	-0.004	0.002	-2.14	0.03

Analysis of Variance

Source	**DF**	**Sum of Squares**	**Mean Square**	**F Ratio**
Model	4	12.795	3.19874	43.97
Error	158	11.494	0.07274	**Prob>F**
C Total	162	24.289		<.0001

Residual analysis of this model shows no particular anomalies (a bit of skewness toward positive values).

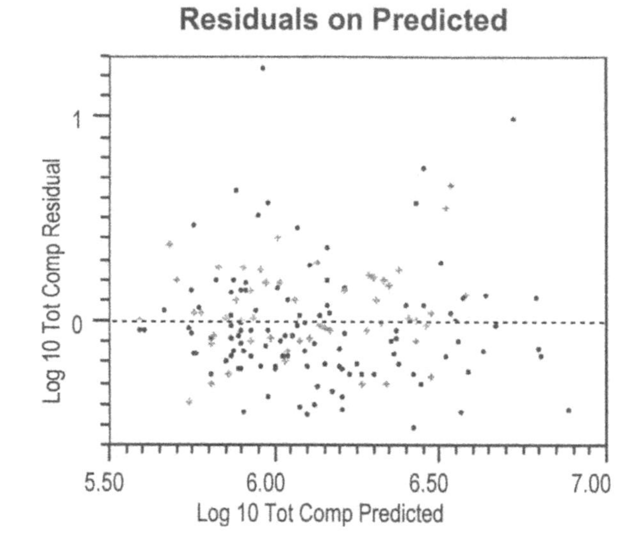

The leverage plot for stock returns shows a very highly leveraged company (First USA).

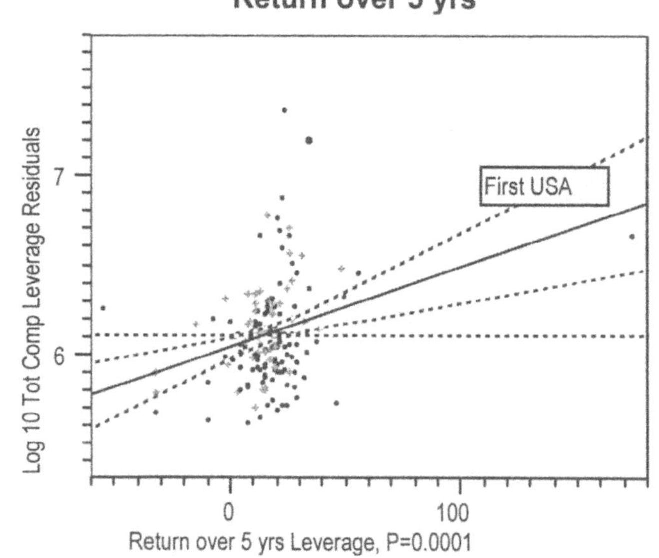

Excluding First USA does not have a dramatic effect upon the fitted model since this observation, though highly leveraged, has a rather small residual and is not influential.

Response: Log 10 Tot Comp (with First USA)

RSquare	0.527
Root Mean Square Error	0.270
Mean of Response	6.113
Observations	163

Term	Estimate	Std Error	t Ratio	Prob>|t|
Intercept	3.864	0.222	17.37	0.00
Log 10 Sales	0.501	0.042	11.96	0.00
Return over 5 yrs	0.005	0.001	3.89	0.00
Age	0.014	0.004	3.59	0.00
Years Firm	-0.004	0.002	-2.14	0.03

Excluding First USA

RSquare	0.525
Root Mean Square Error	0.270
Mean of Response	6.110
Observations	162

Term	Estimate	Std Error	t Ratio	Prob>|t|	VIF
Intercept	3.860	0.223	17.32	<.0001	0.0
Log 10 Sales	0.502	0.042	11.95	<.0001	1.0
Return over 5 yrs	0.005	0.002	3.32	0.0011	1.0
Age	0.014	0.004	3.55	0.0005	1.4
Years Firm	-0.004	0.002	-2.21	0.0288	1.4

At this point in the analysis, in effect we are doing stepwise regression "by hand." Although it is important to be careful and look at the various plots, it takes quite a bit of time and patience, so much so that we have ignored issues of potential interactions among these factors and the possibility of nonlinearities (which do not show up in leverage plots). When used carefully, stepwise regression addresses some of these problems.

JMP offers a very flexible stepwise regression tool via the *Fit Model* platform. You use the same dialog as in multiple regression, but change the "personality" of the modeling process by using the lower right pop-up dialog to change from "Standard Least Squares" to "Stepwise". There is also a trick to picking the variables. In addition to the usual set of predictors, we also want to check for nonlinearities and interactions. JMP will build all of the needed variables for us if we're careful. First, pick the relevant set of variables to use as predictors, starting with the four in the previous regression plus the MBA indicator. Add the other variables in the data set to these five, skipping factors like state or city of birth. To enter them into the model, use the *Effect Macros* button and choose the *Response surface* option; this choice adds the chosen collection of factors, squared factors, and the pairwise interactions to the model as potential predictors. A click on the *Run Model* button opens a large dialog that controls the stepwise regression process.

We'll try two variations on the stepwise algorithm and compare results, using the 166 financial companies with compensation data. First, let's start with nothing (the initial screen) and let the stepwise procedure add variables to the model (so-called "forward" stepwise regression). With a total of 44 predictors, we need to set the p-value for inclusion to 0.05/44 $\approx$ 0.001. Enter this value in the "Prob to Enter" and "Prob to Leave" fields, replacing the default 0.250 and 0.100. Now click the *Go* button and sit back! You do not have to wait very long since only two factors, *Log Sales* and *Return over 5 years*, pass this very strict test. Alternatively, we can start from a model with all of the variables (use the *Enter All* button) and use backward elimination to whittle the model down. It takes a bit longer, but arrives at the same model. Finally, as a third variation, we can start from the model that our deliberate searching had found with *Log Sales, Return over 5 years, Age,* and *Years Firm*. We can force these into the model and make them stay by clicking on the "Lock" item for each. Stepwise regression again removes all of the unlocked factors and will not expand the model.

When you are finished with stepwise selection and want to go back to the usual regression modeling process, just click the *Make Model* button to get a new regression dialog which has the predictors chosen by the stepwise algorithm. (Note, this analysis was done without the two previously identified outliers, First USA and Citizen's Bancorp.)

At this point, it is hard to argue that an MBA degree has a significant effect. If we force feed this variable into the model, we obtain the following fit. The estimate lies near the edge of significance, but at least has the expected sign!

Response: Log 10 Tot Comp

RSquare	0.532
Root Mean Square Error	0.269
Mean of Response	6.110
Observations	162

Expanded Estimates

Term	Estimate	Std Error	t Ratio	Prob>|t|
Intercept	3.822	0.224	17.10	<.0001
Log 10 Sales	0.499	0.042	11.89	<.0001
Return over 5 yrs	0.005	0.002	3.36	0.0010
Age	0.015	0.004	3.78	0.0002
Years Firm	-0.004	0.002	-2.19	0.0298
MBA?[0]	-0.034	0.023	-1.47	0.1429
MBA?[1]	0.034	0.023	1.47	0.1429

Analysis of Variance

Source	DF	Sum of Squares	Mean Square	F Ratio
Model	5	12.843488	2.56870	35.45
Error	156	11.301975	0.07245	**Prob>F**
C Total	161	24.145462		<.0001

The normal quantile plot from this last fit shows statistically significant deviation from normality, but it is hard to judge what it might mean. Perhaps the data have some clusters; clustering can produce a histogram with thicker tails (more extreme values) than the normal. Maybe we need to divide the companies in this collection into yet more well-defined groups, such as separating regional banks from Wall Street firms or using information about the school granting the MBA degree. Who are those in the group of outliers that have the higher than predicted salaries?

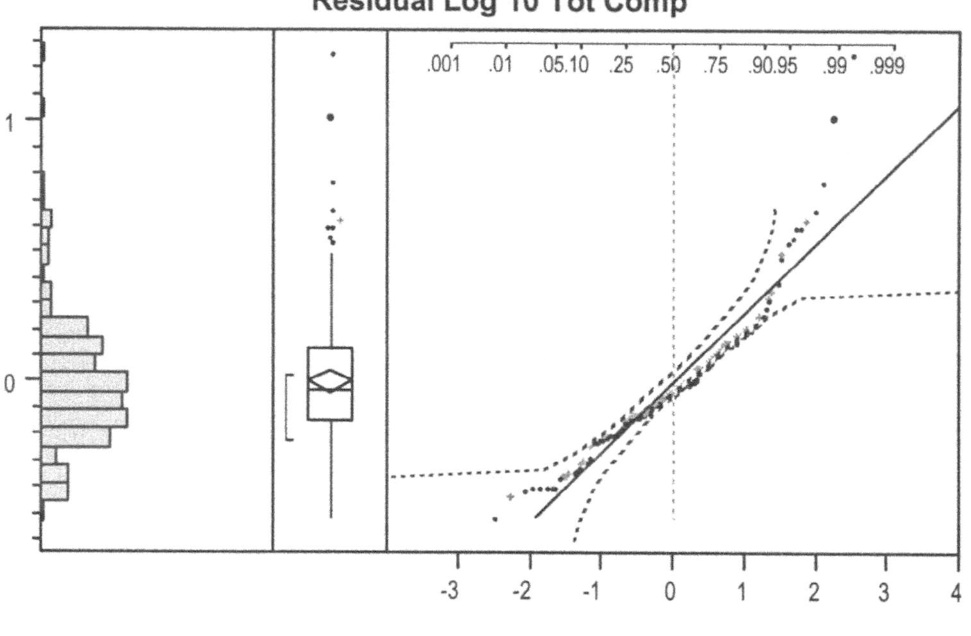

Highly publicized salaries for CEO's in the US. have generated sustained interest in the factors related to total compensation. Does it matter how long the executive has been at the firm? Do CEO's pay themselves less if they have a large stake in the stock of the company? And, does having an MBA increase executive salaries?

Back to our earlier question: How does the MBA degree fit into all of this, now that we have improved the fit of our equation? The effect is in the "right" direction, but at best "marginally" statistically significant. Nonetheless, the difference in compensation obtained by MBA's suggested by this fit is rather large.

To illustrate how to go further (admitting a lack of honest significance), the fit implies that the comparison between having and not having an MBA is *twice* the fitted coefficient, giving

$$\begin{aligned} 0.068 &= \log_{10} \text{Total Comp with MBA} - \log_{10} \text{Total Comp without MBA} \\ &= \log_{10} \frac{\text{Total Comp with MBA}}{\text{Total Comp without MBA}} \end{aligned}$$

Raising 10 to the power from this equation gets rid of the logarithm and implies

$$10^{.068} \approx 1.17 = \frac{\text{Total Comp with MBA}}{\text{Total Comp without MBA}}$$

In words, CEOs in the financial segment who have an MBA are earning about 17% more than their peers who lack an MBA. How much money is this? Since the median total compensation is about $2,000,000, a 17% difference is $340,000 per year. Also be aware that this analysis does not account for the high proportion of MBAs who get this far!

Using Stepwise Regression for Prediction

Nineties.jmp

Some of our previous analyses of stock market returns suggest that these series fluctuate randomly. So, why not use models that possess similar random variation to predict the market returns?

Is it possible to build a model that uses random noise to predict the stock market?

The data for this example is a subset of the financial markets data studied in *Basic Business Statistics* (FinMark.jmp). This subset has the monthly value-weighted returns for the 48 months in 1990-1993. The goal is to try to predict the returns for the months of 1994. Here is a plot of the 48 months of value weighted returns.

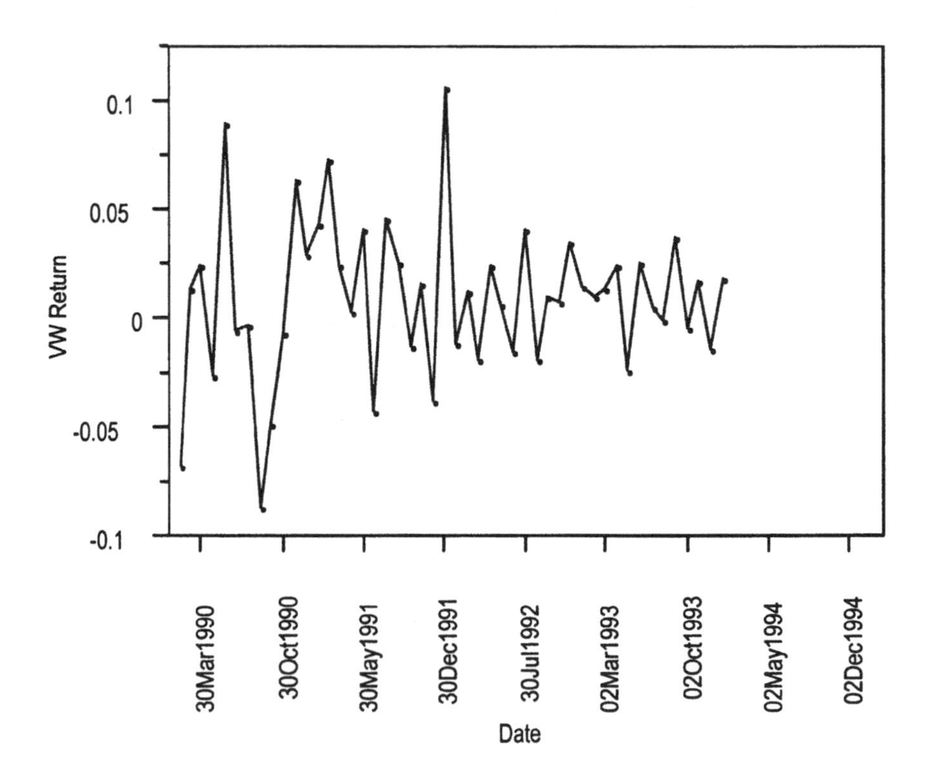

The returns are more volatile in the early 1990s, but they seem to have settled down near the end of the series. The goal is to predict the next 12 values for this series — the empty part of the plot — using 10 predictors included in the data set.

For building the predictions of this series, we have created 10 columns in the spreadsheet labeled X1 – X10. Each column is defined by a formula; and all of the formulas are the same

?normal

Each column is a sample of 60 observations from a standard normal population. (48 months for the observation period, plus another 12 for the prediction period.)

We will use stepwise regression to build a superficially impressive regression model that will explain over half of the variation in the returns. To begin, open the Fit Model dialog and select as response *VW Return*. For predictors, choose the block X1–X10, then enter them into the model using the *Response Surface* item given by pressing the *Macros* button. This adds the variables X1–X10, $X1^2 – X10^2$, and all of the 45 cross-product terms to the set of predictors. Before you click the *Run Model* button, select the Stepwise Regression option just above this button. Now when you click *Run Model*, you get JMP's handy stepwise regression tool. Leave the options alone, and simply click on the *Go* button on the left.

The default options lead to a rather big regression model with a rather impressive $R^2 = 0.94$[1]. To get a more complete set of output using the predictors chosen by stepwise regression, click the *Make Model* button to get the standard *Fit Model* dialog, pre-filled with the variables chosen by the stepwise procedure. Some of the considerable output from fitting that model appears on the page that follows. You will notice that quite a few of the variables that were included are not significant at the 0.05 level in the fit. Some are needed and some are not. As we discussed with categorical predictors, it's a good idea to keep the components of an interaction in the model even if they alone are not significant. So we'll keep terms like X5 since it is part of a significant interaction with X9. We'll get rid of the other interactions, though, that don't appear useful. You probably want to do this one at a time to make sure that once you eliminate something, a correlated factor might appear useful that does not seem so now (e.g., $X1^2$). The ones that are removed in the end are italicized in the output.

[1] Some versions of JMP may give a slightly different fit for these data.

Although some of the individual coefficients are not significantly different from zero, the overall fit of the model is quite significant (judging from the overall F statistic's p-value).

Response VW Return

RSquare	0.941
Root Mean Square Error	0.015
Observations	48

Analysis of Variance

Source	DF	Sum of Squares	Mean Square	F Ratio
Model	30	0.057589	0.001920	9.0536
Error	17	0.003605	0.000212	Prob > F
C. Total	47	0.061193		<.0001

Parameter Estimates

Term	Estimate	Std Error	t Ratio	Prob>|t|
Intercept	-0.0225	0.0084	-2.67	0.0161
X1	-0.0195	0.0051	-3.81	0.0014
X2	0.0053	0.0024	2.19	0.0425
X3	-0.0027	0.0038	-0.70	0.4907
X4	-0.0147	0.0047	-3.16	0.0057
X5	-0.0203	0.0038	-5.31	<.0001
X6	0.0108	0.0040	2.70	0.0151
X7	0.0033	0.0040	0.84	0.4102
X9	0.0013	0.0043	0.30	0.7698
X10	-0.0006	0.0059	-0.11	0.9136
(X1+0.18072)*(X1+0.18072)	0.0045	0.0027	1.68	0.1120
(X3-0.17051)*(X1+0.18072)	0.0306	0.0075	4.09	0.0008
(X3-0.17051)*(X3-0.17051)	-0.0046	0.0031	-1.51	0.1498
(X4+0.21857)*(X2+0.23514)	0.0109	0.0040	2.76	0.0134
(X5+0.0469)*(X2+0.23514)	0.0085	0.0027	3.14	0.0060
(X6-0.04737)*(X1+0.18072)	0.0120	0.0048	2.51	0.0223
(X6-0.04737)*(X3-0.17051)	-0.0225	0.0075	-3.01	0.0078
(X6-0.04737)*(X6-0.04737)	0.0066	0.0036	1.85	0.0822
(X7-0.27522)*(X1+0.18072)	-0.0273	0.0056	-4.88	0.0001
(X7-0.27522)*(X4+0.21857)	-0.0237	0.0073	-3.25	0.0047
(X7-0.27522)*(X5+0.0469)	-0.0158	0.0043	-3.64	0.0020
(X7-0.27522)*(X6-0.04737)	0.0132	0.0060	2.20	0.0420
(X7-0.27522)*(X7-0.27522)	0.0052	0.0039	1.35	0.1933
(X9-0.0028)*(X1+0.18072)	0.0217	0.0044	4.88	0.0001
(X9-0.0028)*(X4+0.21857)	0.0081	0.0048	1.68	0.1115
(X9-0.0028)*(X7-0.27522)	0.0117	0.0056	2.10	0.0509
(X9-0.0028)*(X9-0.0028)	0.0051	0.0021	2.42	0.0268
(X10+0.02929)*(X1+0.18072)	0.0088	0.0046	1.91	0.0729
(X10+0.02929)*(X3-0.17051)	-0.0072	0.0051	-1.40	0.1784
(X10+0.02929)*(X4+0.21857)	0.0194	0.0044	4.36	0.0004
(X10+0.02929)*(X9-0.0028)	-0.0018	0.0041	-0.43	0.6720

Here is the output having deleted the unhelpful interactions. All of the terms that remain are either significant (in the sense of having p-values less than 0.05) or are part of a significant interaction. The deletion of the 13 other predictors has caused R^2 to drop to 0.822 from 0.941. The partial F-statistic for this drop is [(0.941–0.822)13)]/[(1-0.941)/17] = 2.64 is close to, but not quite significant. The overall F-statistic remains significant, is a little smaller (8 versus 9).

Response VW Return

RSquare	0.822
Root Mean Square Error	0.019
Mean of Response	0.009
Observations	48

Analysis of Variance

Source	DF	Sum of Squares	Mean Square	F Ratio
Model	17	0.050280	0.002958	8.1302
Error	30	0.010913	0.000364	Prob > F
C. Total	47	0.061193		<.0001

Parameter Estimates

Term	Estimate	Std Error	t Ratio	Prob>|t|
Intercept	0.0014	0.0036	0.41	0.6880
X1	-0.0126	0.0039	-3.24	0.0029
X2	0.0066	0.0025	2.67	0.0121
X3	-0.0052	0.0036	-1.47	0.1527
X4	-0.0117	0.0041	-2.88	0.0073
X5	-0.0137	0.0038	-3.61	0.0011
X7	0.0004	0.0040	0.09	0.9277
X9	-0.0064	0.0033	-1.93	0.0625
X10	-0.0103	0.0034	-3.06	0.0046
(X3-0.17051)*(X1+0.18072)	0.0137	0.0044	3.14	0.0037
(X4+0.21857)*(X2+0.23514)	0.0135	0.0033	4.07	0.0003
(X5+0.0469)*(X2+0.23514)	0.0092	0.0025	3.75	0.0008
(X7-0.27522)*(X1+0.18072)	-0.0180	0.0046	-3.89	0.0005
(X7-0.27522)*(X5+0.0469)	-0.0173	0.0052	-3.32	0.0024
(X9-0.0028)*(X1+0.18072)	0.0189	0.0039	4.90	<.0001
(X9-0.0028)*(X4+0.21857)	0.0103	0.0038	2.68	0.0119
(X10+0.02929)*(X1+0.18072)	0.0129	0.0033	3.88	0.0005
(X10+0.02929)*(X4+0.21857)	0.0149	0.0034	4.34	0.0001

Even after removing the other factors, this fit looks quite impressive. It seems that you can model the stock market using a collection of random predictors, letting stepwise regression figure out how to combine them to get a good fit. Here are graphical evidence of the prediction ability and a residual plot.

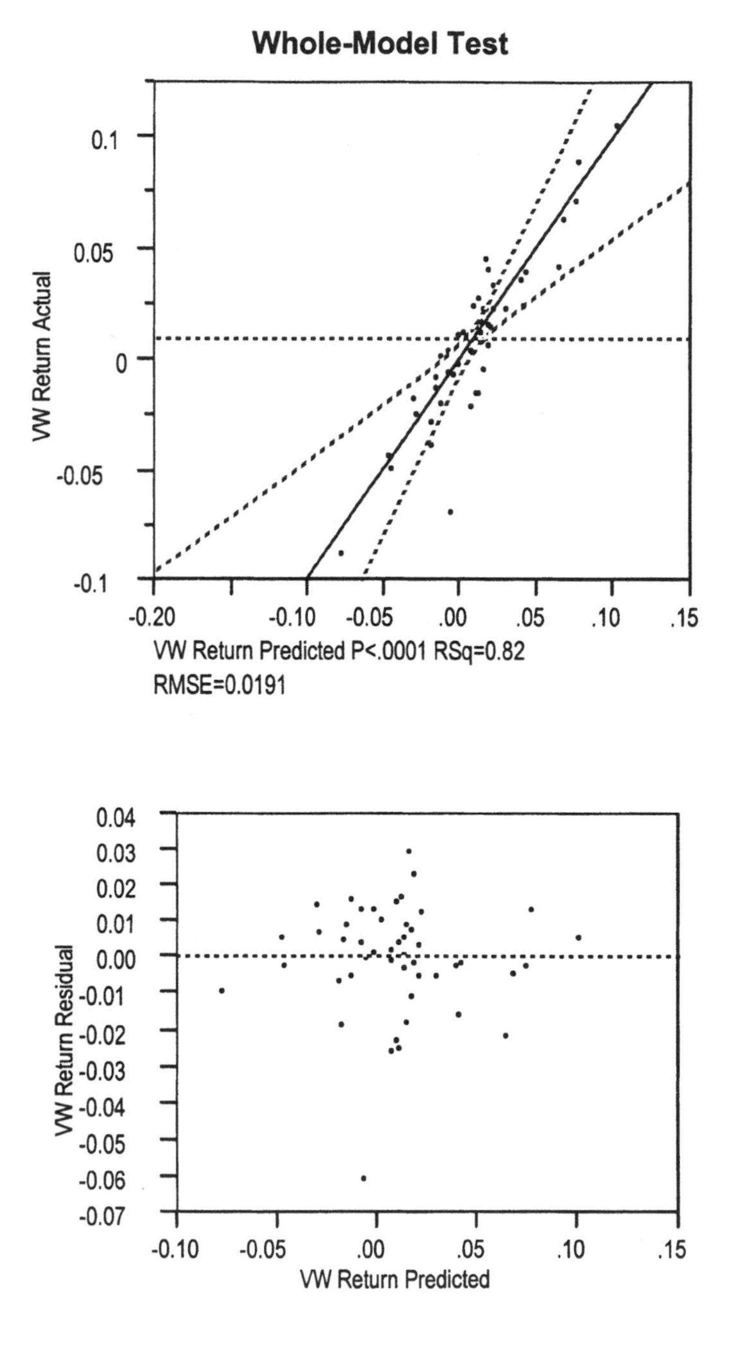

So, how well does this model predict future returns? *Complete Return* has the full set of 60 value-weighted returns. The predictions track the returns well, until the forecast period begins. Then the prediction errors appear to go "out of control".

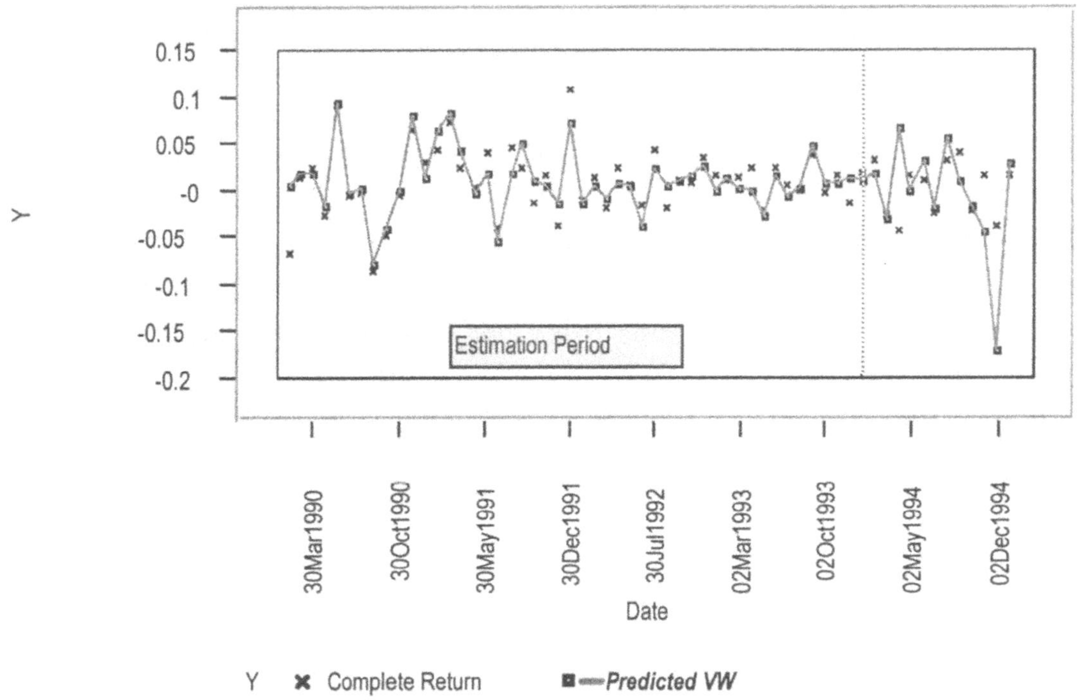

Y

0.15
0.1
0.05
-0
-0.05
-0.1
-0.15
-0.2

Estimation Period

30Mar1990 30Oct1990 30May1991 30Dec1991 30Jul1992 02Mar1993 02Oct1993 02May1994 02Dec1994

Date

Y × Complete Return ■ Predicted VW

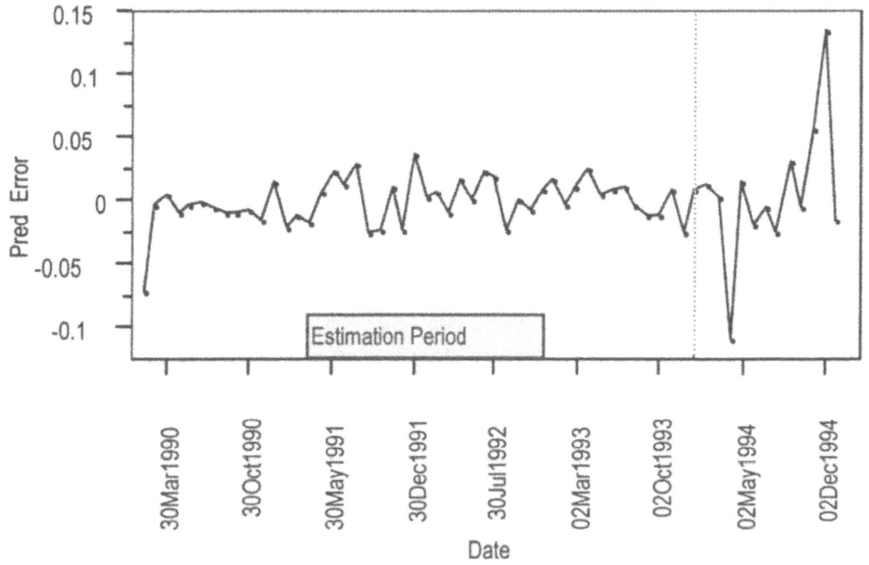

Pred Error

0.15
0.1
0.05
0
-0.05
-0.1

Estimation Period

30Mar1990 30Oct1990 30May1991 30Dec1991 30Jul1992 02Mar1993 02Oct1993 02May1994 02Dec1994

Date

Some of our previous analyses of stock market returns suggest that these series fluctuate randomly. So, why not use models that possess similar random variation to predict the market returns?

Is it possible to build a model that uses random noise to predict the stock market?

The stepwise model fits well. It has a significant overall F-statistic, and numerous coefficients are quite significant. The out-of-sample predictions, though, are quite poor. Predicting all of the future returns to be zero is more accurate in the sense of having less variation than this model for 1994. The SD of the actual returns is 0.03 whereas that of the prediction errors is 0.06. And when this model goes wrong, it can be really far off.

Complete Return **Pred Error**

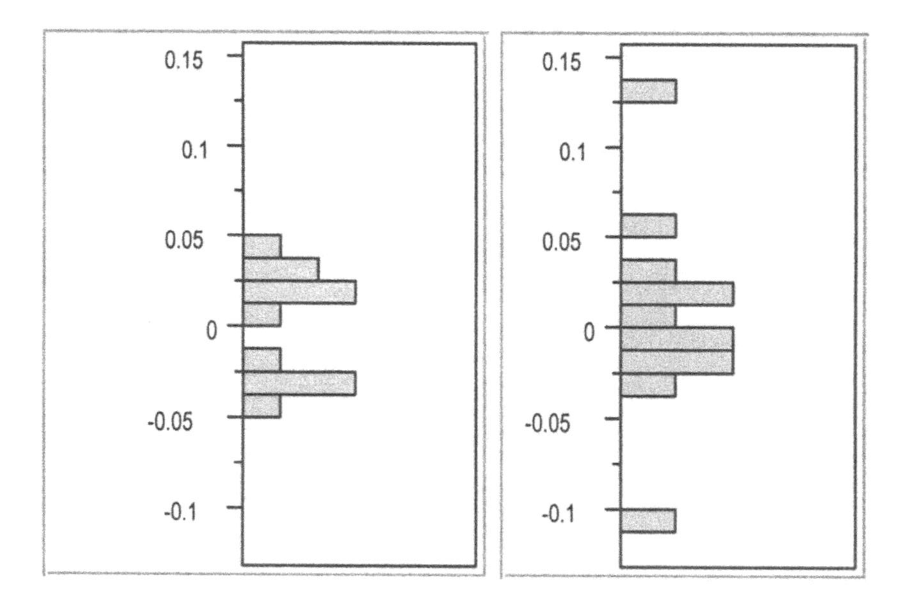

Moments

	VW Returns	Prediction Error
Mean	0.00031	0.00596
Std Dev	0.02927	0.05678
Std Err Mean	0.00845	0.01639
upper 95% Mean	0.01891	0.04203
lower 95% Mean	-0.01829	-0.03012
N	12	12

This example is typical of what happens when one uses a method like stepwise regression to *overfit* the data, capitalizing on random variation to make things appear significant by chance alone.

So, what might you have done to avoid this pitfall of stepwise regression? A better method is to use a harsher criterion for building the model in the first place. We have 65 = 10 + 10 + 45 variables to be considered by stepwise regression. Applying the adjustment from the Bonferroni inequality suggests that the *p*-value for entry to the model ought to be much smaller than the JMP default value of 0.25. If we want to keep our 5% overall criterion, then the Bonferroni inequality suggests reducing the p-value for entry to

$$p\text{-value for entry} = \frac{\text{standard p-value}}{\text{number predictors}} = \frac{0.05}{65} = 0.0008$$

Alas, JMP won't let you set the *p*-value for entry quite so small, so set it to the smallest available, 0.001. With this choice, nothing enters the model in the stepwise regression and you are led to the right conclusion: none of these predictors, including their interactions, is a useful predictor.

Class 9. Comparing Many Mean Values

One-way analysis of variance generalizes two-sample comparison of means (as developed in *Basic Business Statistics*). Rather than be limited to two groups, a one-way analysis of variance (anova) allows the comparison of several groups, identifying which group (if any) has the highest or lowest mean value.

Two methods identify the best (or worst) group and a third handles all possible pairwise comparisons. These procedures adjust the test statistics to reflect the number of comparisons being made. Without such an adjustment, one will eventually find statistically significant differences by chance alone, just as the stepwise regression in Class 9 finds structure in noise. Though our focus is the comparison of many mean values, these ideas are important in regression analysis as well.

Topics

Experimentation; treatment factor and response
Two-sample *t*-test
One-way anova
Multiple comparison and error rates
Assumptions and diagnostics (independence, equal variance and normality)

Examples

1. Selecting the best vendor (comparison with best)
2. Headache pain relief (all possible comparisons)
3. Analysis of variance and tests for linearity (a diagnostic procedure)

Key application

Selecting an incentive program. An oil company wants to determine which incentive program increases profits the most. It has three schemes available: one, the proverbial "free glass," two, a "free window wash," and three, a "determined attempt to keep the gas station environment ultra clean." How should the company go about deciding which is the most effective scheme? One way to think of this problem is to treat the three incentives as three different groups and then ask which group generates the highest mean increase in profits. This question is answered through a technique called the *analysis of variance*, the subject of this and the next class.

Definitions

Hsu comparison. A method of multiple comparisons designed to allow comparisons to either the smallest or largest mean value. Use this procedure if your interest is in finding the lowest or highest group and determining if this group is significantly different from the others.

LSD comparison. A method of multiple comparisons that is the same as doing the collection of pairwise t-statistics. Not recommended unless you are only interested in one comparison to begin with (in which case, why gather all of that other expensive data?).

Tukey-Kramer comparison. A method of multiple comparisons designed to allow all pairwise comparisons of the means. The number of possible pairwise comparison of k means is $k(k-1)/2$. For example, with $k=10$ groups, there are $10 \times 9/2 = 45$ pairwise comparisons.

Concepts

Analysis of variance (comparing the means of many groups, anova). In *Basic Business Statistics*, the two-sample t-test was introduced as a method for determining if a difference exists between the means of two groups. The analysis of variance is the extension of this technique to situations in which you have to compare more than two groups. In the gas station example above, there are three groups to compare. Just as in the two-sample t-test, confidence intervals are a very informative way of presenting results, but in the analysis of variance these confidence intervals become more complex because the problem we are dealing with is also more complex.

Multiple comparisons and error rates. It may be tempting to think that a straightforward way of treating this problem would be to compare the means by carrying out a whole series of two-sample t-tests. In the example above with three incentives, label them A, B and C. We could compare A with B, A with C and then B with C, performing three pairwise comparisons.

Recall that every hypothesis test has a Type I error rate, which is the chance that you reject the null hypothesis when in fact the null is true. For example, in the case of comparing two means, a Type I error occurs when in reality the means are the same, but you incorrectly conclude that there is a difference. A popular choice for the Type I error rate is 5% (i.e., reject the null hypothesis if the *p*-value is less than 0.05, or equivalently use a 95% confidence interval). With this choice, you will incorrectly declare a difference in means 5% of the time when in fact they are the same.

Now imagine performing a whole series of two-sample *t*-tests, each using the same Type I error rate of 5%. To make things specific, consider the gas station example above where we have three groups and do three two-sample *t*-tests for the three pairwise comparisons. Suppose that the null hypothesis is in reality true (i.e., there is no difference among the three population means; they are the same). What is the Type I error rate after performing three two-sample *t*-tests? That is, what is the chance that we have made at least one incorrect decision?

The chance of making at least one incorrect statement is one minus the probability of making *no* incorrect statements. "No incorrect statements" says we have to get the first one correct *and* second one correct *and* the third one correct. To compute the probability of events joined by "*AND*", multiply the separate probabilities together *so long as* the events are independent. Clearly, in this example, since the same groups are involved, these comparisons are *not* independent. We can still get an idea of the probability by using the Bonferroni inequality,

$$\Pr(E_1 \text{ OR } E_2 \text{ OR} \ldots \text{ OR } E_k) \leq \Pr(E_1) + \Pr(E_2) + \ldots \Pr(E_k).$$

In words, the probability that at least one of the k events E_1 through E_k occurs is no more than sum of the probabilities for all (draw a Venn diagram if it helps). In this example, an event is "making the wrong decision" on each of our three comparisons,

$$\Pr(\text{Error in Comparison}_1 \text{ OR Error in Comparison}_2 \text{ OR Error on Comparison}_3)$$
$$\leq 3 \times 0.05 = 0.15.$$

The point is that this upper limit is much higher than the 5% error rate used for each individual test. What this translates to in practice is that you are quite likely to make at least one error if you make enough comparisons. In the illustration above, it could be as large as one chance in seven – almost like rolling a fair die. It's like going fishing: if you do enough comparisons, then just by chance alone you will find something statistically significant in the data even though there's no real difference among the population means. This explains in part how stepwise regression can go awry; doing lots of 5% level tests produces many false positives. The special comparison procedures of this class keep the overall error rate at the chosen amount when working with designed experiments.

Heuristics

Bonferroni procedure. This is the quick fix for the error rate problem in making multiple comparisons. In the calculation carried out above, the chance of making an error on one of the three comparisons is bounded by $3 \times 0.05 = 0.15$, and in general

chance for error overall $\leq$ (number of comparisons) $\times$ (error rate for each).

If we were forced to do a series of three two sample t-tests, we can get an *overall error rate* of 5% by reducing the error rate of each to $0.05/3 = 0.017$. In this case, the chance for some error is less than

$$3 \times 0.05/3 = 0.05.$$

Another example: say we wanted to compare the means of five groups with two sample t-tests and we wanted an overall Type I error rate of 10%. With five groups there are 10 possible comparisons (i.e., $\binom{5}{2} = 10$). If we used individual error rates of $0.1/10 = 0.01$, then the overall error rate would be less than $10 \times 0.01 = 0.1$, as required.

Selecting the Best Vendor

Repairs.jmp

A large university would like to contract the maintenance of the personal computers in its labs. The university would like to make sure that the vendor chosen is both reliable and inexpensive. It has the costs charged by 10 possible vendors for essentially the same repair. The university would like to retain several vendors unless the data indicate excessive costs.

Which company should the university choose (or avoid), or is there no difference?

The data gathered by the university are the costs for a making a common type of repair to the machines in its computer labs. It has gathered data from its records for the costs of making this repair by each of 10 vendors. Using the inventory of records for each vendor, the university selected 10 visits for each, making a total sample of 100 repair visits to the university labs.

Before we move into the example, it is important to recognize that the assumptions of regression also apply to the analysis of variance. In many ways, the analysis of variance *is* regression, but using only categorical factors. Thus, the assumptions needed for the one-way analysis of variance method are

1. Independent observations, both within and between groups,
2. Equal group variances, and
3. Normality within each group.

The last two assumptions are easily checked from the model residuals. As usual, the assumption of independence is difficult to verify from the data alone and depends on the context. In this example, the random sampling of vendor records makes independence seem more appropriate. Had we collected records from, say, 10 consecutive repairs, the assumption would be more questionable.

We begin by using the *Fit Model* platform so that we get the now-familiar regression output. The R^2 measure of the goodness-of-fit is pretty low and the corresponding F-test (F=1.04) is not significant (the differences in means are not large when compared to the within-vendor variation). The vendor terms have equal standard errors since we have the same number of observations for each. The t-statistic for one of the vendors is significant, with $p = 0.03$ implying that costs for this vendor are significantly lower than the overall average. Or are they?

Response: RepairCost

RSquare	0.094
Root Mean Square Error	11.217
Mean of Response	150.390
Observations	100

Expanded Estimates

Term	Estimate	Std Error	t Ratio	Prob>|t|
Intercept	150.39	1.12	134.07	<.0001
Vendor[1]	*-7.39*	*3.37*	*-2.20*	*0.0307*
Vendor[2]	3.11	3.37	0.92	0.3579
Vendor[3]	-4.69	3.37	-1.39	0.1668
Vendor[4]	0.01	3.37	0.00	0.9976
Vendor[5]	0.91	3.37	0.27	0.7875
Vendor[6]	-1.09	3.37	-0.32	0.7468
Vendor[7]	2.11	3.37	0.63	0.5322
Vendor[8]	4.21	3.37	1.25	0.2142
Vendor[9]	2.61	3.37	0.78	0.4400
Vendor[10]	0.21	3.37	0.06	0.9504

In *Basic Business Statistics*, we discussed how repeated testing in quality control increases the chance for a false positive (declaring the process out of control when it's not). A similar issue arises here. With so many t-statistics, what is the probability of one being significant by chance alone? Since the results are not independent (each is compared to an overall average, and the data are pooled to estimate a common scale), we can't multiply probabilities, but we can bound the probability using the Bonferroni inequality.

$$\begin{aligned} &\Pr(\text{At least one } p\text{-value} < 0.05 \text{ by chance alone}) \\ &= \Pr(p_1 < 0.05 \;\; OR \;\; p_2 < 0.05 \;\; OR \ldots OR \; p_{10} < 0.05) \\ &\le \Pr(p_1 < 0.05) + \Pr(p_2 < 0.05) + \ldots + \Pr(p_{10} < 0.05) = 10 \times 0.05 = 0.5 \end{aligned}$$

The probability of a significant value by chance alone could be as large as 50%. What can we do? One approach is to use a stricter rule for p-values. If we use $0.05/10 = 0.005$ as a cutoff, then the same calculation as above shows that the chance is then at most 5%. With this smaller threshold for p-values, p=0.03 is too large, and no term is significant.

Because multiple comparisons are so common in anova problems, JMP offers a variety of special tools when using the *Fit Y by X* platform. Since the variable *Vendor* is declared to be nominal, JMP automatically provides the proper analysis. The *Means, Std Dev, Std Err* command on the analysis button produces the following output. From the output, Vendor #1 has the lowest average cost, with Vendor #3 not far behind. The SD's vary, but (with the exception of Vendor #6) are reasonably comparable.

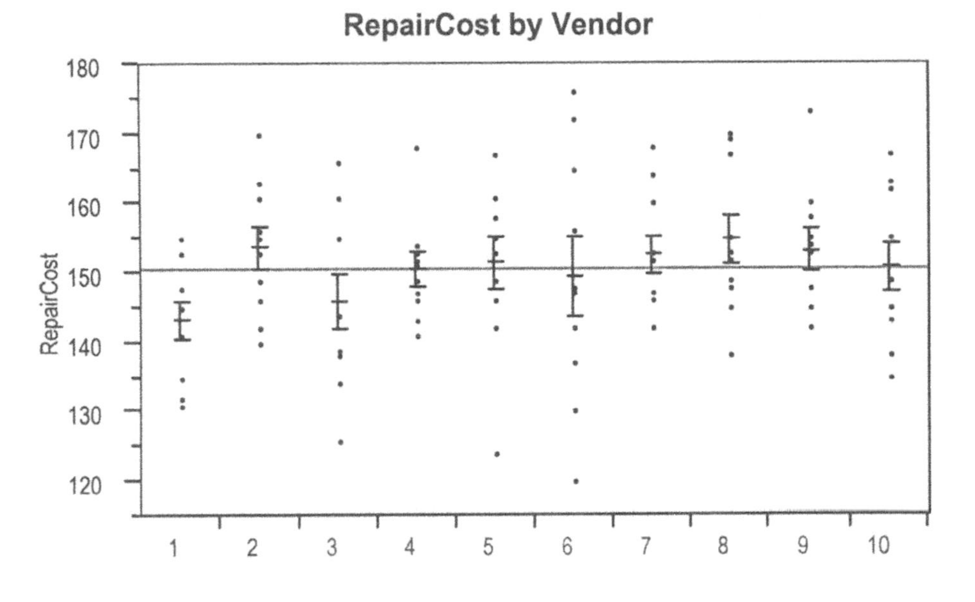

Means and Std Deviations

Level	Number	Mean	Std Dev	Std Err Mean
1	10	143.0	8.26	2.61
2	10	153.5	9.54	3.02
3	10	145.7	12.89	4.08
4	10	150.4	7.52	2.38
5	10	151.3	12.07	3.82
6	10	149.3	18.14	5.74
7	10	152.5	8.72	2.76
8	10	154.6	10.80	3.42
9	10	153.0	9.49	3.00
10	10	150.6	10.90	3.45

Since we are trying to find the best vendor, the one with the lowest cost, we need to use a special comparison procedure. The procedure recognizes that we are letting the data generate the hypothesis and constructs the appropriate intervals. The comparisons we need in this case are generated by the *Compare means* command provided by the pop-up menu at the red triangle at the left of the title. This command augments the plot with comparison circles and adds a table to the output. JMP offers several ways to compare the mean values, and here we'll use Hsu's method.

The comparison circles created by this procedure identify significantly different vendors. Clicking on a circle highlights the vendors that differ significantly according to the selection comparison method. In this case, *none* of the vendors differs significantly from the minimum.

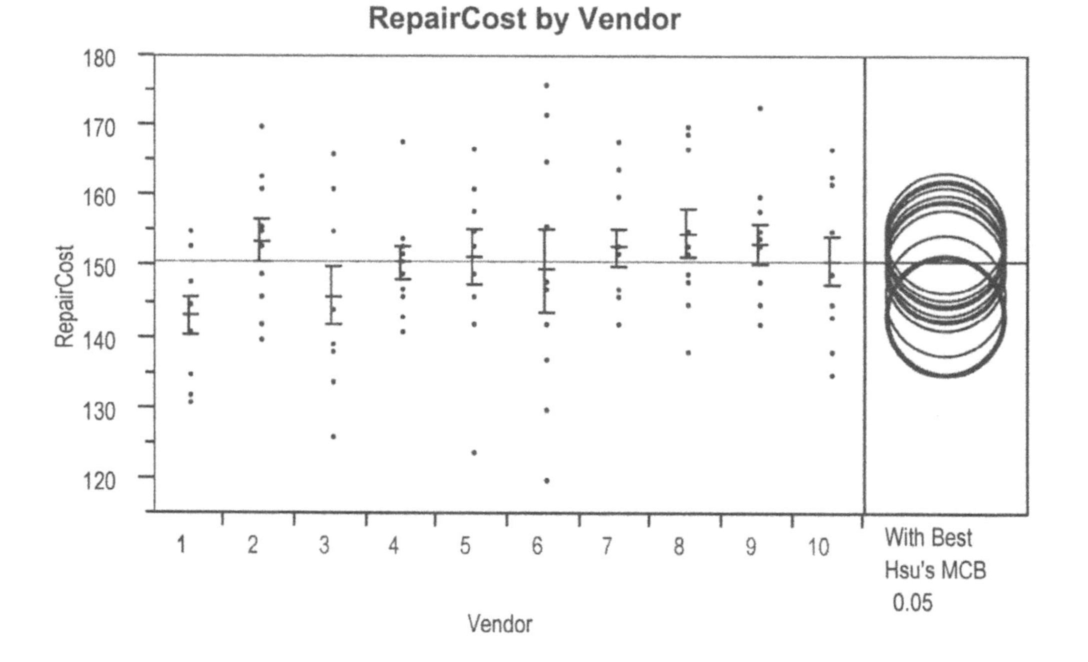

Two tables accompany the graphical output. The table of means comparisons shows the pairwise differences in the averages. The columns (and rows) are arranged in order of the mean values so that the most similar groups are adjacent to each other. Since Vendor #8 and Vendor #1 are most different, they are at the extremes of the table.

Means Comparisons

Dif=Mean[i]-Mean[j]	8	2	9	7	5	10	4	6	3	1
8	0.0	1.1	1.6	2.1	3.3	4.0	4.2	5.3	8.9	11.6
2	-1.1	0.0	0.5	1.0	2.2	2.9	3.1	4.2	7.8	10.5
9	-1.6	-0.5	0.0	0.5	1.7	2.4	2.6	3.7	7.3	10.0
7	-2.1	-1.0	-0.5	0.0	1.2	1.9	2.1	3.2	6.8	9.5
5	-3.3	-2.2	-1.7	-1.2	0.0	0.7	0.9	2.0	5.6	8.3
10	-4.0	-2.9	-2.4	-1.9	-0.7	0.0	0.2	1.3	4.9	7.6
4	-4.2	-3.1	-2.6	-2.1	-0.9	-0.2	0.0	1.1	4.7	7.4
6	-5.3	-4.2	-3.7	-3.2	-2.0	-1.3	-1.1	0.0	3.6	6.3
3	-8.9	-7.8	-7.3	-6.8	-5.6	-4.9	-4.7	-3.6	0.0	2.7
1	-11.6	-10.5	-10	-9.5	-8.3	-7.6	-7.4	-6.3	-2.7	0.0

Below the table of differences of means is a comparison table. For every pair of means, the comparison table below shows *one endpoint of a confidence interval* for the difference in population means. In the two-sample case, the 95% interval for $\mu_1 - \mu_2$ is approximately

$$[(\bar{X}_1 - \bar{X}_2) - 2\,\mathrm{SE(diff)}, (\bar{X}_1 - \bar{X}_2) + 2\,\mathrm{SE(diff)}]\ .$$

If this interval includes zero, then we cannot reject the null hypothesis $\mu_1 = \mu_2$. If the upper endpoint $(\bar{X}_1 - \bar{X}_2) + 2\,\mathrm{SE(diff)}$ is negative, then the interval does not include zero and we can reject the null and conclude a difference exists. For the comparison table, JMP shows the upper interval endpoints from a special interval. If the tabled value is negative, then those means are different. The bottom row gives the upper endpoints for $\bar{X}_1 - \bar{X}_j$. Since all of the endpoints are positive, none of the differences is significant.

Comparisons with the best using Hsu's MCB

Mean[i]-Mean[j]+LSD	8	2	9	7	5	10	4	6	3	1
8	12	13	14	14	16	16	17	18	21	24
2	11	12	13	13	15	15	15	17	20	23
9	11	12	12	13	14	15	15	16	20	22
7	10	11	12	12	14	14	14	16	19	22
5	9	10	11	11	12	13	13	14	18	21
10	8	9	10	10	12	12	13	14	17	20
4	8	9	10	10	11	12	12	13	17	20
6	7	8	9	9	10	11	11	12	16	19
3	3	5	5	6	7	7	8	9	12	15
1	*1*	*2*	*2*	*3*	*4*	*5*	*5*	*6*	*10*	*12*

If a column has any negative values, the mean is significantly greater than the min.

Rather than see whether any are significantly larger than the minimum, we might also be interested in whether *any* of the pairwise comparisons are significant. Again, the *Analysis* button provides the needed summary. From the *Compare all pairs* command, we see that none of the pairwise comparisons is significant. The comparison circles are wider for this method than the comparison to the best. Rather than just look at each mean compared to the minimum, this procedure considers all 45 pairwise comparisons.

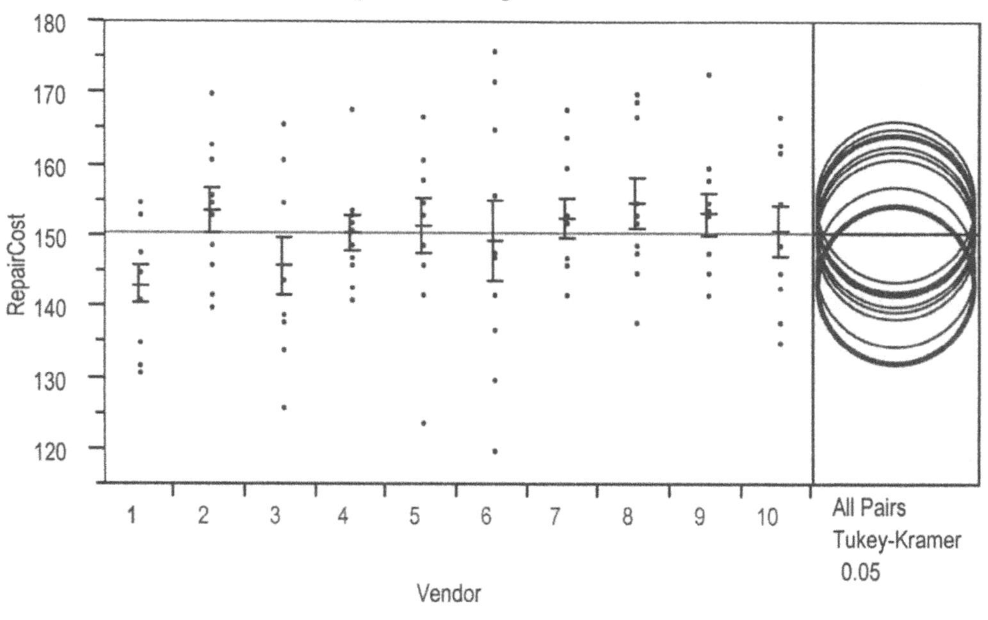

For Tukey-Kramer comparisons, JMP gives the *lower endpoint of a confidence interval* for the absolute value of the difference of the means $|\bar{X}_i - \bar{X}_j|$, denoted by **Abs(Dif)** in the labeling of the table on the next page. You can tell that JMP is giving a lower endpoint since it *subtracts* the LSD value, denoted **-LSD**, from the difference of the means. (LSD is an abbreviation for what would be $2SE(\bar{X}_1 - \bar{X}_2)$ in the case of two means; a value larger than 2 is needed for handling multiple comparisons.) Since $|\bar{X}_i - \bar{X}_j| > 0$, a positive lower endpoint implies the interval does not include zero and that the means are significantly different. Since all are negative in this example, we conclude that no two means are different.

Comparisons for all pairs using Tukey-Kramer HSD

Abs(Dif)-LSD	8	2	9	7	5	10	4	6	3	1
8	-16	-15	-15	-14	-13	-12	-12	-11	-7	-5
2	-15	-16	-16	-15	-14	-13	-13	-12	-8	-6
9	-15	-16	-16	-16	-15	-14	-14	-13	-9	-6
7	-14	-15	-16	-16	-15	-14	-14	-13	-9	-7
5	-13	-14	-15	-15	-16	-16	-15	-14	-11	-8
10	-12	-13	-14	-14	-16	-16	-16	-15	-11	-9
4	-12	-13	-14	-14	-15	-16	-16	-15	-12	-9
6	-11	-12	-13	-13	-14	-15	-15	-16	-13	-10
3	-7	-8	-9	-9	-11	-11	-12	-13	-16	-14
1	-5	-6	-6	-7	-8	-9	-9	-10	-14	-16

Positive values show pairs of means that are significantly different.

The more comparisons that we are willing to entertain, the wider the comparison circles become. A t-test assumes that we are interested in only one mean comparison and that we had chosen that comparison prior to peeking at the data. Hsu's comparison is valid for comparisons to either the minimum or maximum mean. Tukey-Kramer comparisons allow for snooping around at all of the comparisons. To accommodate this flexibility, the width of the comparison circles grows as we increase the number of possible comparisons.

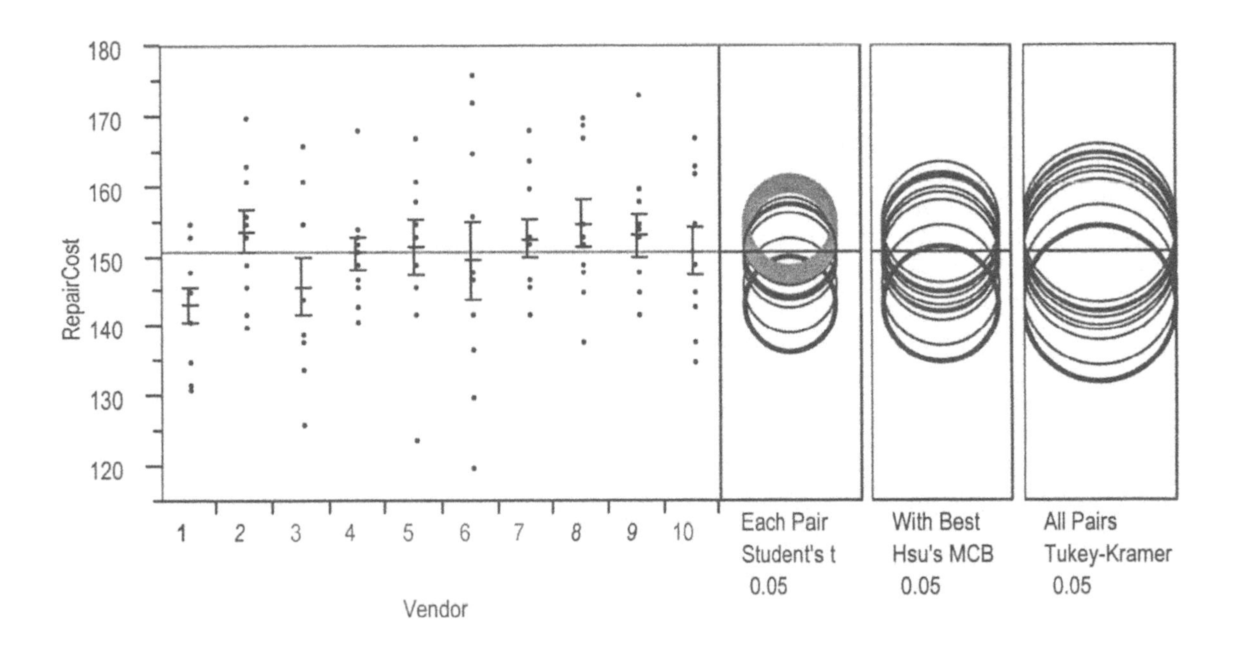

Before accepting the results provided by the analysis of variance, we need to check the assumptions of constant variance and normality. Three plots are most useful for investigating these assumptions: residuals plotted on predicted (fitted) values or comparison boxplots for checking the constant variance assumption, and quantile plots of residuals for checking the normality assumption. As in other examples, it makes most sense to check for constant variance before checking for normality. Often, data that lack constant variance will also deviate from normality — after all, they possess different levels of variation. If you check normality first, you might not learn the source of nonnormality, namely that the group variances are different.

Here is the plot of the residuals on the fitted values, which are just the averages for the 10 vendors. The dots fall on vertical lines, one for the mean of each group. The spread in the "column" above each mean suggests the variance for that group. There do not appear to be meaningful differences in variation. (You can also check the constant variance assumption by saving the residuals and using the *Fit Y by X* platform to get comparison boxplots.)

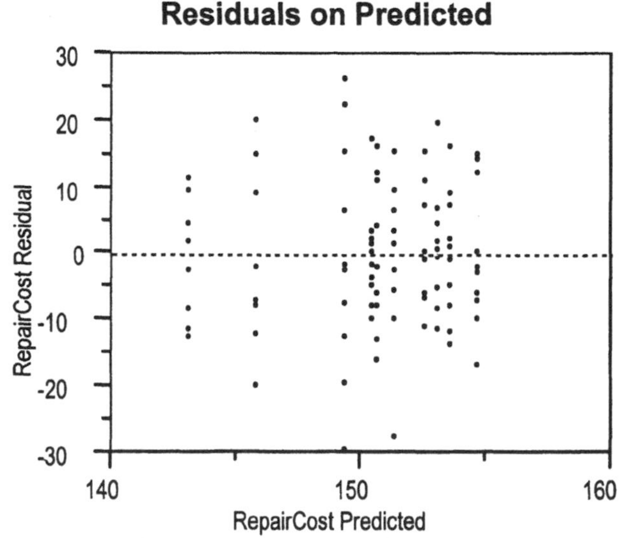

The normal quantile plot is also fine.

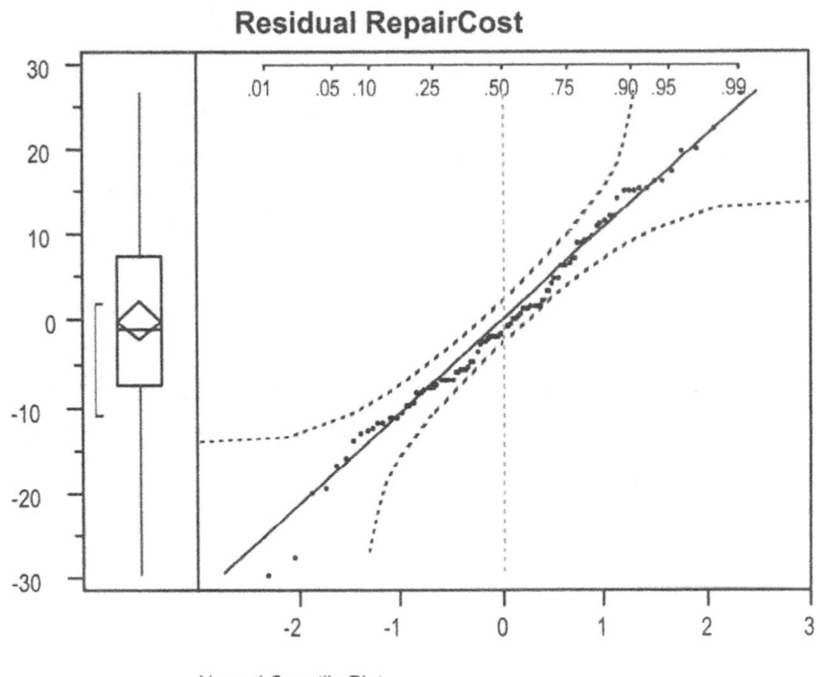

A large university would like to contract the maintenance of the personal computers in its labs. The university would like to make sure that the vendor chosen is both reliable and inexpensive. It has the costs charged by 10 possible vendors for essentially the same repair. The university would like to retain several vendors unless the data indicate excessive costs.

Which company should the university choose (or avoid), or is there no difference?

Although the sample means differ, the differences among the means are not large relative to the variation in the repair costs. Hence, we find no significant differences among the vendors. All of the needed assumptions seem valid.

Caveat emptor

The initial t-statistics of the regression of *Repair Cost* on *Vendor* showed a significant effect. It would be easy to claim that "I was only interested in that vendor, so this result is meaningful." The interpretation of a p-value is dependent upon how the analysis was done. Certainly the meaning of a p-value is lost when we add enough variables to a regression to virtually guarantee something will be significant. The Bonferroni approach is a reasonable insurance policy, though it's seldom used in practical applications (people *like* to find significant effects).

This sort of deception can also be more subtle. One can simply hide the other groups, suggesting no need for the protection offered by Bonferroni. For example, the simple pairwise two-sample t-tests would declare that the average repair costs of vendors 2, 8, and 9 are larger than those of vendor 1. If you don't recall the two-sample t-tests, they're introduced in Class 7 of *Basic Business Statistics*.

Headache Pain Relief

Headache.jmp

A pharmaceutical manufacturer would like to be able to claim that its new headache relief medication is better than those of rivals. Also, it has two methods for formulating its product, and it would like to compare these as well.

From the small experiment, what claims can the marketers offer?

In the data set from this small experiment, the categories (in the column *Drug*) are

(1) active compound #1,
(2) active compound #2,
(3) rival product, and
(4) control group (aspirin).

The response is a pain relief score which lies in the range of 0 (no relief) to 50 (complete relief). Subjects were randomly assigned to the four groups using a "double-blind" procedure so that neither the physician nor the patient knows who is getting which drug. The drugs are packaged in identical containers that are distinguished only by serial numbers; even the tablets themselves are manufactured to be indistinguishable.

The data analysis begins by inspecting the comparison plot of the relief supplied by the four drugs. A particularly large outlier is evident, and a look at the data confirms it to be a coding error. The sign was reversed — all of the scores are supposed to be positive.

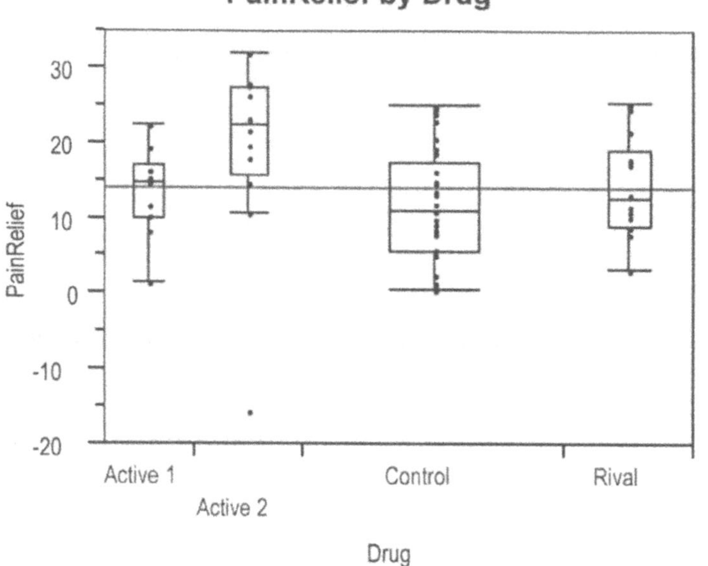

Means and Std Deviations

Level	Number	Mean	Std Dev	Std Err Mean
Active 1	10	13.4	5.9	1.87
Active 2	12	19.1	12.4	3.59
Control	29	11.5	7.7	1.43
Rival	14	14.2	6.6	1.77

Since the manufacturer is interested in various comparisons from this one set of data, we will use the Tukey-Kramer procedure. With the negative value corrected by reversing the sign (after a check of the source data, of course), the Tukey-Kramer comparison obtained by using the *Compare all pairs* command indicates significant differences. In the figure on the next page, the *Active 2* group is selected and found to be different from the other three. Positive values in the comparison table (which tables *lower* endpoints of intervals for the absolute mean differences) confirm that the only meaningful differences are between *Active 2* and the others. You will reach a different conclusion if the outlier is left in the data, uncorrected.

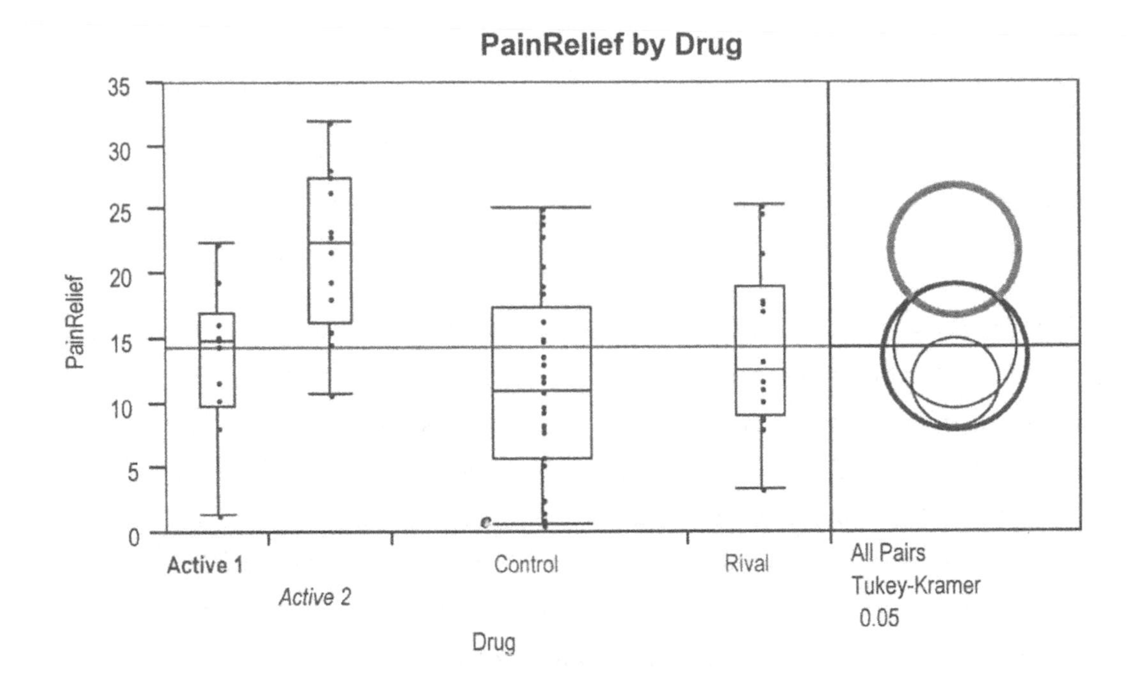

Means and Std Deviations

Level	Number	Mean	Std Dev	Std Err Mean
Active 1	10	13.4	5.9	1.87
Active 2	12	21.7	6.3	1.82
Control	29	11.5	7.7	1.43
Rival	14	14.2	6.6	1.77

Comparisons for all pairs using Tukey-Kramer HSD

Abs(Dif)-LSD	Active 2	Rival	Active 1	Control
Active 2	-7.53	0.19	0.43	3.90
Rival	0.19	-6.97	-6.75	-3.21
Active 1	0.43	-6.75	-8.24	-4.85
Control	3.90	-3.21	-4.85	-4.84

Positive values show pairs of means that are significantly different.

Further residual analysis (not shown) indicates no violation of assumptions. The variances are comparable in the four groups, and the pooled data is close to normality, as seen in a quantile plot.

A pharmaceutical manufacturer would like to be able to claim that its new headache relief medication is better than those of rivals. Also, it has two methods for formulating its product.

From the small experiment, what claims can the marketers offer?

After correcting an evident coding error, the second active compound (Active 2) is the clear winner. The other three seem comparable with no differences among them. Hence advertising can claim that we're better than aspirin or that other stuff you've been taking (the rival product). Suppose the manufacturer wanted only to identify the best and not make other comparisons. Would Hsu's procedure also indicate that *Active 2* was best?

The table of output showing the comparison of means offered by the *Fit Y by X* platform can be confusing since it shows just one endpoint of the adjusted confidence interval for the differences of pairs of means. You can get a more complete table of Tukey-Kramer comparisons by using the more elaborate *Fit model* platform. For this example, use *Fit model* to regress *PainRelief* on the single categorical predictor *Drug*. A red, triangular button near the leverage plot for *Drug* offers the option "LSMeans Tukey HSD". This command generates a more complete, color-coded table of comparisons.

LSMeans Differences Tukey HSD

Alpha= 0.050 Q= 2.64 LSMean[i] By LSMean[j]

Mean[i]-Mean[j] Std Err Dif Lower CL Dif Upper CL Dif	Active 1	Active 2	Control	Rival
Active 1	0 0 0 0	-8.3217 2.9883 -16.215 -0.4287	1.90793 2.55939 -4.8522 8.66803	-0.88 2.88965 -8.5124 6.75241
Active 2	8.32167 2.9883 0.42869 16.2146	0 0 0 0	10.2296 2.39555 3.90224 16.557	7.44167 2.74559 0.18976 14.6936
Control	-1.9079 2.55939 -8.668 4.85217	-10.23 2.39555 -16.557 -3.9022	0 0 0 0	-2.7879 2.2713 -8.7871 3.21124
Rival	0.88 2.88965 -6.7524 8.51241	-7.4417 2.74559 -14.694 -0.1898	2.78793 2.2713 -3.2112 8.7871	0 0 0 0

Analysis of Variance and Tests for Linearity

Cleaning.jmp

> Our initial evaluation of the performance of cleaning crews accepted the linearity of the relationship between number of crews and number of rooms cleaned.
>
> Does the analysis of variance suggest more?

Recall the data on the performance of cleaning crews from Classes 1 and 2. When data have repeated Y values for each value of the predictor X, it is possible to compute some additional diagnostics.

RoomsClean by Number of Crews

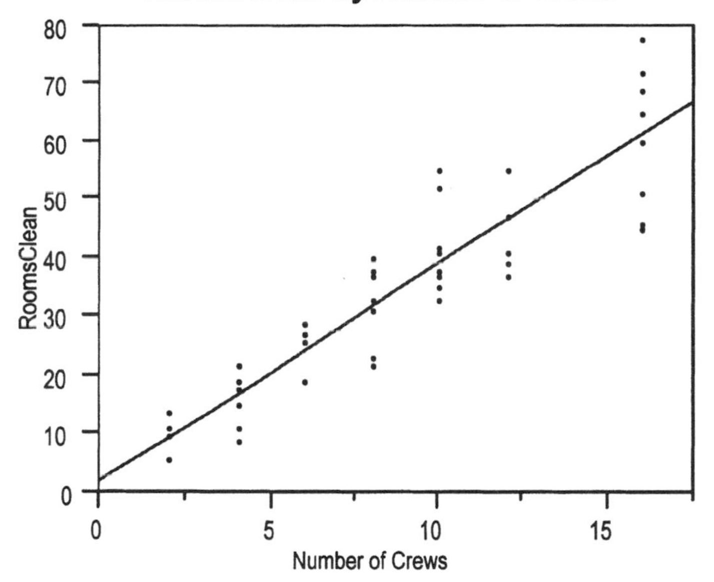

RoomsClean = 1.78 + 3.70 Number of Crews

RSquare	0.86
Root Mean Square Error	7.34
Observations	53

Term	Estimate	Std Error	t Ratio	Prob>|t|
Intercept	1.78	2.10	0.9	0.40
Number of Crews	3.70	0.21	17.5	0.00

In this data set, we observe several values of the number of rooms cleaned (Y) on days with the same number of crews (X). In this lucky circumstance with repeated values of the response, we can use anova to see if there is evidence of a systematic departure from linearity. The idea is simple. Just collect the residuals for each value of the predictor (here, the number of crews), and run an anova. If the means are different from zero, then we have evidence that the linear fit is inadequate. Although JMP does this procedure automatically, we'll first see what happens doing it "by hand."

First, save the residuals from the model. Then relabel *Number of Crews* as a ordinal (or nominal) variable and use *Fit Y by X* to compute the analysis of variance. The results are on the next page. Notice that the fit is not significant. Alternatively, JMP automatically does *almost* the same thing as part of the *Fit Model* command. The output box labeled "Lack of Fit" gives the following summary.

Lack of Fit

Source	DF	Sum of Squares	Mean Square	F Ratio
Lack of Fit	5	97.76	19.55	0.34
Pure Error	46	2647.04	57.54	**Prob>F**
Total Error	51	2744.80		0.8862
				Max RSq
				0.8620

This summary is quite similar to the anova on the next page, with one important exception. The DF value for Lack of Fit has 5 shown above, but 6 in the anova we just did. Why are they different? We can count 7 groups, so it seems like 6 is right (the number of groups minus one). The difference is that we are working with residuals that are the product of a model, not the original data. In the end, though, neither method shows any evidence for a lack of linearity. The value for the maximum R^2 shown in the table above indicates the best possible R^2 using any transformation of X. It's not much larger than the value that we obtained with the initial fit, 0.857.

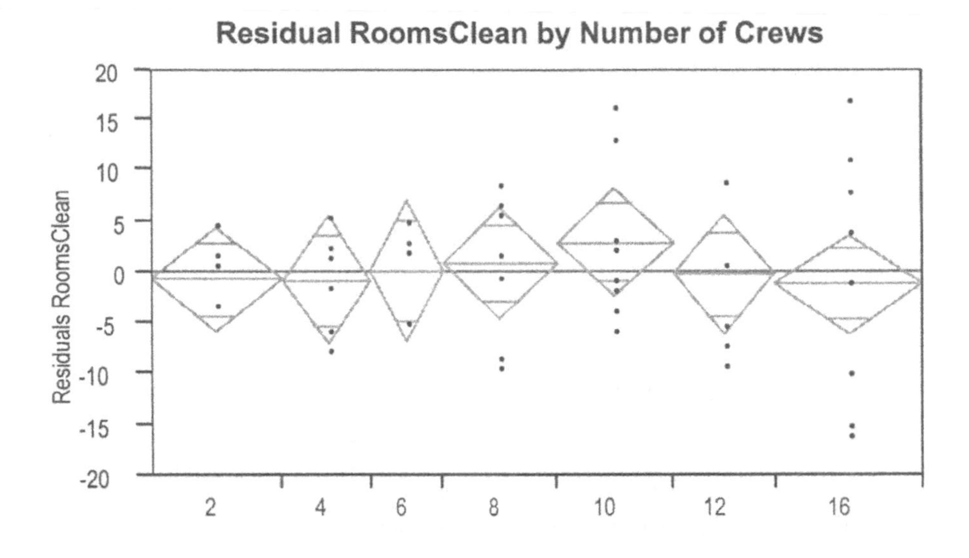

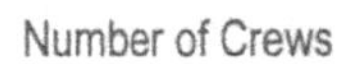

Mean of Response	3.55e-15	ZERO!
Observations	53	

Source	DF	Sum of Squares	Mean Square	F Ratio
Model	6	97.76	16.29	0.28
Error	46	2647.04	57.54	**Prob>F**
C Total	52	2744.80	52.78	0.94

Means for Oneway anova

Level	Number	Mean	Std Error
2	9	-0.85	2.5
4	6	-0.92	3.1
6	5	0.01	3.4
8	8	0.73	2.7
10	8	2.83	2.7
12	7	-0.34	2.9
16	10	-1.30	2.4

Std Error uses a pooled estimate of error variance

Our initial evaluation of the performance of cleaning crews accepted the linearity of the relationship between number of crews and number of rooms cleaned.

Does the analysis of variance suggest more?

When data have natural groups, one can use anova to help check for possible nonlinearities in a regression model. JMP makes the calculations easy by doing the needed anova automatically. JMP also adjusts the calculations for the use of residuals, something that we cannot do easily when performing the anova "by hand."

Of course, this diagnostic procedure has some assumptions associated with it. In particular, it expects the variability of the data to be constant. As we noted back in Class 2, it's not. The variability gets larger when more rooms are cleaned. Since this assumption is violated, we have to treat the results in this illustration with some skepticism. Nonetheless, it remains a quick and easy check for linearity.

It may be useful to see briefly a second example. The data set used to model the dependence of wine sales on display feet (Display.jmp) also has repeated values of the response for each value of the predictor. Here is the JMP summary of for lack of fit.

Source	DF	Sum of Squares	Mean Square	F Ratio
Lack of Fit	5	59898.62	11979.7	8.0030
Pure Error	40	59875.86	1496.9	**Prob>F**
Total Error	45	119774.47		<.0001
				Max RSq
				0.8560

Unlike the what we found with the cleaning crews, the relationship between wine sales and display feet, as found in Class 1, is clearly nonlinear.

Class 10. Analysis of Variance with Two Factors

Often we have several factors that we need to simultaneously compare. For example, both the color and style of a package might influence potential customers, and we would like to make sure that we choose the right combination. The problem of choosing the right combination is made more difficult because such factors often interact. In this example, the best color for one style of package might be the worst color for a different style. Two-way analysis of variance with interactions permits one to resolve these issues when only two grouping factors are involved. More complicated analysis methods generalize this approach to problems with three or more factors. Indeed, regression analysis can be used to obtain the results for any type of anova problem, albeit the results will be tedious to sort out.

Topics

Two-way analysis of variance with interaction
Comparison of additive models versus models with interaction
Profile plots (interaction plot)
Multiple comparisons in two-factor studies
Relationship of anova to regression with dummy variables

Examples

1. Package design experiment (no interaction)
2. Evaluating employee time schedules (with interaction)

Key application

Two-way anova and interaction. In the introduction to Class 9, we discussed a one-way anova for incentive programs at gasoline stations. The aim was to determine which of three incentive schemes was the most effective in increasing profits.

Analysts at the oil company believe that the effectiveness of the incentive scheme may also depend on the geographic location of the gas station. For example, a clean gas station might be more valued in a city than in the country. The question naturally arises how to determine whether or not such differences really exist.

The methodology used to perform such an analysis is called a two-way anova, because the model has two variables (or factors), "incentive scheme" and "geographic location." Two-way anova allows us to investigate how the mean increase in profits depends on both factors, finding the right combination.

Definitions

*Two-way anova **without** interaction.* A method for determining how the mean value of the response (*Y*) differs according to the levels of two different factors, under the assumption that the impact of one factor is unrelated to the impact of the other.

*Two-way anova **with** interaction.* A method for determining how the mean value of the response (*Y*) differs according to the levels of two different factors, allowing for the possibility that the impact of one factor may depend on the level of the other factor.

Concepts

Two-way anova and interaction. The gas station example was originally analyzed using a one-way anova because *one* factor was used to explain variability in the profits at gas stations. (Profit is the response (*Y*-variable) and "incentive scheme" is the predictor *X*). Even though there are three levels to the predictor ("free glass", "window wash" and "clean fuel station"), the procedure is a one-way anova because this factor occupies a single column in the spreadsheet. It's one variable.

Now consider the inclusion of the other factor "geographic region" which is defined as either "Urban" or "Rural." Clearly, we have a spreadsheet with two predictors (*X*-variables), and hence the name "two-way anova." The question that we would like to ask is "How does the mean increase in profits depend on the type of incentive scheme and the location of the gas station?"

The critical aspect to realize when investigating this question is to recognize that there may not be a simple answer! The reason for this is that there could be "interaction" between the two independent variables. Go back to the discussion on indicator variables in regression from Classes 6 and 7 and read again what we had to say about interaction. The idea remains the same in two-way anova. An interaction requires (at least) three variables, one response (*Y* variable) and (at least) two predictors (*X* variables). To say that interaction exists in the gas station example is to believe that the impact of incentive scheme on mean profits depends on the geographic location, or generically

the impact of X_1 *on* Y *depends on the level of* X_2.

When you do a two-way anova, you will have the option not to include an interaction. If you choose to leave it out, then you must believe that the two independent variables act separately. In the sense of regression seen in Classes 6 and 7, you are saying that the lines are parallel. In the gas station example, the lack of an interaction means you believe that the impact of incentive scheme on profits does *not* depend on the geographic location.

The *advantage of having no interaction* is that you are free to look at the independent variables in isolation, one at a time. Essentially you can answer the questions like "Which location has the higher profits?" and "Which incentive scheme gives the highest profits?" separately. This can simplify the decision making and implementation process. In this example the absence of interaction would translate to implementing the same incentive scheme in both rural and urban locations. If there is, however, interaction in the model then this might not be a wise decision because the best incentive scheme may depend on which location you are talking about.

Since interaction is obviously an important (and practically meaningful) concept, how do you decide if there should be interaction in the model? Fortunately graphical methods and statistical tests discussed in this class help you identify interactions. One word of caution though: whenever you look at the output of a statistical test you must recognize that it doesn't tell you how *practically* important a result is. As always, there may be significant interaction, but statistical significance does not necessarily translate to importance. It is up to the analyst (you) to gauge whether or not the interaction you have identified is of any practical importance. Statistics can tell you whether or not it's there, but not how you should deal with it.

Heuristics

Communicating results. It's a sad fact that, even though you are now learning a lot of new and important concepts that translate to better decision making, these concepts invariably prove barren unless you can get them across to other people. Interaction is one such concept. When you've got it, it's transparent. But people without quantitative training often miss the point entirely. Here's another way of putting interaction into words.

Consider a marketing company that has two separate product design groups: one with expertise in lettering and the other with expertise in coloring. There is a new product to be designed. Should you allow the two groups to choose independently or should you compel them to work together? The answer depends on whether or not there is an interaction between lettering and color. If you answer the question "Which lettering should we use?" by responding "It depends on the color," then you are implicitly saying that there is interaction between the two variables.

From the managerial perspective, interaction says that you have to get your product design groups together in the same room when the design and implementation decisions are made. Believing that there is no interaction says that you can have each of the groups go off on their own, doing the best job that they can, and you can be sure that the end result will be the best possible.

Package Design Experiment

Design.jmp

As part of a brand management experiment, several new types of cereal packages were designed. Two colors and two styles of lettering were considered. Each combination was used to produce a package, and each of these combinations was test marketed in 12 comparable stores. At the end of the evaluation period, sales in the 48 stores were compared.

What is the right combination of color and type style, or does it matter?

When approaching this problem, you must think about the presence of *interaction*. In particular, if we fill in the average values for three of the four combinations, what would you guess for the fourth?

	Block	Script
Red	148	161
Green	119	?

The shift of *Red* to *Green* in the *Block* column suggests that the value that goes in the missing cell ought to be 161 – (148-119) = 132. Alternatively, the change from *Block* to *Script* in the *Red* row suggests that the missing cell ought to be 119 + (161-148) = 132.

Each of these natural approaches (which lead to the same answer) presumes the absence of interaction. Interaction implies that the effect of changing from red to green packages depends upon the type of lettering. (Or that the effect of changing from block to script lettering depends upon the color of the package.)

This data set is used in exercises 10.62-10.64 of Hildebrand and Ott, *Statistics for Managers*.

The results of a two-way analysis of variance appear in the following summary. This output is generated by the *Fit Model* command, using a complete factorial design. Since both of the predictors *Color* and *TypeStyle* are nominal columns, JMP automatically uses the appropriate modeling technique (a two-way analysis of variance). Make sure that the cross-product term *TypeStyle*Color* also appears. It is this last term that captures the presence of any interaction. As an aside, you will notice that there is *no collinearity* in this problem. Since we have a "balanced" experiment, one with equal numbers of stores for each combination. Because of this balance, all of the standard errors are the same. Plus, we avoid the collinearity issues that often complicate the interpretation and choice of a regression model.

Expanded Estimates

Term	Estimate	Std Error	t Ratio	Prob>|t|
Intercept	144.92	5.52	26.26	<.0001
Color[Green]	-9.83	5.52	-1.78	0.0816
Color[Red]	9.83	5.52	1.78	0.0816
TypeStyle[Block]	-11.17	5.52	-2.02	0.0491
TypeStyle[Script]	11.17	5.52	2.02	0.0491
TypeStyle[Block]*Color[Green]	-4.50	5.52	-0.82	0.4191
TypeStyle[Block]*Color[Red]	4.50	5.52	0.82	0.4191
TypeStyle[Script]*Color[Green]	4.50	5.52	0.82	0.4191
TypeStyle[Script]*Color[Red]	-4.50	5.52	-0.82	0.4191

Effect Test

Source	Nparm	DF	Sum of Squares	F Ratio	Prob>F
Color	1	1	4641.3333	3.1762	0.0816
TypeStyle	1	1	5985.3333	4.0959	0.0491
TypeStyle*Color	1	1	972.0000	0.6652	0.4191

Analysis of Variance

Source	DF	Sum of Squares	Mean Square	F Ratio
Model	3	11598.667	3866.22	2.6457
Error	44	64297.000	1461.30	**Prob>F**
C Total	47	75895.667		0.0608

As before, this sort of regression model is most easy to understand if we write out the equation for the different groups. For example, the fitted model for green package with block lettering is

$$
\begin{aligned}
\text{Mean(Green, Block)} &= 144.9 + (-9.8) + (-11.2) + (-4.5) \\
&= 144.9 - 25.5 \\
&= 119.4
\end{aligned}
$$

whereas the model for the green package with script lettering is

$$
\begin{aligned}
\text{Mean(Green, Script)} &= 144.9 + (-9.8) + (11.2) + (4.5) \\
&= 150.8
\end{aligned}
$$

Before going further with interactions, we note that the residual plot suggests no problems. The variance appears constant with no unusual outliers. A quick check for normality (not shown) also indicates no problems at this point.

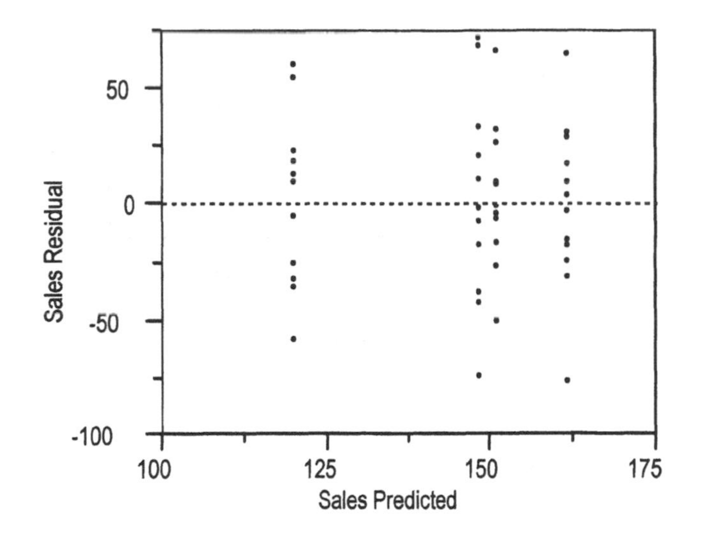

The associated *profile plots* (or LSMeans Plots as called by JMP) generated by the button in the title of the leverage plots show the means of the cells defined by each of the terms in the model. The almost parallel lines in the final profile plot confirm the absence of any significant interaction.

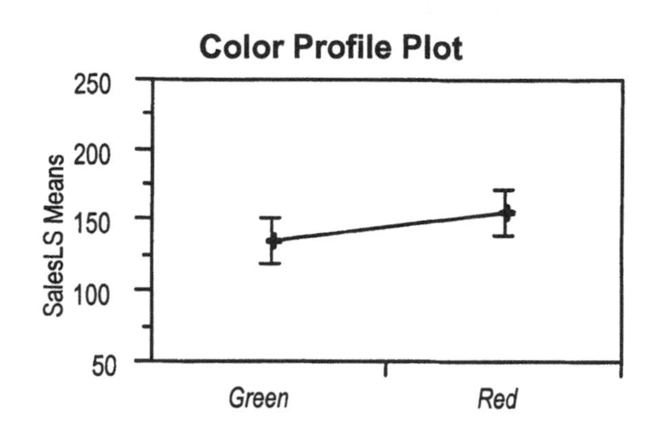

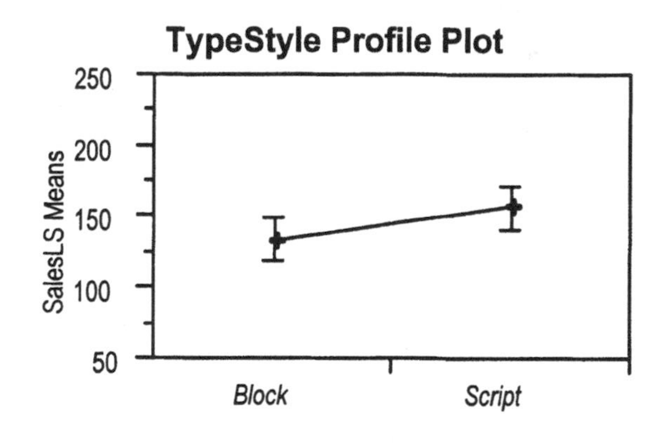

Which combination works best? Since there is little interaction, it's safe to choose the best row and combine it with the best column, in this case leading to red script packages. Is it significantly better (perhaps it costs more)? To answer this more challenging question, we return to a one-way analysis of variance using the four combinations as the single explanatory factor. We find that although the red package with script lettering generated the most sales, this combination was not significantly better than green with script or red with block lettering. It is significantly better than the green/block combination.

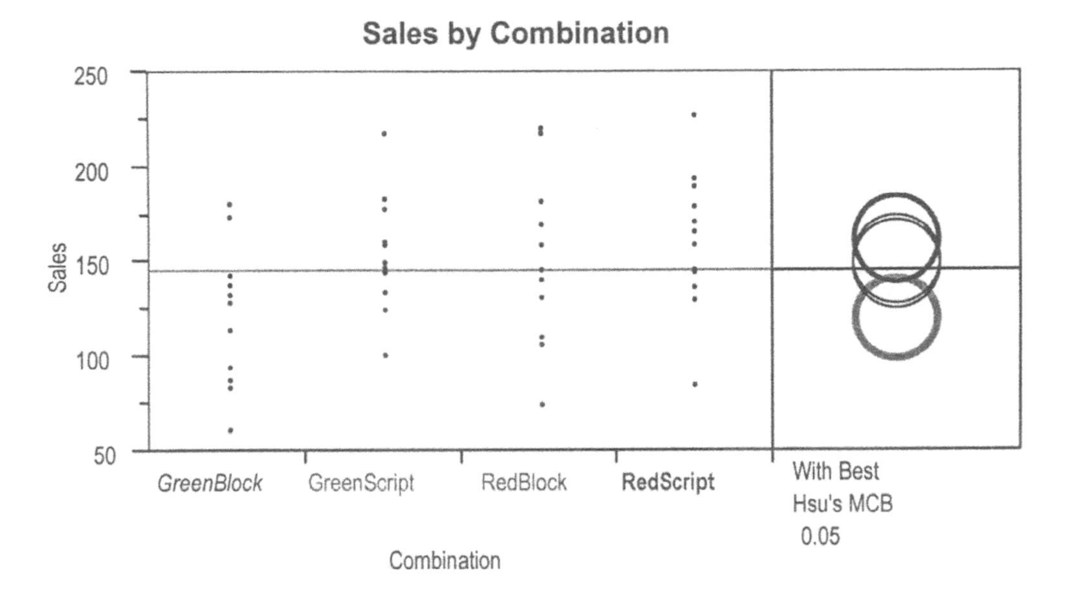

Means and Std Deviations

Level	Number	Mean	Std Dev	Std Err Mean
GreenBlock	12	119.42	37.49	10.82
GreenScript	12	150.75	33.51	9.67
RedBlock	12	148.08	44.85	12.95
RedScript	12	161.42	36.13	10.43

Comparisons with the best using Hsu's MCB

Mean[i]-Mean[j]-LSD	RedScrip	GreenScr	RedBlock	GreenBlo
RedScript	-33.1	-22.4	-19.7	8.9
GreenScript	-43.7	-33.1	-30.4	-1.7
RedBlock	-46.4	-35.7	-33.1	-4.4
GreenBlock	-75.1	-64.4	-61.7	-33.1

If a column has any positive values, the mean is significantly less than the max.

Before completing the analysis, we need to inspect the residuals further, checking as usual for constant variance and normality. The residuals are generated from the *Fit Model* platform.

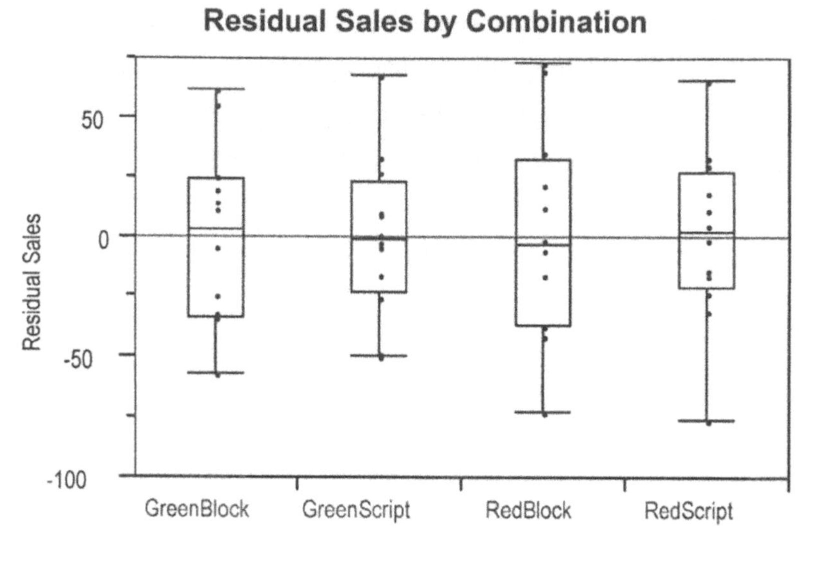

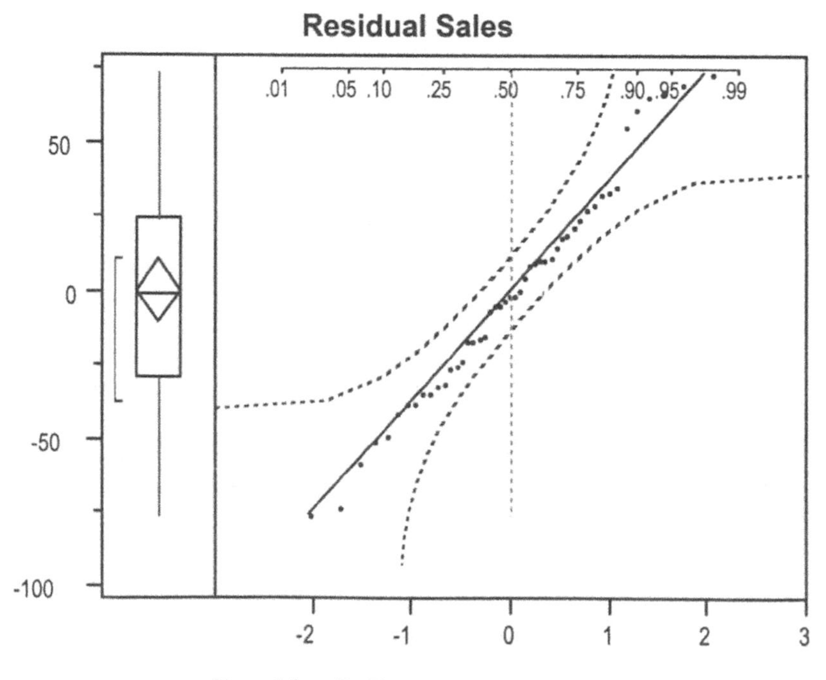

As part of a brand management experiment, several new types of cereal packages were designed. Two colors and two styles of lettering were considered. Each combination was used to produce a package, and each of these combinations was test marketed in 12 stores. At the end of the evaluation period, sales in the 48 stores were compared.

What is the right combination of color and type style, or does it matter?

Although the red, script-lettered package type was preferred (in that it had the highest mean value), it was not significantly better than two other package styles. Only the green, block combination is to be avoided.

In addition, we learned that there is essentially no interaction between the choice of color and lettering style, a result that might be useful in designing new types of packages. The absence of interaction implies that we are free to choose the style of lettering independently of the color of the letters. That is, we can choose the best combination by using the "marginal" information on color and lettering, without having to investigate combinations.

As with any analysis, there are some obvious questions to consider regarding the source of the data. For example, were the stores randomized to the different package and lettering styles? Are the stores in comparable neighborhoods? Are the packages displayed in the same manner in all of the stores? Anova works best when the results are determined by the treatment effects, not other extraneous sources of variation. Although randomization protects from the effects of these other factors, we obtain a better estimate of the group differences if we can remove these other sources of variation and focus on the differences of interest.

Evaluating Employee Time Schedules

Flextime.jmp

Should the clerical employees of a large insurance company be switched to a four-day week, allowed to use flextime schedules, or kept to the usual 9-to-5 workday?

The data measure percentage efficiency gains over a four-week trial period. After several preliminary one-way looks at the data, the analysis will use a two-way analysis of variance including interactions. The two factors are the department and the condition of the schedule. For reference, the levels of these factors are

Department	Condition
1. Claims	1. Flextime
2. Data Processing	2. Four-day week
3. Investments	3. Regular hours

This data set is the subject of a case study in Hildebrand and Ott, *Statistics for Managers* (pages 404–406).

Let's start by ignoring the department and do a one-way analysis of variance by type of schedule. Here are the results from the *Compare means* procedure that we used in doing a one-way analysis of variance.

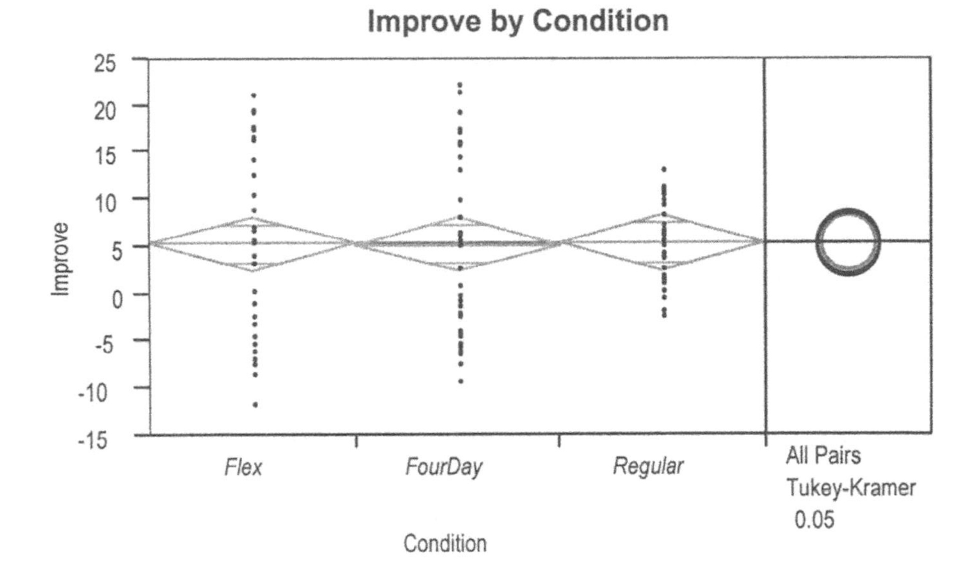

Oneway anova

Source	DF	Sum of Squares	Mean Square	F Ratio
Model	2	0.5	0.3	0.0040
Error	96	6603.4	68.8	**Prob>F**
C Total	98	6603.9	67.4	0.9960

Level	Number	Mean	Std Error
Flex	33	5.19	1.44
FourDay	33	5.08	1.44
Regular	33	5.27	1.44

All three scheduling methods appear equivalent. Does this comparison hold up when we look within a department? You can get a quick hint of what is going to happen by color coding the points shown in this plot by the "hidden" variable, *Department*.

Here are the results if we focus our attention on the data processing division, again using a one-way analysis of variance to compare the scheduling methods. The differences are very significant, with a clear preference for the flextime schedule.

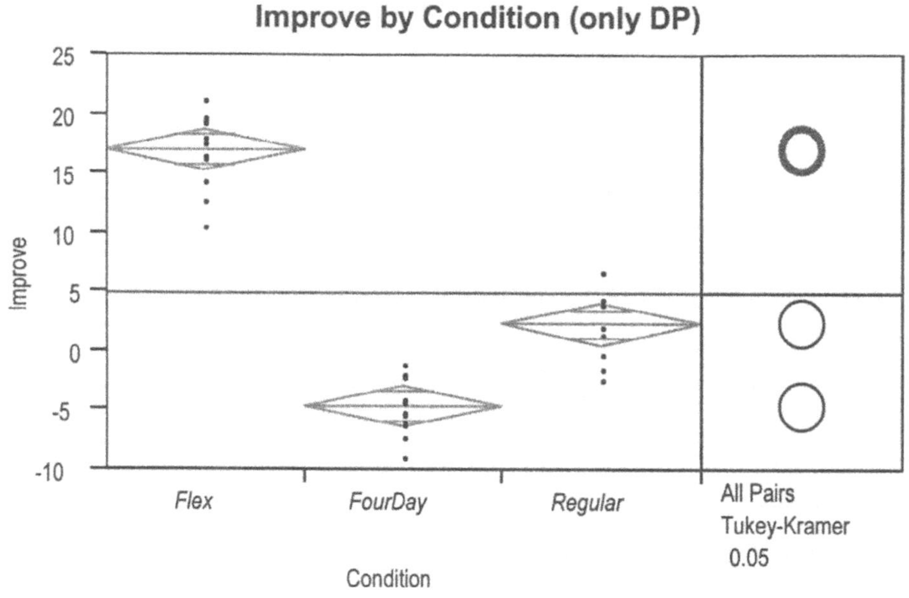

Improve by Condition (only DP)

Oneway anova

Source	DF	Sum of Squares	Mean Square	F Ratio
Model	2	2683.7	1341.84	167.6933
Error	30	240.1	8.00	**Prob>F**
C Total	32	2923.7	91.37	<.0001

Level	Number	Mean	Std Error
Flex	11	16.89	0.85
FourDay	11	-4.75	0.85
Regular	11	2.22	0.85

For the investment division, significant differences are again apparent, but not the same preferences as found in the DP group. For the investments group, the four-day week is preferred, and flextime (the favorite of DP) is worst.

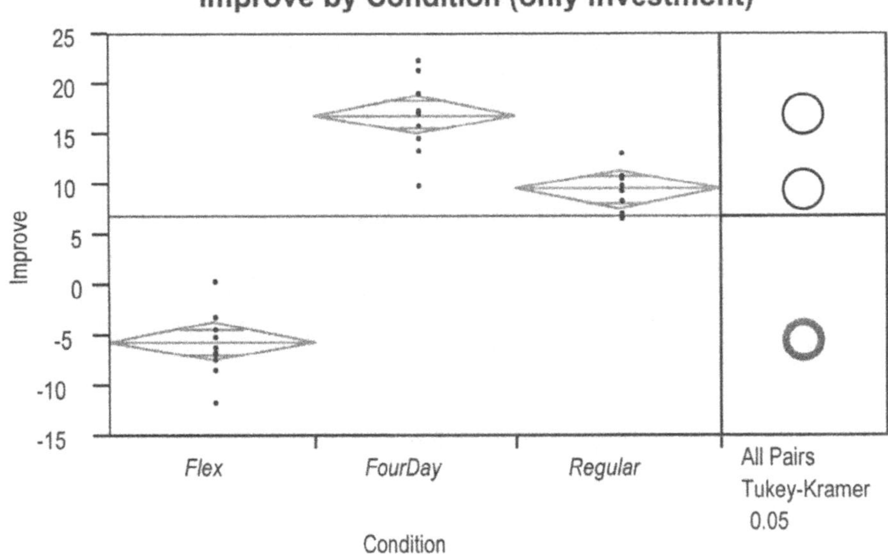

Oneway anova

Source	DF	Sum of Squares	Mean Square	F Ratio
Model	2	2895.5	1447.75	168.3921
Error	30	257.9	8.60	**Prob>F**
C Total	32	3153.4	98.54	<.0001

Level	Number	Mean	Std Error
Flex	11	-5.65	0.88
FourDay	11	16.87	0.88
Regular	11	9.38	0.88

Clearly, our initial one-way anova ignoring the department missed a lot of structure. We need a two-way anova with an interaction.

As suggested by our results within the two departments, the two-way analysis of variance indicates substantial interaction, making it pointless to consider the effects of department and condition alone. In contrast to the packaging example, we have to examine the combinations of the factors to see which is best.

Effect Test: Improve

Source	Nparm	DF	Sum of Squares	F Ratio	Prob>F
Department	2	2	154.3087	8.07	0.0006
Condition	2	2	0.5487	0.03	0.9717
Condition*Department	4	4	5588.2004	146.06	<.0001

Analysis of Variance

Source	DF	Sum of Squares	Mean Square	F Ratio
Model	8	5743.0578	717.882	75.0520
Error	90	860.8618	9.565	**Prob>F**
C Total	98	6603.9196		<.0001

Residuals on Predicted

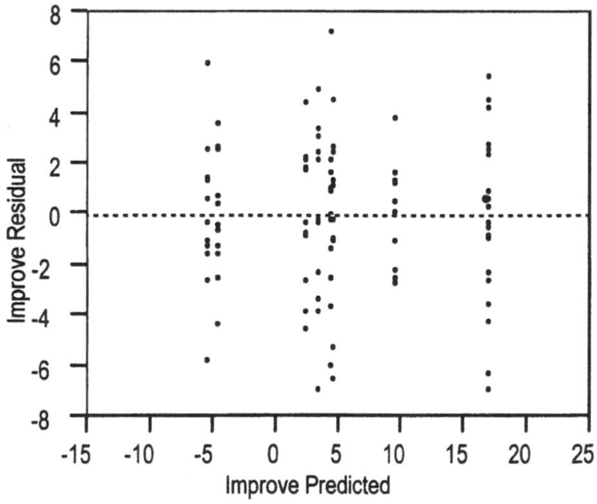

Improve Residual

Improve Predicted

The profile plot (LS Means plot in JMP terminology) for the interaction makes it most clear what is going on in this data set. The profile plot shows the means for the nine cells of the two-way design. Lines join the means for the different methods of scheduling.

Condition*Department Profile Plot

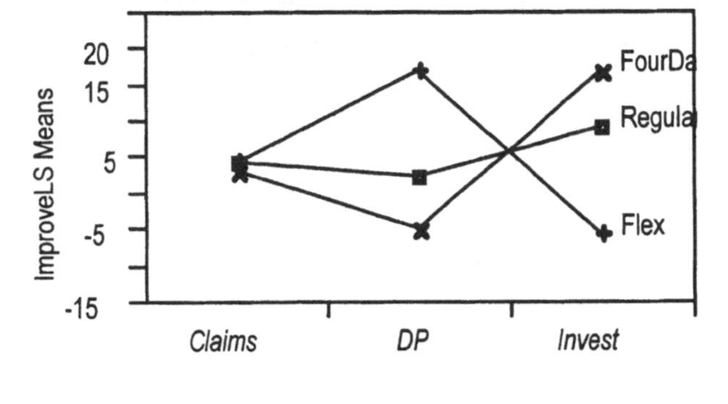

The profile plots for the separate factors give the misleading impression that each of these is not important, when in fact a strong interaction exists.

Department Profile Plot

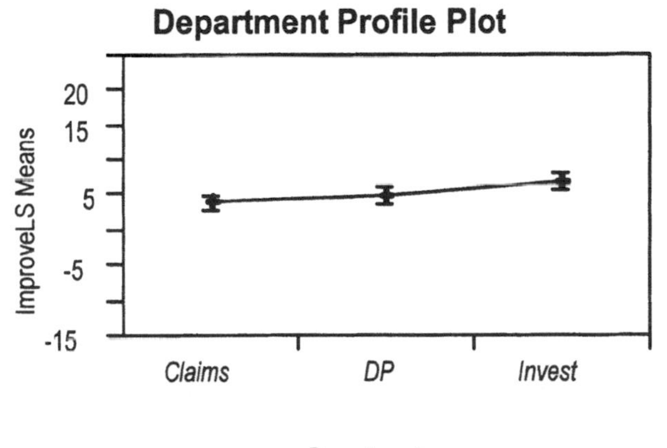

Condition Profile Plot

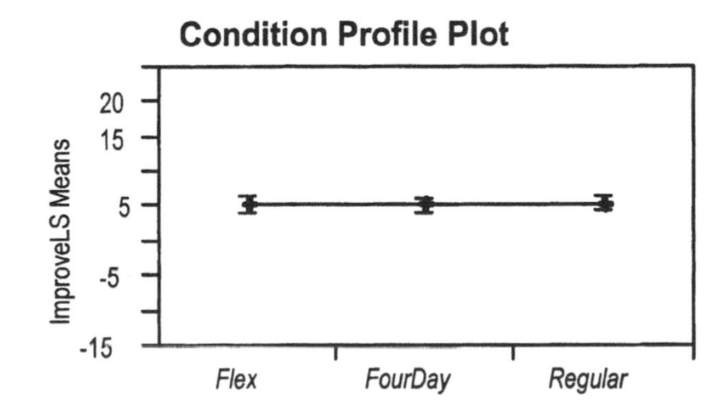

Condition

Clearly each department prefers one schedule or another, but the preferences are not consistent from one department to another. To see if the differences shown in the profile plot for the interaction are meaningful, we can do a one-way analysis with each of nine combinations of *Condition* and *Department* forming the factor. The only essential difference in the analyses is that JMP will do the multiple comparisons that we seek.

The results show that the means for the claims department (*Dept* = 1) do not differ in a meaningful way. The observed differences within the other two departments, DP and Investments, include significant comparisons, with flextime the best in DP and four-day weeks winning in investments.

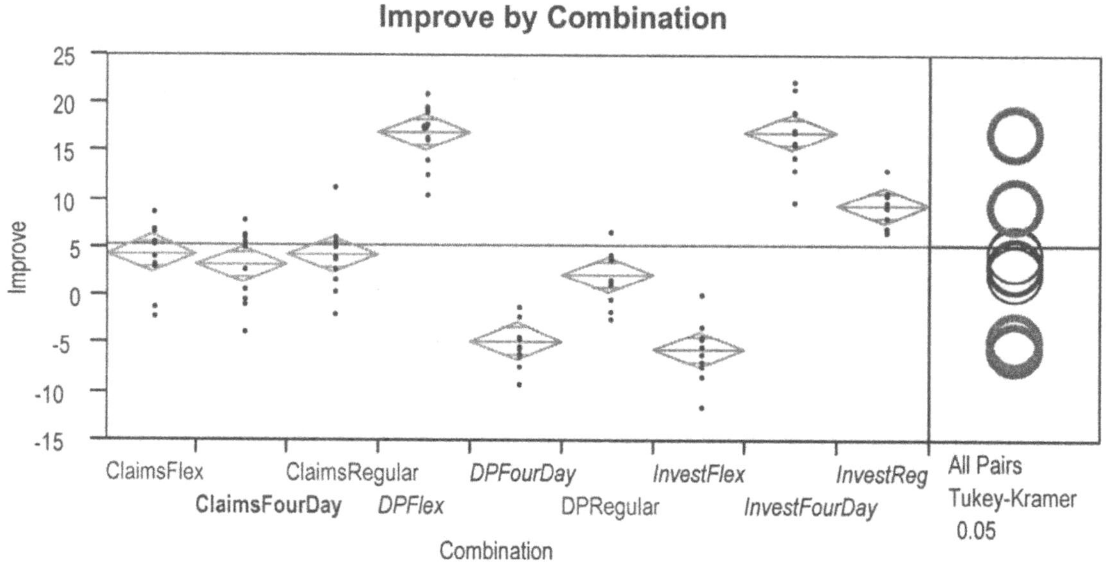

Means and Std Deviations

Level	Number	Mean	Std Dev	Std Err Mean
ClaimsFlex	11	4.3	3.32	1.00
ClaimsFourDay	11	3.1	3.66	1.10
ClaimsRegular	11	4.2	3.44	1.04
DPFlex	11	16.9	3.23	0.97
DPFourDay	11	-4.7	2.38	0.72
DPRegular	11	2.2	2.82	0.85
InvestFlex	11	-5.7	3.04	0.92
InvestFourDay	11	16.9	3.52	1.06
InvestRegular	11	9.4	2.04	0.61

Here is the comparison table for the Tukey-Kramer comparisons shown on the previous page. As before, this table gives the lower endpoints for confidence intervals for the absolute difference in two means. Since this difference must be positive (that's the job of the absolute value), the associated intervals are centered on a positive value. If the lower endpoint is less than zero, then the interval includes zero and the difference is not significant. However, if the lower endpoint is positive, zero lies outside the interval for the difference in means and the difference is significant.

Comparisons for all pairs using Tukey-Kramer HSD

Abs(Dif)-LSDDP/F	**In/4**	**In/R**	**Cl/F**	**Cl/R**	**Cl/4**	**DP/R**	**DP/4**	**In/F**	
DPFlex	-4.2	-4.2	3.3	8.4	8.5	9.6	10.5	17.4	18.4
InvestFourDay-4.2	-4.2	3.3	8.4	8.5	9.6	10.5	17.4	18.3	
InvestRegular3.3	3.3	-4.2	0.9	1.0	2.1	3.0	9.9	10.8	
ClaimsFlex	8.4	8.4	0.9	-4.2	-4.1	-3.0	-2.1	4.9	5.8
ClaimsRegular8.5	8.5	1.0	-4.1	-4.2	-3.1	-2.2	4.8	5.7	
ClaimsFourDay9.6	9.6	2.1	-3.0	-3.1	-4.2	-3.3	3.7	4.6	
DPRegular	10.5	10.5	3.0	-2.1	-2.2	-3.3	-4.2	2.8	3.7
DPFourDay	17.4	17.4	9.9	4.9	4.8	3.7	2.8	-4.2	-3.3
InvestFlex	18.4	18.3	10.8	5.8	5.7	4.6	3.7	-3.3	-4.2

Positive values show pairs of means that are significantly different.

The residual plots for this last model do not suggest deviations from the crucial assumptions of constant variance and approximate normality. Do you believe the assumption of independence in this setting?

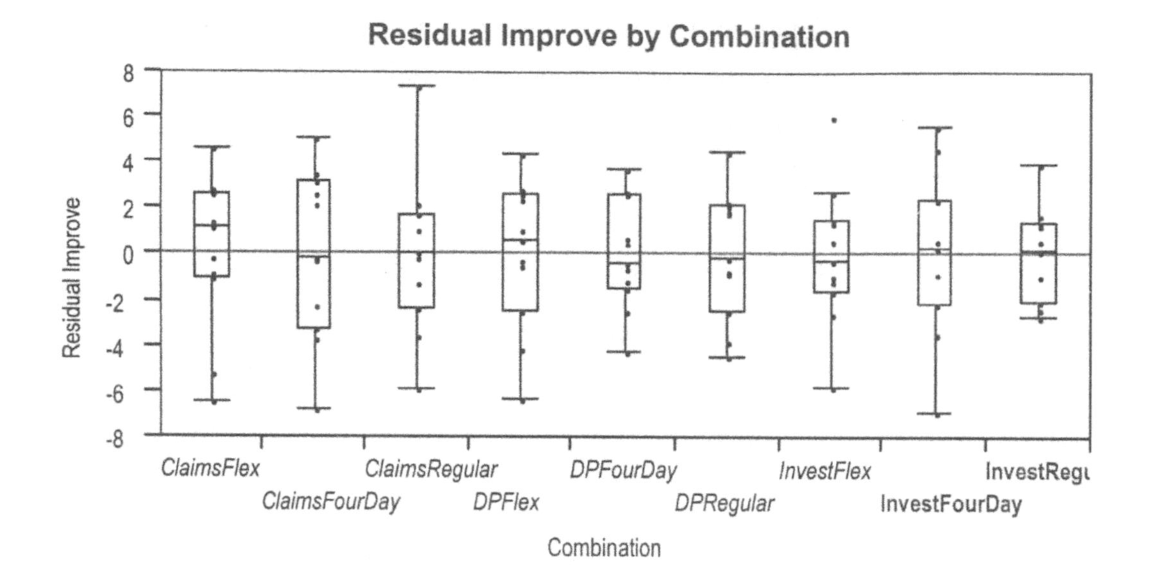

Should the clerical employees of a large insurance company be switched to a four-day week, allowed to use flextime schedules, or kept to the usual 9-to-5 workday?

The interaction found in the analysis implies that the choice of the best system of scheduling depends upon the department. No one schedule will be best for the entire firm. As is often the case, interaction implies the need for a more complicated solution than would be possible were there no interaction. If there were no interaction, it might be the case that one schedule would be best throughout the firm. Here that's not the case, and one will have to customize the schedule if we hope to reap any of the suggested gains. Of course, would different schedules work in different groups that had to interact?

As to the specific departments, in the claims department, it does not matter which schedule is used. All are equally (in)effective. In data processing, flextime was preferred. In investments, the four-day week was best.

Class 11. Modeling a Categorical Response

We have seen how categorical variables are useful as explanatory variables in a regression model. This class considers a modeling technique which is needed when the response itself is a categorical variable. Though more general methods exist, we will consider only cases in which the response is dichotomous, and use a method called logistic regression.

Logistic regression builds a model for the conditional probability that the response is in one category or the other. Rather than directly predicting the response, this model estimates the probabilities of group membership. Logistic regression resembles least squares regression in some ways; for example, a logistic regression has slopes and an intercept and also various measures of goodness-of-fit and significance. In some cases, the results from logistic regression resemble those of a least squares regression. However, the interpretations of the slopes and intercept of a logistic regression differ from those in a least squares regression analysis.

Topics

- Logistic function
- Likelihood and goodness of fit
- Odds ratio

Examples

1. The Challenger disaster (one predictor)
2. Marketing orange juice (multiple predictors)

Key application

Modeling discrete choices.

Marketing departments typically need to identify those attributes of consumers which make them more or less likely to purchase a particular product. How does sex, income, or past purchase behavior influence future purchase behavior? This is the basic idea of market segmentation. A useful method for answering these questions is logistic regression.

Definitions

Odds. The ratio of the probability p of an event to one minus that probability, $\frac{p}{1-p}$.

Logit. The (natural) log of the odds, $\log_e \frac{p}{1-p}$.

Concepts

Logistic regression. Every analysis which we have done up to this point has used a continuous variable as the response. We have used categorical variables in regression, but always as predictors. In the marketing problem outlined above, it is clear that it would be useful to have a categorical response variable, in this case either "buy" or "don't buy." These alternatives could be coded as either zero or one in the spreadsheet, so that it would be possible to run a standard regression on this data. However, this regression could be meaningless because points on the Y-axis between zero and one have no meaning when the response is categorical.

The way that this problem is handled is by *modeling the probabilities* of purchase rather than the purchase event itself. Our interest lies in finding out which market segments have a high probability of purchase and which have a low probability. The fact that probabilities lie between zero and one means that a thoughtless application of standard regression will get into a lot of trouble because it totally disregards these bounds.

Logistic regression is tailor-made for dealing with categorical response variables. We can arrive at the idea of a logistic regression by the following chain of thought. Least squares regression assumes that the average response at a given set of X values is a linear function, or weighted average, of the X values. The weights are the slopes.

Average(Y at given X's) = constant + weighted combination of X's.

If the response Y is a categorical variable that has been coded as 0 for "no" and 1 for "yes," then the average of Y given some set of predictors is just the probability that Y is 1 (that the response is "yes"). So, the above statement becomes

Probability($Y = 1$ at given X's) = constant + weighted combination of X's.

Because a probability must fall between zero and one, it may be crazy to model it as a weighted sum of the X's because, like a line, such combinations are unbounded and will not respect the zero–one bounds.

To deal with this problem, *logistic regression models transformed probabilities* rather than the raw probabilities. A logistic regression fits the model

Transformed[Prob(Y=1 at given X)] = constant + weighted combination of *X*'s.

On the right hand side, we still have a weighted combination of the *X*'s just as in standard regression, but on the left-hand side we have a transformed version of the probabilities. The particular transformation of the probabilities is called the "logistic transform" and hence the name logistic regression. In summary, logistic regression models a transformed version of the probabilities in terms of a linear combination of *X*-variables.

The key to understanding logistic regression is to recognize the issues that arise when working with a transformed response variable. Essentially there are three worlds to work in when using logistic regression. Firstly, there is the world of the logit, that is, the world of the transformed probabilities. The logit of a probability p is defined as follows

$$\text{logit}(p) = \log\left(\frac{p}{1-p}\right).$$

The helpful thing about the logit is that, unlike a probability which is constrained to values between 0 and 1, the logit can take on any value. This means that fitting a line to the logits is more sensible than fitting a line to the probabilities.

Secondly, there is the world of the "odds." To go from a probability p to its odds requires the formula

$$\text{odds} = \frac{p}{1-p}.$$

Alternatively, you could take the logit and "undo" the log transform, that is, exponentiate it,

$$\text{odds} = \exp(\text{logit}) = e^{\text{logit}}.$$

Though the world of the "odds" is not intuitive to everyone, many people, especially those who like to gamble, feel far more comfortable interpreting odds than probabilities.

Finally, there is the world of the probabilities. Given the odds, you can recover the probability by calculating

$$p = \frac{\text{odds}}{1 + \text{odds}},$$

and, given a logit, you can recover the probability by calculating

$$p = \frac{\exp(\text{logit})}{1 + \exp(\text{logit})} = \frac{1}{1 + \exp(-\text{logit})}$$

At this point things can get quite confusing and it's natural to ask "Why bother with these three worlds?" "Why not just work with the probabilities?" It turns out that each world has some advantage. The world of the logit is where the regression model is fit. It's mathematically a good place to do the model fitting. The world of the odds is a good place to interpret the regression coefficients that come out of the logistic regression model. The world of the probabilities is a great place to work when you want to convey exactly what you have done to other people because 95% of people are going to have a much better feel for probabilities than the other two worlds. Understanding logistic regression has a lot to do with becoming comfortable in moving between the three worlds, and that is what the class examples should facilitate.

Interpretation of coefficients. Since we fit a linear model on the logit scale, the interpretation of the regression coefficients is again on the logit scale. In a univariate (one predictor) logistic regression, the interpretation of the slope, call it β_1 is generically "for every one unit change in X_1, the logit of the probability changes by β_1."

Very few people have a good feeling for logits and prefer to interpret the coefficient in the odds world. Recall that we go from logits to odds by using the exponential transform. The model on the logit scale is linear with the parts added together, and exponentiating something that is additive gives rise to a multiplicative model because

$$\exp(a + b) = \exp(a) \times \exp(b) \quad \text{or} \quad e^{a+b} = e^a e^b$$

The result is that the regression coefficients have a multiplicative interpretation on the odds scale, so that a generic interpretation of slope β_1 is "for every one unit change in X_1, the odds change by a multiplicative factor of $\exp(\beta_1)$."

Try this for a logistic regression coefficient of zero: "for every one unit change in X_1, the odds change by a multiplicative factor of $\exp(0) = 1$." If we multiply anything by 1, it does not change. Hence a logistic regression coefficient of 0 corresponds to no relationship between the response and the x-variable, just as in the usual regression model.

The Challenger Disaster

Orings.jmp

Might the Challenger explosion in 1986 have been avoided?

Seconds after liftoff on January 28, 1986, the space shuttle Challenger exploded in a ball of fire, killing all seven crew members aboard. The reasons for the disaster have been extensively investigated (by the Rogers commission) and may be attributed to a breakdown in communication among the management structures within NASA.

The immediate cause of the accident was a leaking joint in one of the two solid rocket boosters that propel the shuttle into space. The leaking joint had failed to contain the hot gases from the burning solid fuel propellant. Just after liftoff, hot gases began to burn through the seal joint in the right solid rocket booster (SRB). Seconds later the right SRB broke away from the vehicle, rupturing the external fuel tank. Massive, "almost explosive" burning ensued.

The leaking joint was a result of the failure of the secondary O-rings to seal. The shuttle boosters use two seals to guard against this type of accident. Prior to the accident there had been discussions as to whether or not this failure to seal was related to the temperature in the joint, which itself was affected by the temperature at the launch site. In his testimony before the Rogers Commission, Lawrence Mulloy, the Solid Rocket Booster program manager at the Marshall Space Flight Center, said, referring to the failure of the O-rings:

> "This is a condition that we have had since STS-2, that has been accepted; that blow-by of the O-rings cannot be correlated to the temperature by these data. STS-61A had blow-by at 75°. Soot blow-by of the primary O-rings has occurred on more than one occasion, independent of temperature. This is the nature of challenges. Think about this, think about your data."

Instinctively, Mulloy appears to have considered the following data which shows the number of incidents of O-ring thermal distress (burning/blow-by of the first of the two rings) for the seven previous shuttle flights that had displayed such problems.

FLIGHT	JOINT TEMPERATURE	#INCIDENTS
STS 51-5	52	3
41B	56	1
61C	58	1
41C	63	1
41D	70	1
STS-2	70	1
61A	75	2

The data set appears to corroborate Mulloy's claim of no significant correlation. At the time that the recommendation to launch the shuttle was sent from the O-ring manufacturers to NASA, the overnight low was predicted to be 18°. The temperature at the time of launch was 31°.

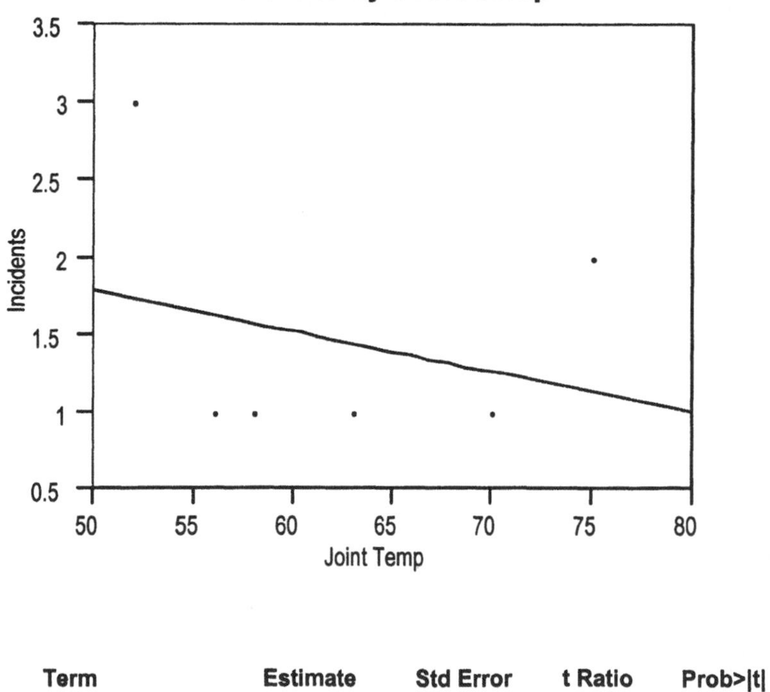

Term	Estimate	Std Error	t Ratio	Prob>|t|
Intercept	3.071	2.533	1.21	0.28
Joint Temp	-0.026	0.040	-0.65	0.54

Perhaps Mulloy should have thought more about the data. In particular he should have considered *all the previous shuttle flights*, not just those that had at least one O-ring incident. The data in Orings.jmp describes the 24 shuttle launches prior to the Challenger disaster. A regression analysis of these data is subject to criticisms on account of the discreteness and limited range of the dependent variable. However, it does show a clear negative trend that is significant. These data reveal a propensity for O-ring incidents to occur at lower temperatures. In particular all of the flights below 65° suffered at least one incident, whereas only 3 out of 20 flights launched above 65° suffered such failures. A point is partially obscured by the dashed bands at 52° and 3 failures. (The recent interpretation offered by E. Tufte in *Visual Explanations* (1997) gives a careful account of the way NASA handled the information that was available.)

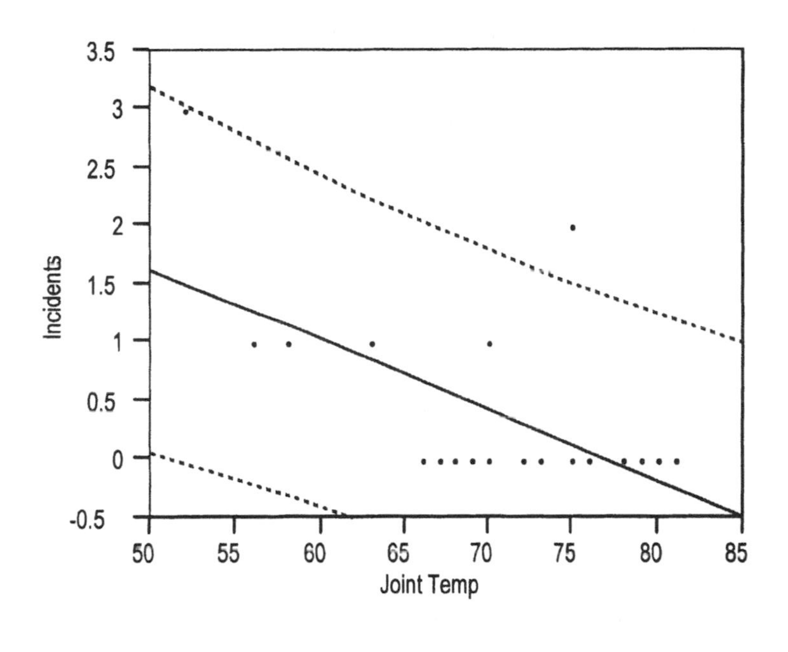

RSquare	0.332
Root Mean Square Error	0.648
Mean of Response	0.417
Observations	24

Term	**Estimate**	**Std Error**	**t Ratio**	**Prob>|t|**
Intercept	4.637	1.282	3.62	0.002
Joint Temp	-0.060	0.018	-3.31	0.003

Logistic regression offers an alternative analysis of these data. Rather than fit a linear model to the number of failures, a logistic regression builds a model for the probability of a failure (such a model does not distinguish flights with multiple failures from those with a single failure). The fitted curve in this figure is the probability of a leak in at least one joint as a function of temperature. The curve is described by a logistic function which has the form

$$\text{Fitted Probability} = \frac{\exp(a+bx)}{1+\exp(a+bx)} = \frac{1}{1+\exp(-a-b\,x)}$$

The chi-square column in the table of parameter estimates is computed as the square of the *t*-ratio that you would expect in a regression model. That is, the chi-square column is simply (Estimate/Std Error)^2. You can interpret the Prob>ChiSq column just as in regression (it's the p-value). In this case, we see that the effect of temperature is very significant.

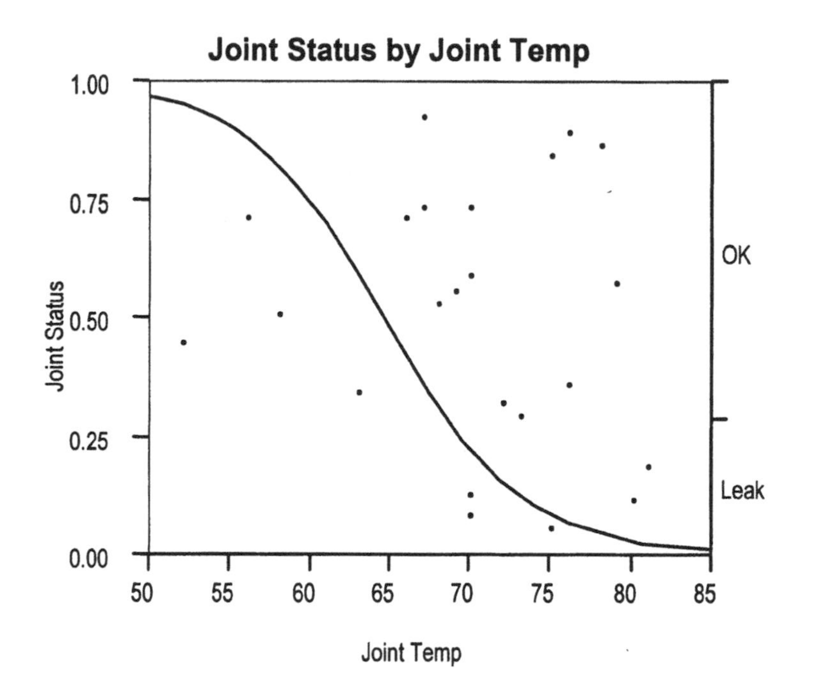

Term	Estimate	Std Error	ChiSquare	Prob>ChiSq
Intercept	15.05	7.28	4.28	0.04
Joint Temp	-0.23	0.11	4.76	0.03

To interpret the coefficient –0.23 of joint temperature, we need to convert this value into one that can be interpreted using "odds". The odds of a chance event is the ratio of the probability that it does occur to the probability that it does not,

$$\text{Odds (Event)} = \frac{\text{Prob of Event}}{1 - \text{Prob of Event}}\ .$$

For example, the odds of a fair coin showing heads after a toss are one to one. The odds that a roll of a fair die shows one dot on top are one to five.

The coefficient of joint temperature has a similar interpretation. The fitted logistic equation for the shuttle booster data is

$$\begin{aligned}\text{Pr(Failure at Temp } t) &= \frac{\exp(a+b\,t)}{1+\exp(a+b\,t)} \\ &= \frac{1}{1+\exp(-a-b\,t)} = \frac{1}{1+\exp(-15.05+0.23\,t)}\end{aligned}$$

To interpret the "slope" 0.23, rearrange the equation as the odds of failure and cancel the common term in the product,

$$\begin{aligned}\text{Odds of Failure at Temp } t &= \frac{\text{Pr(Failure at temp } t)}{\text{Pr(OK at temp } t)} \\ &= \frac{1}{1+\exp(-15.05+0.23\,t)} \times \frac{1+\exp(-15.05+0.23\,t)}{\exp(-15.05+0.23\,t)} \\ &= \exp(15.05-0.23\,t)\end{aligned}$$

Now take the log of the odds to get the logit, and we discover the linear equation underlying logistic regression

$$\text{Logit of Failure at Temp } t = 15.05 - 0.23\,t$$

Thus, the slope indicates how fast the log of the odds of failure changes as temperature increases. Alternatively, by exponentiating the slope,

$$e^{-0.23} = 0.79$$

we see that the odds of failure become 21% smaller for each increase of one degree (it's 79% of what it was). Thus if the odds of a failure at 65° are one to one (i.e., a 50% chance), then at 75° the odds are roughly $0.8^{10} \approx 0.1$.

Might the Challenger explosion in 1986 have been avoided?

What's the probability of failure at 31°, the launch temperature? From the previous calculations

$$\Pr(\text{Failure at } 31^\circ) = \frac{\exp(15.05-0.23\times31)}{1+\exp(15.05-0.23\times31)} = \frac{1}{1+\exp(-15.05+0.23\times31)} = 0.99965$$

Clearly, there was a high probability of a joint leaking, though as noted the failure of a single O-ring does not cause an explosion. Is this a valid prediction? There was no recorded data at this low a temperature, so this probability represents an extrapolation far outside the range of experience, with the consequent uncertainties. Nonetheless, there is strong evidence that O-rings would fail even if we do not believe every digit in this probability estimate.

In retrospect, had someone done the above analysis, the incident might have been avoided. This example emphasizes the following points:

1. Use all the data for the analysis. Do not select observations based on the size of the dependent variable.
2. Regression is not the right tool for all problems. Discreteness of the response and the fact that it has to be greater than or equal to zero limit the usefulness of predictions based on a simple linear regression model.
3. Even though regression may not be "technically" correct, it provides a qualitative description. It indicated the inverse relationship between O-ring failure and temperature, and it flagged a possible outlier.

A Simple Alternative

A simple analysis of the data provides the following table:

	At least one failure	No failures
Temp < 65	4	0
Temp > 65	3	17

Based on this table one can calculate the p-value associated with the null hypothesis of no relationship between temperature and failure (Fisher's exact test). The *p*-value for the null hypothesis of no relationship is 0.0033, and the hypothesis must be rejected. There was plenty of evidence that could have saved Challenger.

Marketing Orange Juice

Juice.jmp

What factors affect customer purchase of orange juice? Is it brand loyalty, price, or a discount coupon?

The data for this example are from a marketing study of consumer purchases of orange juice at five supermarkets. The data consist of 1070 purchase occasions, and the dependent variable is the type of orange juice purchased: Minute Maid or Citrus Hill. The data consider only occasions on which one or the other of these two brands of orange juice was purchased.

This is a large data set with many factors. It is a good idea to familiarize yourself with the information before diving into a logistic regression analysis. The data in the JMP spreadsheet are sorted by household and include the week of purchase. The following analysis is quite involved; here's the plan for what we are going to do. We

(1) begin with some exploratory graphical analysis of the data;

(2) fit marginal logistic regression models for several predictors (that is, one predictor at a time);

(3) build a multiple logistic regression model;

(4) interpret the coefficients of the multiple logistic regression;

(5) deal with collinearity; and

(6) conclude with our usual summary.

We begin with some exploratory graphical analysis of the data. The majority of the purchases are of Citrus Hill, a brand that is no longer sold.

Purchase

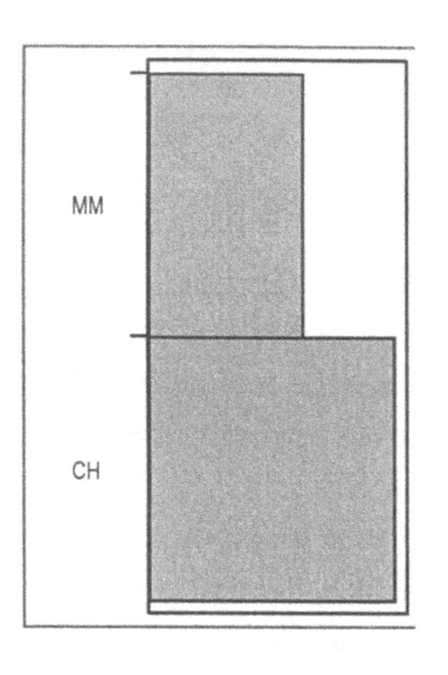

Level	Count	Probability	Cum Prob
CH	653	0.61	0.61
MM	417	0.39	1.00
Total	1070		

Store #7 has the largest share of purchases.

Store ID

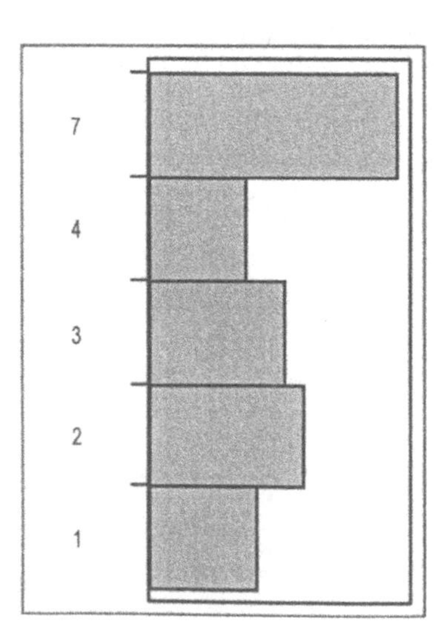

Level	Count	Probability	Cum Prob
1	157	0.15	0.15
2	222	0.21	0.35
3	196	0.18	0.54
4	139	0.13	0.67
7	356	0.33	1.00
Total	1070		

There seem to be large differences in purchase frequency across the stores. For example, Minute Maid has a 62% share in store #3, but only a 23% share in store #7.

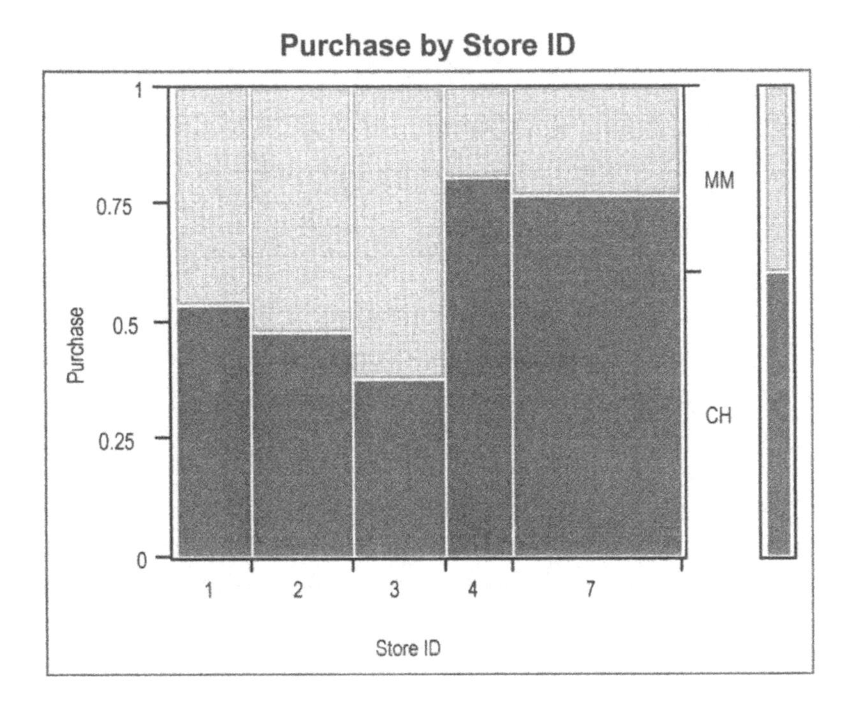

Crosstabs

Count Col %	Purchase Store ID 1	2	3	4	7	
CH	85 54.14	107 48.20	75 38.27	112 80.58	274 76.97	653
MM	72 45.86	115 51.80	121 61.73	27 19.42	82 23.03	417
	157	222	196	139	356	1070

Store #7 offers more specials and also has the lowest median sale price for Citrus Hill.

Special CH by Store ID

Special CH

1 0.8 0.6 0.4 0.2 0

1 2 3 4 7

Store ID

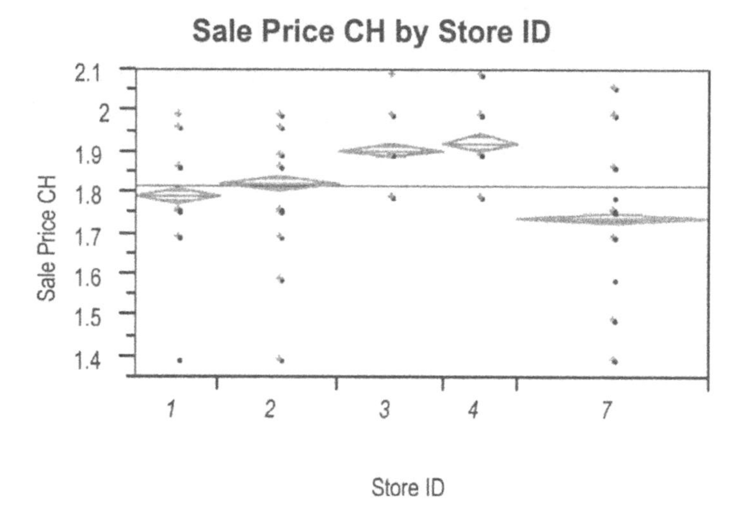

Have the lower prices in store #7 produced the higher sales? Or, have higher sales led to volume discounts and consequently lower prices? Since the data are collected as part of an observational study, we cannot answer this question. Such issues plague observational studies. To get a reliable answer, we would need to manipulate prices as part of an experiment to obtain data for answering such questions.

Price is an important factor. Discounts are also offered. The variables *Sale Price MM* and *Sale Price CH* are the actual prices available on that occasion, having subtracted off the discount available. The variable

Price Diff = *Sale Price MM* – *Sale Price CH*

is the difference in sale prices available for the two brands. Minute Maid is usually more expensive. The 95% confidence interval for mean price difference is [0.13, 0.16]. From the boxplot, Minute Maid was the more expensive brand on about 75% of the purchase occasions.

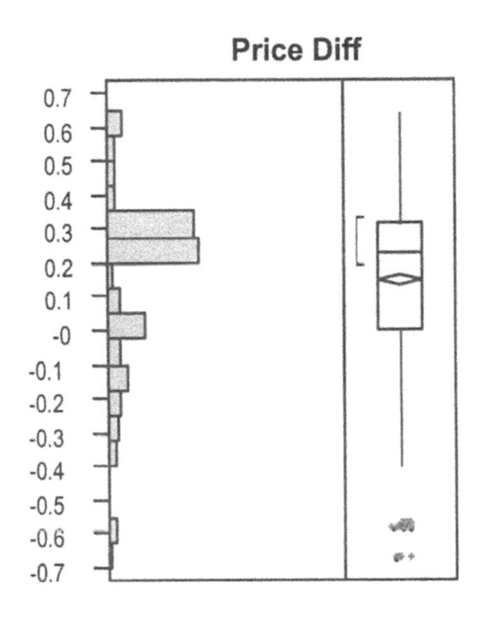

Moments

Mean	0.146
Std Dev	0.272
Std Error Mean	0.008
Upper 95% Mean	0.163
Lower 95% Mean	0.130
N	1070

For this part of our analysis, we will consider logistic regression models for several predictors, taken one at a time. The data includes two redundant measures of past brand preferences, termed *Loyalty CH* and *Loyalty MM* (the two sum to one). The loyalty index is a measure of the recent brand preferences of the households in the study. This index is the most important predictor of purchase behavior in this data.

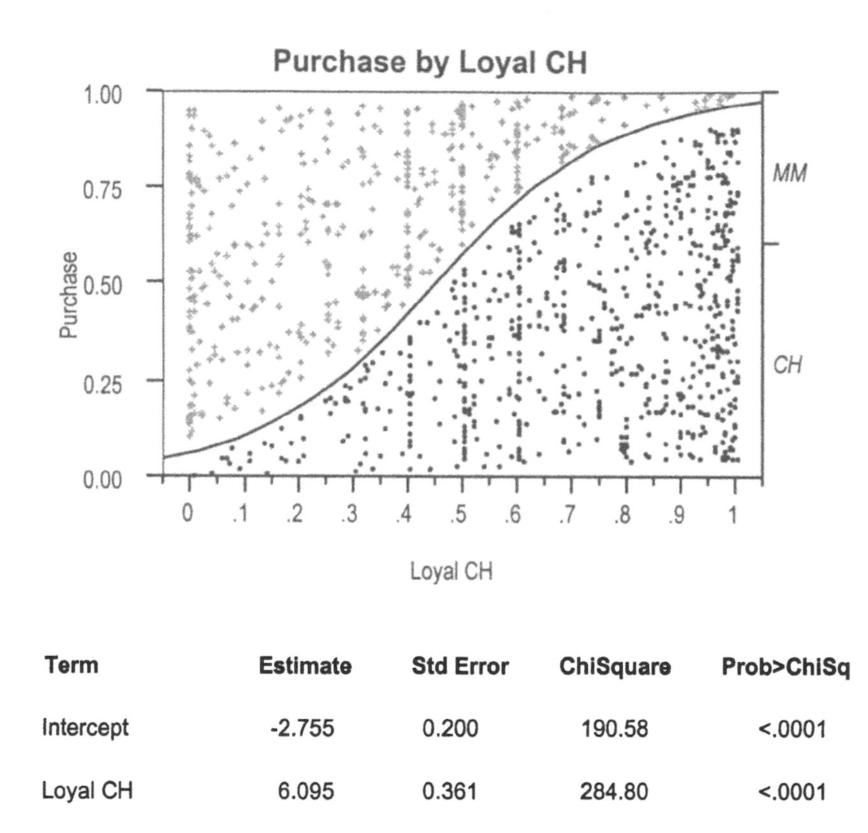

Term	Estimate	Std Error	ChiSquare	Prob>ChiSq
Intercept	-2.755	0.200	190.58	<.0001
Loyal CH	6.095	0.361	284.80	<.0001

JMP, like other SAS programs, uses an unusual convention for logistic regression. When interpreting a plot such as this, think of the logistic curve shown as a dividing line, separating Minute Maid purchases on the top from Citrus Hill on the bottom. As the loyalty to Citrus Hill increases, its "share" of the vertical axis increases (the curve increases). When the loyalty to Citrus Hill is small, Minute Maid has the larger share (the curve is near the bottom of the figure). As the loyalty of a household to Citrus Hill increases, the curve grows and Minute Maid's share shrinks.

Both price factors are also individually predictive of purchase choice. For example, here is a logistic regression of *Purchase* on *Sale Price MM*. The summary of the fitted model indicates that this price factor is significant (albeit not as good a predictor as loyalty).

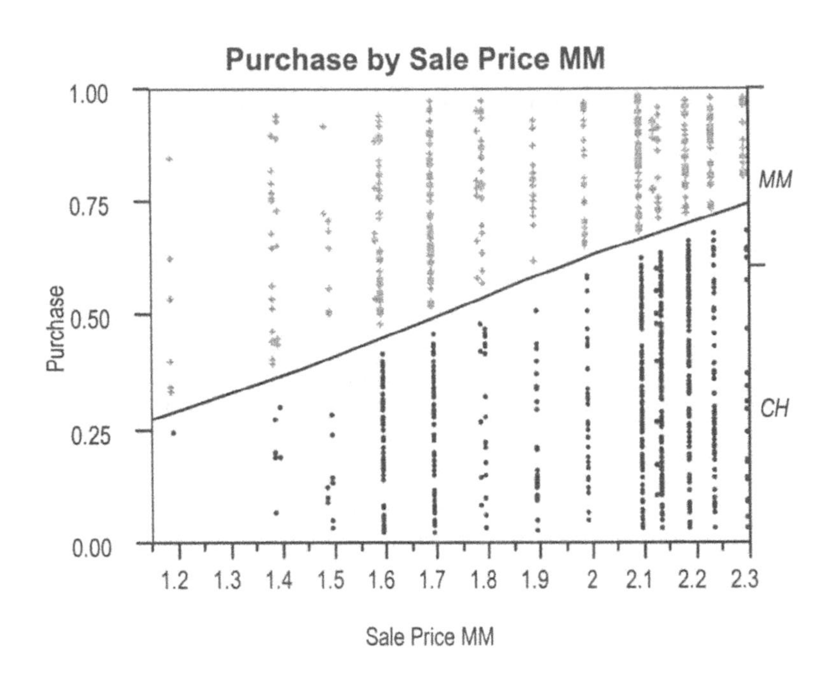

Term	Estimate	Std Error	ChiSquare	Prob>ChiSq
Intercept	-3.056	0.498	37.63	<.0001
Sale Price MM	1.795	0.254	50.00	<.0001

As you would expect, the logistic curve decreases as the sale price of Citrus Hill increases. The more expensive Citrus Hill becomes, the less likely is a customer to purchase it. Once again, this price is a significant predictor of purchase preference.

Term	Estimate	Std Error	ChiSquare	Prob>ChiSq
Intercept	4.078	0.854	22.78	<.0001
Sale Price CH	-1.993	0.467	18.25	<.0001

The difference in price has an even better fit, with a more distinctly curved separation of the two types of purchases. Note the increased level of significance of the fitted slope coefficients (ChiSquare is now 74, versus 50 and 18 in the previous models).

Term	Estimate	Std Error	ChiSquare	Prob>ChiSq
Intercept	0.156	0.073	4.58	0.0324
Price Diff	2.169	0.252	74.12	<.0001

Is a price advantage enough to overcome past preferences? Both loyalty and price difference are important predictors of purchase preference when considered separately, but what happens when we combine these predictors into one model?

Let's try to combine some of these factors in a multiple logistic regression model, starting from the obvious choices, loyalty and price difference. With the price of Citrus Hill also included, we see that the price difference, not the actual price, is the key factor. After all, these data are restricted to occasions on which a purchase occurred. Keep in mind that positive coefficients imply an increasing preference for Citrus Hill.

Response: Purchase

Parameter Estimates

Term	Estimate	Std Error	ChiSquare	Prob>ChiSq
Intercept	-2.804	1.532	3.35	0.0673
Loyal CH	6.407	0.388	272.83	<.0001
Price CH	-0.244	0.826	0.09	0.7677
Price Diff	2.856	0.343	69.21	<.0001

The presence of special promotions adds a little, but not much. The chi-square probability for this term is not significant.

Term	Estimate	Std Error	ChiSquare	Prob>ChiSq
Intercept	-3.297	0.227	211.64	<.0001
Loyal CH	6.362	0.386	271.06	<.0001
Price Diff	2.883	0.347	68.95	<.0001
Special CH	0.431	0.257	2.82	0.0931

Some differences among the stores also exist, but these seem to be rather small effects in comparison to loyalty and price differential. Store 7 seems to have the largest effect; Store 7 has the most specials and best prices for Citrus Hill. To find the value for Store 7, add the slopes for stores 1–4 and change the sign. JMP does not do the expanded estimates for logistic regression.

Term	Estimate	Std Error	ChiSquare	Prob>ChiSq
Intercept	-3.196	0.236	183.75	<.0001
Loyal CH	6.214	0.402	238.79	<.0001
Price Diff	2.830	0.351	64.89	<.0001
Store ID[1]	-0.143	0.184	0.60	0.4380
Store ID[2]	-0.287	0.166	3.00	0.0833
Store ID[3]	-0.195	0.190	1.05	0.3056
Store ID[4]	0.093	0.225	0.17	0.6789

Effect Test

Source	Nparm	DF	Wald ChiSquare	Prob>ChiSq
Loyal CH	1	1	238.786	0.0000
Price Diff	1	1	64.893	0.0000
Store ID	4	4	14.333	0.0063

Should we pursue the analysis and focus on store #7, or have we just found this one store by chance alone (think back to the Hsu and Tukey-Kramer comparisons of Class 9). Unfortunately, JMP (like all other packages) does not provide these protections against trying to read too much from a collection of p-values in its logistic regression output. Fortunately, we can still use the Bonferroni idea. In focusing on the five stores, if we are working at a 95% level of confidence, rather than use 0.05 as the threshold for a p-value, the Bonferroni rule suggests that we use 0.05/5 = 0.01 instead.

Since JMP does not give the explicit estimate for store #7, we have to work a little harder than otherwise. To get JMP to print the value (and its standard error) for store #7, its value is recoded as 0 (zero) in the variable *STORE*. The analysis is repeated with this variable replacing *Store ID*. Here are the parameter estimates.

Response: Purchase

Parameter Estimates

Term	Estimate	Std Error	ChiSquare	Prob>ChiSq	
Intercept	-3.196	0.236	183.75	<.0001	
Loyal CH	6.214	0.402	238.79	<.0001	
Price Diff	2.830	0.351	64.89	<.0001	
STORE[0]	0.531	0.157	11.42	0.0007	STORE #7
STORE[1]	-0.143	0.184	0.60	0.4380	
STORE[2]	-0.287	0.166	3.00	0.0833	
STORE[3]	-0.195	0.190	1.05	0.3056	

The intercept and slopes for loyalty and price difference are the same as in the prior output. Since JMP uses the first k–1 of k categories in its output, store #7 now appears under the label for store #0.

The p-value shown for store #7 is 0.0007. Clearly, this store stands out from the rest. The others are not so distinct. To keep the rest of this analysis more simple, we'll drop the others and just use store #7 in what follows.

These first analyses suggest a model with loyalty, price difference, and a dummy variable for store #7 (distinguishing among the other stores is not significant). All of the terms in the model shown below are significant.

Response: Purchase

Model	-LogLikelihood	DF	ChiSquare	Prob>ChiSq
Difference	301.79	3	603.5806	<.0001
Full	413.64			
Reduced	715.43			
	RSquare (U)		0.4218	
	Observations		1070	

Term	Estimate	Std Error	ChiSquare	Prob>ChiSq
Intercept	-3.06	0.23	178.80	<.0001
Loyal CH	6.32	0.39	263.45	<.0001
Price Diff	2.82	0.35	64.92	<.0001
Store#7?[No]	-0.35	0.10	12.77	0.0004

How are we to interpret these coefficients? Odds ratios help. The odds ratio represents the odds for purchasing Citrus Hill *based on the extremes of the indicated predictor*. For example, the odds ratio for *Loyal CH* means that

$$\frac{\text{Odds(purchase CH} \mid \text{highest } \textit{Loyal CH})}{\text{Odds(purchase CH} \mid \text{lowest } \textit{Loyal CH})} = 553 .$$

JMP will add these to the parameter summary when using the *Fit Model* platform.

Term	Estimate	Std Error	ChiSquare	Prob>ChiSq	Odds Ratio
Intercept	-3.06	0.23	178.80	<.0001	0.05
Loyal CH	6.32	0.39	263.45	<.0001	553.48
Price Diff	2.82	0.35	64.92	<.0001	40.04
Store#7?[No]	-0.35	0.10	12.77	0.0004	0.50

Thinking back to the analysis of variance, we should check for interactions. Since store #7 is different, the loyalty and price difference effects might differ there from the others. You can try it, but the estimates (not shown) are not significant.

What about discounting? The model to this point only considers price differential — ignoring how that price was reached. Are customers more impressed by a discount than a cheaper list price? Adding both discount measures does not help.

Term	Estimate	Std Error	ChiSquare	Prob>ChiSq
Intercept	-3.19	0.30	111.39	0.00
Loyal CH	6.30	0.39	262.00	0.00
Price Diff	3.15	0.81	15.10	0.00
Store#7?[No]	-0.32	0.10	9.53	0.00
Disc CH	*0.37*	*1.14*	*0.11*	*0.74*
Disc MM	*0.54*	*0.94*	*0.33*	*0.57*

Nor does adding both discounts measured as percentages.

Term	Estimate	Std Error	ChiSquare	Prob>ChiSq
Intercept	-3.25	0.31	109.31	<.0001
Loyal CH	6.30	0.39	262.16	<.0001
Price Diff	3.34	0.83	16.00	<.0001
Store#7?[No]	-0.31	0.10	8.92	0.0028
Pct Disc MM	*1.64*	*2.03*	*0.65*	*0.4211*
Pct Disc CH	*0.48*	*2.17*	*0.05*	*0.8261*

However, combining the factors does lead to a significant improvement.

Term	Estimate	Std Error	ChiSquare	Prob>ChiSq
Intercept	-3.43	0.31	122.73	0.00
Loyal CH	6.39	0.39	262.71	0.00
Price Diff	3.71	0.68	29.65	0.00
Store#7?[No]	-0.26	0.11	6.02	0.01
Disc MM	-16.98	7.97	4.53	0.03
Pct Disc MM	38.03	17.04	4.98	0.03

Why is this? Note the very large, relatively unstable coefficients for the price discount and percentage price discount. Quite a bit of collinearity is present in these last two terms.

In contrast to all of the plots it offers for least squares regression, JMP offers little diagnostic help with logistic regression. Though it is not quite appropriate, we can use the least squares regression diagnostics to view what is happening in this model. Simply replace the response by the dummy variable *Buy*. Here is the multiple regression summary. The VIFs make the collinearity effect apparent (as do the leverage plots which are not shown here). The effects are significant, but the coefficient estimates seem very unreliable.

Response: Buy

RSquare	0.461
Root Mean Square Error	0.359
Mean of Response	0.610
Observations	1070

Term	Estimate	Std Error	t Ratio	Prob>|t|	VIF
Intercept	-0.00	0.03	-0.14	0.89	0.00
Loyal CH	0.96	0.04	26.24	0.00	1.06
Price Diff	0.44	0.08	5.46	0.00	3.94
Store#7?[No]	-0.03	0.01	-2.06	0.04	1.28
Disc MM	-2.27	1.08	-2.11	0.04	438.93
Pct Disc MM	4.98	2.30	2.16	0.03	454.90

Below is the plot of the last two factors. The discount and percentage discount are almost perfectly correlated. The correlation is 0.9988.

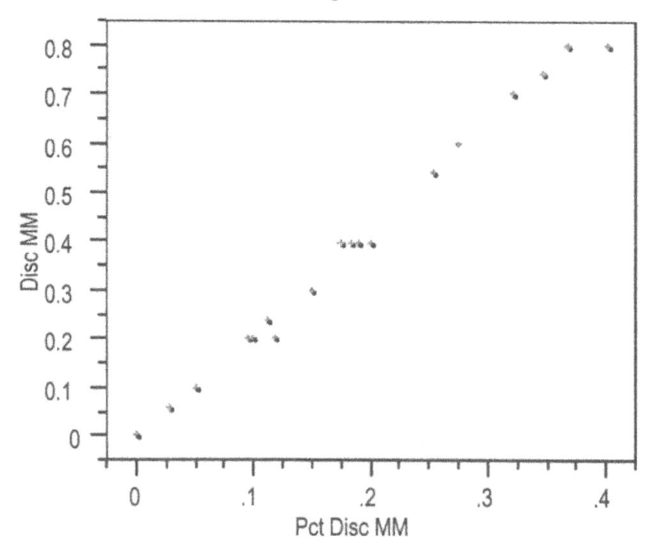

To sort out what is happening, we return to the logistic model and formulate it directly in terms of prices, discounts, and interactions. With the price difference included as in the previous models, interactions are a bit too confusing.

Response: Purchase

Model	-LogLikelihood	DF	ChiSquare	Prob>ChiSq
Difference	306.666	10	613.3313	<.0001
Full	408.760			
Reduced	715.426		RSquare (U)	*0.4286*

Term	Estimate	Std Error	ChiSquare	Prob>ChiSq
Intercept	-5.07	1.87	7.38	0.007
Loyal CH	6.16	0.47	173.68	0.000
Store#7?[No]	-0.28	0.11	6.74	0.009
Price CH	-2.64	1.26	4.36	0.037
Price MM	3.60	0.86	17.65	0.000
Disc CH	22.64	19.69	1.32	0.250
Disc MM	18.58	9.33	3.97	0.046
Price MM*Disc MM	-10.02	4.55	4.85	0.028
Price CH*Disc CH	-7.43	10.19	0.53	0.466
Loyal CH*Disc MM	-0.29	1.87	0.02	0.879
Loyal MM*Disc CH	-8.72	5.62	2.41	0.120

Removing the terms that are not significant makes little difference to the overall fit and the reduction of the collinearity gives better estimates of the remaining parameters (though clearly a lot of collinearity remains — look at the odds ratio for the discount of MM).

Model	-LogLikelihood	DF	ChiSquare	Prob>ChiSq
Difference	306.386	8	612.7713	<.0001
Full	409.040			
Reduced	715.426			
	RSquare (U)	*0.4283*		

Term	Estimate	Std Error	ChiSquare	Prob>ChiSq	Odds Ratio
Intercept	-4.61	1.76	6.89	0.01	0.010
Loyal CH	6.13	0.41	221.61	0.00	459.340
Store#7?[No]	-0.27	0.11	6.63	0.01	0.580
Price CH	-2.95	1.19	6.16	0.01	0.307
Price MM	3.67	0.85	18.52	0.00	9.031
Disc CH	8.60	3.41	6.37	0.01	73.622
Disc MM	18.58	9.24	4.04	0.04	2842722.8
Price MM*Disc MM	-10.08	4.40	5.24	0.02	0.000
Loyal MM*Disc CH	-8.69	5.61	2.40	0.12	0.023

What factors affect customer purchase of orange juice? Is it brand loyalty, price, or a discount coupon?

The major factors determining the brand purchased are past "loyalty" to that brand and the difference in price. One store (#7) seems to be pushing one brand.

Discounting also has a special effect aside from altering the difference in prices between the two. Though the collinearity in these data masks the effect of discounting, we can see that households treat the effect of a discount somewhat differently from just the price differential alone.

An important aspect of these data that we have not discussed is the *longitudinal nature* of the data. The data set consists of varying numbers of purchases by a much smaller set of households. The data set has 1070 purchase occasions, but only 174 households. One household is observed on 42 occasions; whereas many others are seen but once. Is the assumption of independence reasonable in this setting?

Class 12. Modeling Time Series

Data that are gathered over time (time series) often present special problems. In particular, observations within a single time series are seldom independent. Such autocorrelation has a powerful effect on a regression model, leading to a spurious impression of accuracy. In addition, several time series often show high correlation, tending to move up and down together. This correlation between series introduces multicollinearity into models using time series data. Special methods help accommodate such data. Regression analysis is also used to model cross-sectional data in which one has before-after pairs of measurements.

Topics

- Autocorrelation
- Durbin-Watson test
- Differenced and lagged variables
- Stationarity

Examples

1. Predicting cellular phone use, revisited (time trend with lags)
2. Trends in computer sales (differencing)

Key applications

Adjusting regression models for missing factors. Regression is a powerful tool for prediction, but often we omit an important predictive factor. In the sequential context of a time series, the values of such omitted factors often "track" each other. By omitting this factor, we leave its tracking pattern as part of the unexplained residual variation. Left alone, the resulting regression model suffers two key flaws. First, we have not modeled all of the available information and will not obtain as accurate a prediction as is possible. Second, the standard errors associated with the regression are wrong and artificially inflate the precision of our estimates (i.e., the SE's are generally too small and ought to be much larger).

Tracking annual sales. Executives who control divisions or product lines in many companies are held accountable for the sales of their groups. Each year, the company sets targets for the sales level expected of each. As the year rolls by, the executives want to be assured that they are making the expected progress toward their targets.

Clearly, it's seldom advantageous to come in below the target, and in some cases it's also unwise to be too high above the target (pulling future sales from next year, making it harder to meet those ever rising expectations). As the data come in, management wants predictions for the whole year, using patterns from the past to make adjustments. For example, if 50% of annual sales historically occur in December, it may not be surprising to be at 40% of the target going into November.

Modeling financial markets. How will the market react to a future change in interest rates? What sort of return can you expect on your mutual fund next year? Such questions are easy to pose, but very hard to answer. Most statistical models predict the future by extrapolating past trends, but do you expect interest rates to affect the market in the future in the same way they have affected the market in the past? Markets evolve, and the difficult part in time series modeling is trying to accommodate such changes. A consistent theme we have seen is that more data leads to more precise statistical estimates (*i.e.,* smaller standard errors and narrower confidence intervals). In time series analysis, more data can be a curse if the market (or whatever is being studied) has changed during the period of observation.

Definitions

Autocorrelation. The correlation between two measurements of the same thing, taken at different times, such as the correlation between the level of unemployment from one month to the next. We saw autocorrelation in Class 2 where dependent residuals tracked over time, violating a key assumption of regression. Autocorrelation in residuals is often due to a specification error in the regression model and may indicate that we have omitted a useful predictor.

Autoregression. A regression model in which the response is predicted using *lags* of itself as explanatory variables, as in the so-called first-order autoregression $Y_t = \beta_0 + \beta_1 Y_{t-1} + e_t$. Clearly, you can add more lags, with usually diminishing gains. Though autoregressions often fit well (i.e., explain much of the variation in the response), they typically offer little in the way of explanation for why the series behaves as it does.

Differencing. Often, we obtain a simpler model for a time series by working with changes rather than levels. Rather than model the size of an economy via GDP, for example, model period-to-period changes obtained by differencing GDP as $GDP_t - GDP_{t-1}$. Differencing is related to the analysis of stock returns rather than prices.

Durbin-Watson statistic. This statistic measures the presence of first-order autocorrelation in the unobserved errors of a regression. It infers correlation between adjacent errors using adjacent residuals; in seasonal models where the correlation is, say, between errors four quarters apart, the DW statistic may be misleading since the autocorrelation is not between adjacent errors.

Lags, lagged variable. A lagged series looks back in time; for example, Y_{t-1} is the one-period lag of the series Y_t. One often sees lags of the form Y_{t-12} with monthly data, or Y_{t-4} with quarterly data, to capture seasonal patterns.

Stationary, stationarity. A time series is stationary if its statistical properties are stable over the period of observation. In particular, stationarity requires that the *autocorrelations* are fixed over the time of observation.

Time series. Data gathered over time, such as monthly GDP or weekly sales.

Concepts

Effective sample size. Observations that are positively autocorrelated (i.e., the correlation of adjacent values is greater than zero) are not so informative about a mean value as uncorrelated observations. From a sample of n uncorrelated observations with variance σ^2, we know that the standard error of the average decreases as n increases,

$$SE(\bar{Y}) = \frac{\sigma}{\sqrt{n}}.$$

If the observations from a time series $Y_1, \ldots, Y_n$ are autocorrelated with $Corr(Y_t, Y_{t+s}) = \rho^{|s|}$, then the formula for the standard error becomes

$$SE(\bar{Y}) = \frac{\sigma}{\sqrt{n}} \times \sqrt{\frac{1+\rho}{1-\rho}}.$$

For example if $\rho=0.6$, then the $SE(\bar{Y})$ is twice the value for the uncorrelated case since $(1+.6)/(1-.6) = 4$. In this sense, n correlated observations with $\rho=0.6$ give the same standard error (and thus the same length confidence interval) as $n/4$ uncorrelated observations. For $\rho=0.6$, the effective sample size for estimating the mean is $n/4$.

An important implication is that if we are ignorant of the autocorrelation and proceed as though the data are uncorrelated, our claimed confidence intervals for a mean will be much shorter than they should be when $\rho>0$. The situation is typically more extreme in regression because of autocorrelation in the predictors (i.e., the effective sample size is smaller still).

Stationarity. This powerful assumption underlies most of the statistical analysis of time series. Without some form of stationarity, one can do little with time series. In very general terms, stationarity means that the statistical properties — averages, standard deviations, correlations — are stable over time. This stability is crucial because it implies that if we identify a systematic relationship in the observed data, then we can extrapolate this relationship into the future. The assumption of stationarity means that we believe that the observed autocorrelation in historical data between Y_t and Y_{t-1}, for example, will also be present between future observations. Thus if we can predict Y_t from Y_{t-1} in the observed data, we ought to be able to do the same in the future.

More recent modeling techniques for time series try to avoid this assumption as much as possible, but inevitably fall back upon it in some fashion. For example, a popular recent class of time series models, known by the acronym ARCH, allow the standard deviation of a time series to change. Although they allow this flexibility, ARCH models assume that such changes in the variance (or volatility) of a time series themselves obey a model that remains constant over time.

Heuristic

Judging the accuracy of predictions.

Are the predictions of a model useful? A simple way to look at these that is very useful particularly with time series is to compare the size of the prediction interval (the interval designed to hold a single new value of the series) to the variation in the data itself. If the prediction interval is so wide as to capture most of the data, the forecasting effort has produced little gain.

Potential Confuser

Autocorrelation versus collinearity.

Both are problems in regression, and both have to do with correlation. How do I keep them straight? Autocorrelation is correlation between the *rows* in the spreadsheet, typically representing a sequence of values gathered over time. Collinearity is correlation among the *columns* that define the predictors in the model.

Predicting Cellular Phone Use, Revisited

Cellular.jmp

Our previous analysis of this data exposed correlation among the error terms: the residuals track over time. How can we model this feature of the data to obtain more accurate predictions?

A plot of the initial subscriber data (which ends in December, 1995) is shown below. Each point corresponds to the cumulative count at the end of a six-month period spanning the years 1984 through 1995. In this analysis, we'll see how well we can predict this series by comparing models fit to the data shown here to what actually happened during the next five years (at least as claimed by the source of this data).

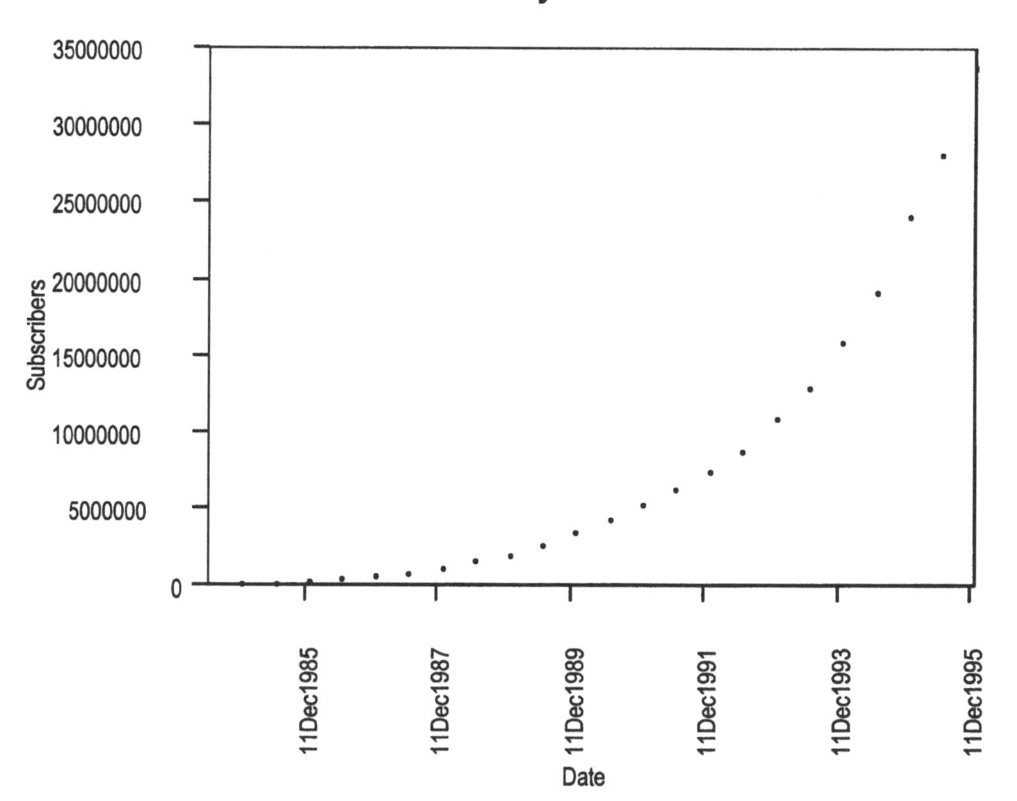

Owing to the nonlinear growth that is evident, we considered several models and (perhaps) settled on the model that uses the quarter-power of the number of subscribers. This model fits extremely well. This plot shows the 95% prediction intervals associated with the fit.

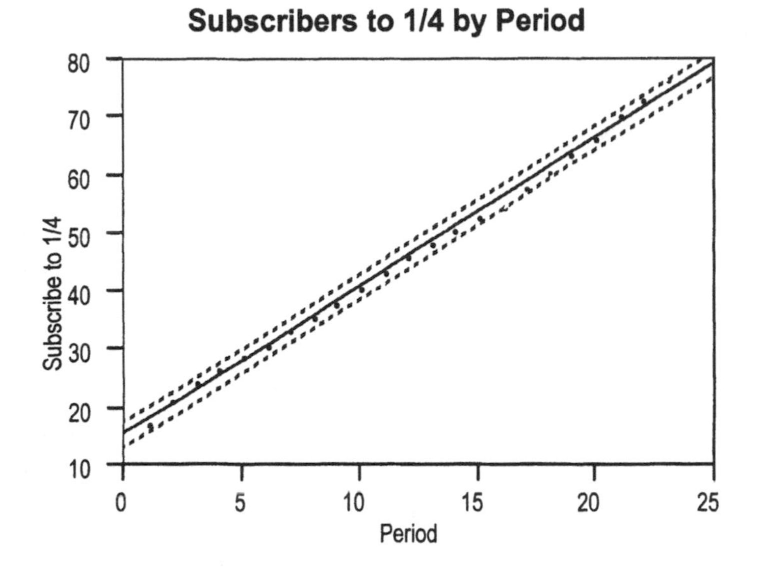

Subscribers to 1/4 = 15.38 + 2.547 Period

Summary of Fit

RSquare	0.997
Root Mean Square Error	0.972
Mean of Response	45.946
Observations	23

Source	**DF**	**Sum of Squares**	**Mean Square**	**F Ratio**
Model	1	6567.2	6567.2	6948.143
Error	21	19.8	0.9	**Prob>F**
C Total	22	6587.0		<.0001

Term	**Estimate**	**Std Error**	**t Ratio**	**Prob>|t|**
Intercept	15.38	0.42	36.70	<.0001
Period	2.55	0.03	83.36	<.0001

The residuals from this model, however well it fits, are not random. Rather they track each other through time.

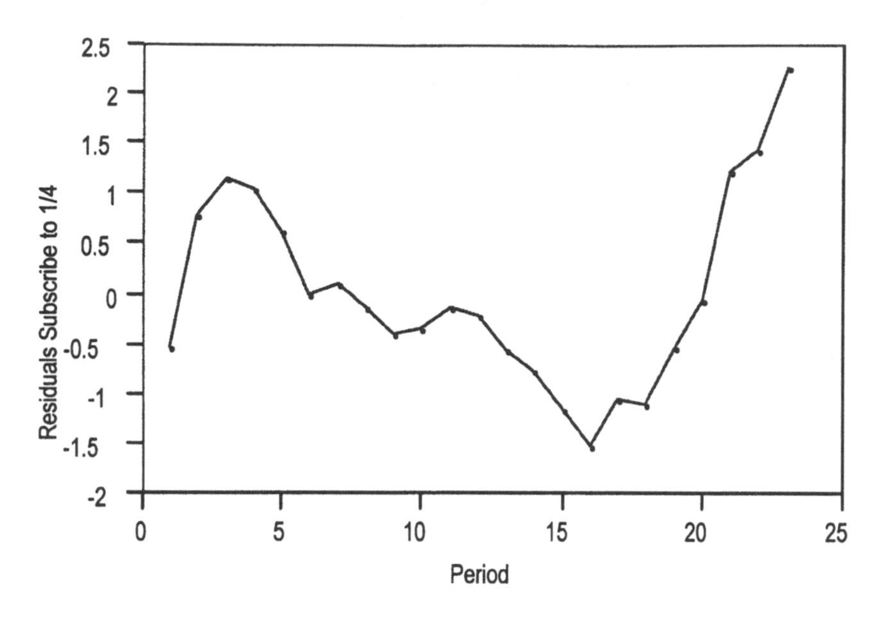

To obtain a better characterization of the degree to which the residuals track through time, we need to look at a plot of each residual on its predecessor. Here is the formula that does the trick, using a the Lag function to obtain the lagged time series,

Lag(Residuals Subscribe to 1/4, 1)

Notice that this operation just shifts the values in the residual column down one row.

The plot of the residuals shows a clear relationship between the series with its lagged values. The correlation is high ($0.84 \approx \sqrt{R^2}$) and the slope is significant. The correlation between adjacent sequential observations is known as the *lag one autocorrelation* of the series. (The time series capabilities of JMP include a special plot that shows a sequence of such autocorrelations for various lags. These tools are outside our scope and are usually covered in a course on time series methods.)

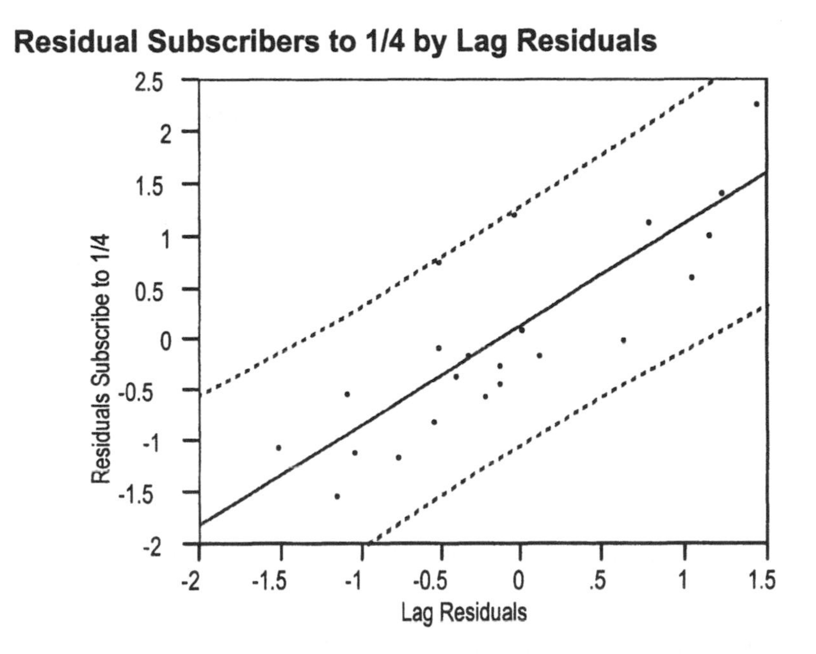

Residuals Subscribers to 1/4 = 0.12 + 0.97 Lag residuals

Summary of Fit

RSquare	0.702
Root Mean Square Error	0.540
Mean of Response	0.024
Observations	22

Term	Estimate	Std Error	t Ratio	Prob>|t|
Intercept	0.12	0.12	1.08	0.30
Lag residuals	0.97	0.14	6.87	0.00

The *Fit Model* platform offers another method for determining the extent of the sequential dependence between residuals. This diagnostic is known as the Durbin-Watson statistic. You can find this command in the group titled "Row diagnostics" in the pop-up menu at the top of the multiple regression output.

The autocorrelation printed here differs from that given by the previous regression (0.71 vs. 0.84). The difference is caused by the way that the ends of the data are handled and is only noticeable with short series like this one.

Response: Subscribers to 1/4

Term	Estimate	Std Error	t Ratio	Prob>|t|
Intercept	15.38	0.42	36.70	<.0001
Period	2.55	0.03	83.36	<.0001

Durbin-Watson

Durbin-Watson	Number of Obs.	Autocorrelation
0.31	23	0.71

The Durbin-Watson statistic lives on an unusual scale, ranging from 0 to 4. Values of this statistic near 2 indicate an absence of autocorrelation or tracking. A DW statistic near 0 suggests substantial positive autocorrelation. Values of DW larger than 2 indicate negative autocorrelation. The DW statistic is related to the size of the autocorrelation between adjacent residuals. In particular, it is approximately the case that

$$\text{DW} \approx 2(1\text{-autocorrelation}) \quad \text{OR} \quad \text{autocorrelation} \approx 1 - \frac{\text{DW}}{2}$$

In this example, $2(1-0.71) = 0.58$, which is not too close. The approximation is poor in this example because of the small sample size and large values near the ends of the observed residuals. It works well with more data.

Although examining the DW statistic is more convenient than looking at either the sequence plot or lag plot, both plots provide much more insight as to why the autocorrelation is present.

Modeling data with autocorrelation requires more sophisticated tools than we have access to in this course. However, we can certainly take the analysis quite a bit further. Our results suggest that the model (with Y = Subscribers$^{1/4}$) is of the form:

$$\begin{aligned} Y_t &= \text{Time Trend} + \text{error}_t \\ &= \text{Time Trend} + \rho\ \text{error}_{t-1} + (\text{smaller error})_t \ . \end{aligned}$$

This simple manipulation suggests that we use the past error to help predict the future error. However, how are we to do that since the errors are not observable? Use the residuals instead. (Better methods are available, but the results that follow give a sense for what can be accomplished using the regression methods we have been studying.)

Response: Subscribe to 1/4

RSquare	0.9991
Root Mean Square Error	0.5318
Mean of Response	47.2437
Observations	22

Term	Estimate	Std Error	t Ratio	Prob>|t|
Intercept	15.22	0.25	60.04	<.0001
Period	2.57	0.02	139.93	<.0001
Lag residuals	1.02	0.14	7.07	<.0001

Durbin-Watson

Durbin-Watson	Number of Obs.	Autocorrelation
1.11	22	0.24

Evidently, some structure remains in the residuals even with this adjustment, although we seem to have captured a good deal of it. A plot clarifies the remaining structure.

Although the fit has significantly improved, the residual plot indicates that structure remains. The appearance of the data continues to show tracking. Evidently, the residuals have more autocorrelation than we first suspected.

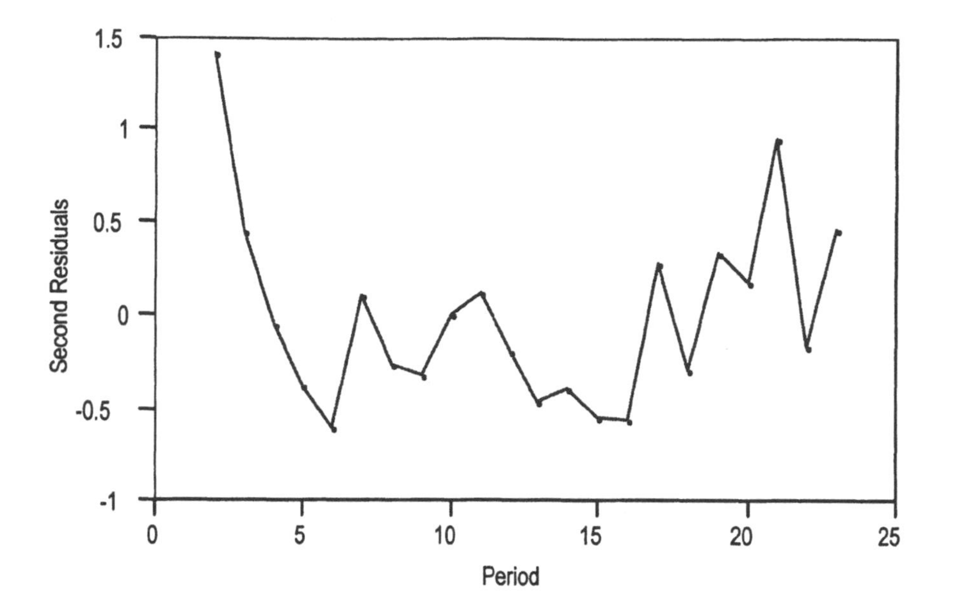

It seems that most of the unexplained pattern happens in the early part of the series. You can get a sense of why this happens by looking at the plot of the raw data on a logarithmic scale (not shown). In that plot, the early years appear to grow differently from most of the data.

A common approach taken in time series analysis in such problems is to use more lags. Here is the fit using two lags of the original residuals. Collinearity is starting to be a problem (the lagged residuals are correlated with each other), and we have lost another initial value, but again the fit seems better. This model seems to have extracted most of the autocorrelation that remained.

Response: Subscribers to 1/4

RSquare	0.9995
Root Mean Square Error	0.3901
Mean of Response	48.4819
Observations	21.0000

Term	Estimate	Std Error	t Ratio	Prob>|t|	VIF
Intercept	14.91	0.21	69.86	0.00	0.00
Period	2.59	0.02	163.52	0.00	1.27
Lag residuals	1.28	0.17	7.37	0.00	2.84
Lag 2 Resids	-0.25	0.20	-1.24	0.23	3.24

Durbin-Watson

Durbin-Watson	Number of Obs.	Autocorrelation
2.54	21	-0.31

Time Series Plot

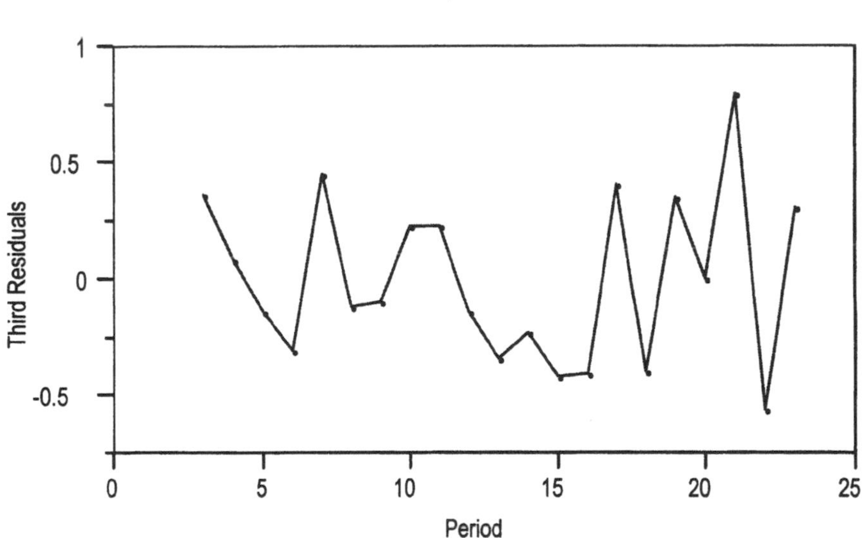

It is easy to lose sight of how well these models are doing. The next plot shows the data along with the predictions from the model with just the period (ignoring any autocorrelation) and the model that uses two lagged residuals. The fits are quite close.

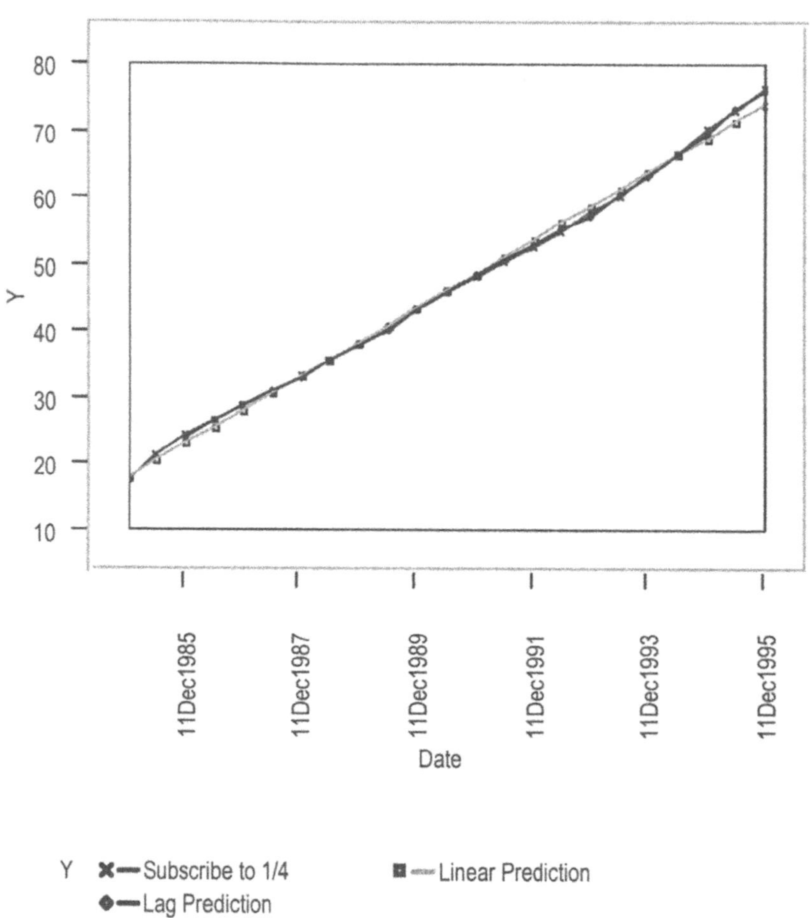

Zooming in on the last few points (using the magnifying glass tool) shows that the models differ by quite a bit near the end of the observed series. The model using the lagged residuals follows the tracking residuals and suggests a much higher prediction for the end of June, 1996.

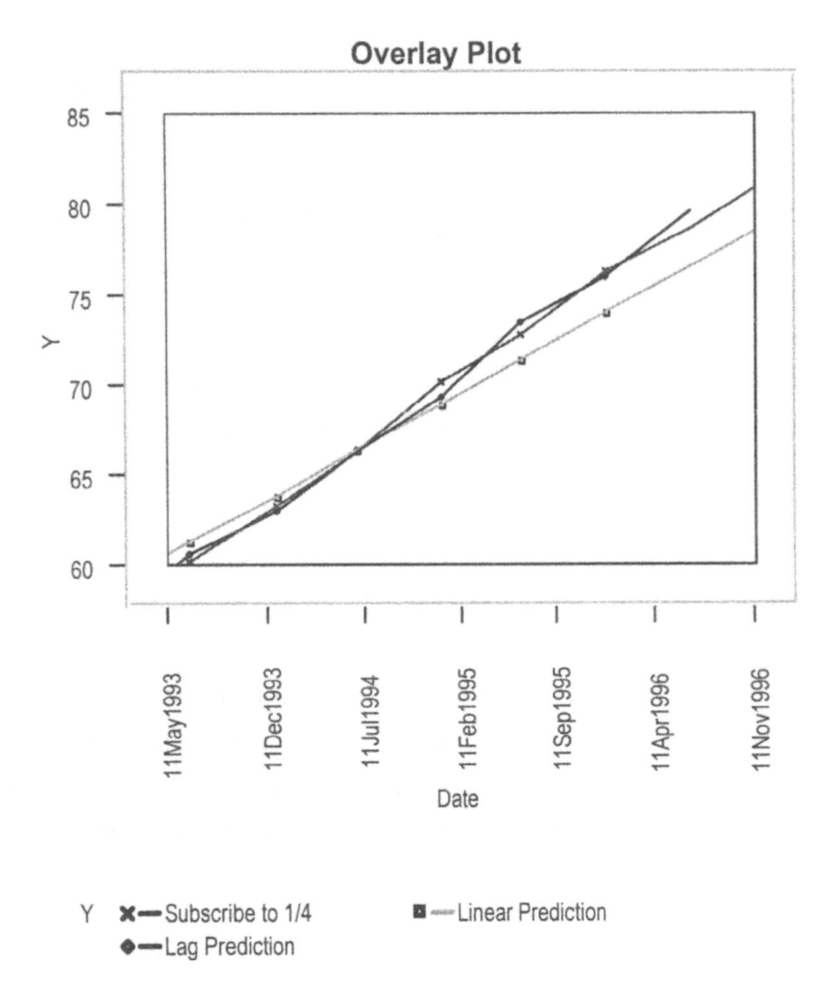

Even though both models fit extremely well, the resultant prediction intervals for the next six-month period are rather wide and almost disjoint. The interval using the lags is more compatible with the alternative analysis presented in Class 2 for this data set than with the first analysis presented in Class 2.

Model	Prediction	Prediction Interval on Subscriber scale
Period alone	76.51	[74.31–78.72] ⇒ [30.5 – 38.4] million
Period and lags	79.56	[78.45–80.68] ⇒ [37.9 – 42.4] million

The predictions using *Period* alone are rather poor, considering that this model has an R^2 of over 99%. In particular, the prediction 76.51 on the quarter-power scale is almost the same as the last data value (for the second half of 1995, 76.24). It's very unlikely that growth would slow so much.

So, what actually happens? Though not used in this analysis, the data file includes values through December 1999. The next value for the series beyond that used in this analysis, the June 1996 value we are predicting with the models shown above, turned out to be (on the quarter power scale) 78.61. This result lies just inside both intervals but about twice as close to the prediction made using lagged residuals. Growth was not so fast as the model with lags predicted, but the lag model did better than a model using period alone.

Our previous analysis of these data exposed correlation among the error terms: the residuals track over time. How can we model this feature of the data to obtain more accurate predictions?

Using lags can help to model residual autocorrelation. Capturing this correlation can explain patterns in the residuals and produce a more accurate forecasting tool. The resulting prediction interval in this example is much higher than and almost disjoint from the interval based on a model which ignores the autocorrelation. The poor prediction of the linear fit to the quarter-power of subscribers shows that large R^2 values do not imply accurate predictions.

Ideally, we would like to use other factors to improve the predictions. Perhaps some other variable can be used to capture the residual pattern that we modeled here using lag residuals. The use of lag residuals improves the fit but does not help to explain how and why the response is changing.

Keep in mind that the methods used in this example are only rough approximations to what is done in a more thorough analysis of these data using specialized tools from time series analysis. Special models, known as ARIMA models and state-space models, are available for modeling and forecasting time series. Both are regression models at heart, but specialized for time series problems. JMP includes tools for building such specialized time series models, but these are outside our scope.

Trends in Computer Sales

CompSale.jmp

Retail sales at a large computer outlet store have been growing steadily in recent years. The management would like to maintain this growth, but is unsure of the source of its own success. They have collected data on sales and what they believe are related factors for 52 two-week periods over the last two years.

What should the management do to sustain this growth? What factors have been most influential in producing this high growth?

The variables used in this problem by and large measure the size of the store:

Sales	Retail sales
Titles	Number of titles of software on display
Footage	Number of feet of display
IBMBase	IBM customer base size
AppBase	Apple customer base size
Period	Integer denoting which two-week period

As a result, we can expect to see substantial collinearity in this example. As in any data analysis, form some opinions about what to expect in this problem *before* looking at the data. For example, anticipate the algebraic signs that the estimated coefficients should possess.

Time series analysis using regression is often complex because one has so many possible approaches that are plausible. In this example, we illustrate two more that use covariates other than simple time trends as part of the model:

(1) Modeling the level of sales directly, and
(2) Modeling the changes (or differences) of sales.

Further discussion and questions about this data set appear in Exercises 12.39-12.42 of Hildebrand and Ott, *Statistics for Managers*.

The analysis begins with inspection of the correlations and scatterplot matrix. The response (dependent variable, *Y*) appears in the first row, with the time index in the last column. (A smaller scatterplot matrix appears on the next page.) The anticipated collinearity is apparent.

Correlations

Variable	Sales	Titles	Footage	IBM Base	AppleBase	Period
Sales	1.00	0.95	0.93	0.93	0.92	0.90
Titles	0.95	1.00	0.96	0.97	0.97	0.97
Footage	0.93	0.96	1.00	0.91	0.94	0.87
IBMbase	0.93	0.97	0.91	1.00	0.95	0.94
AppBase	0.92	0.97	0.94	0.95	1.00	0.94
Period	0.90	0.97	0.87	0.94	0.94	1.00

Notice also that the multivariate analysis reveals a clear outlying period that we will see again later. (Period was omitted when computing this outlier analysis. Otherwise, points near the start and finish of the data are more likely to be identified as outliers.).

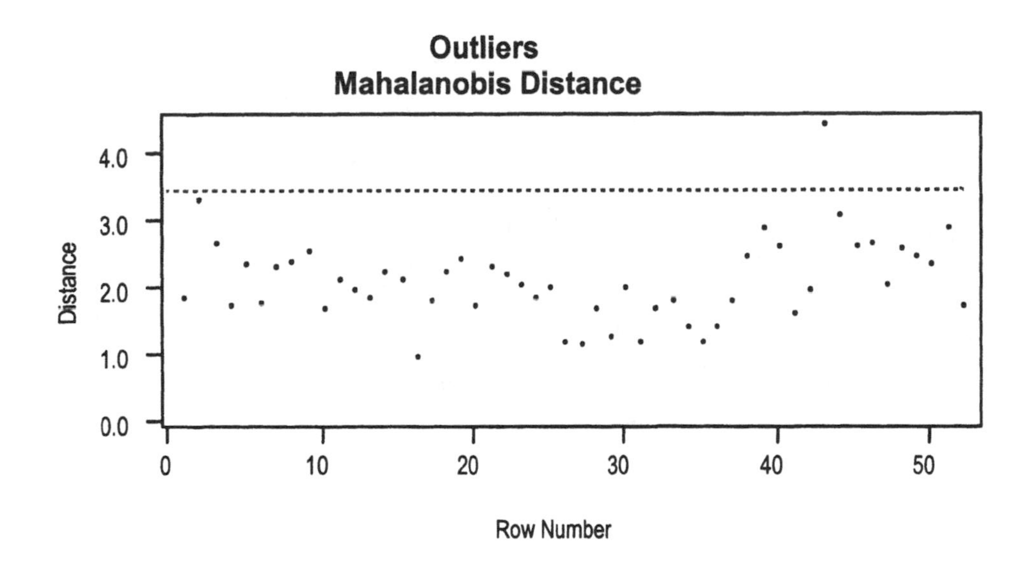

With the response *Sales* as the first variable and the time trend *Period* as the last, we see the scatterplots of the response on each candidate predictor in the top row of the scatterplot matrix. The time series plot of each variable appears in the last column.

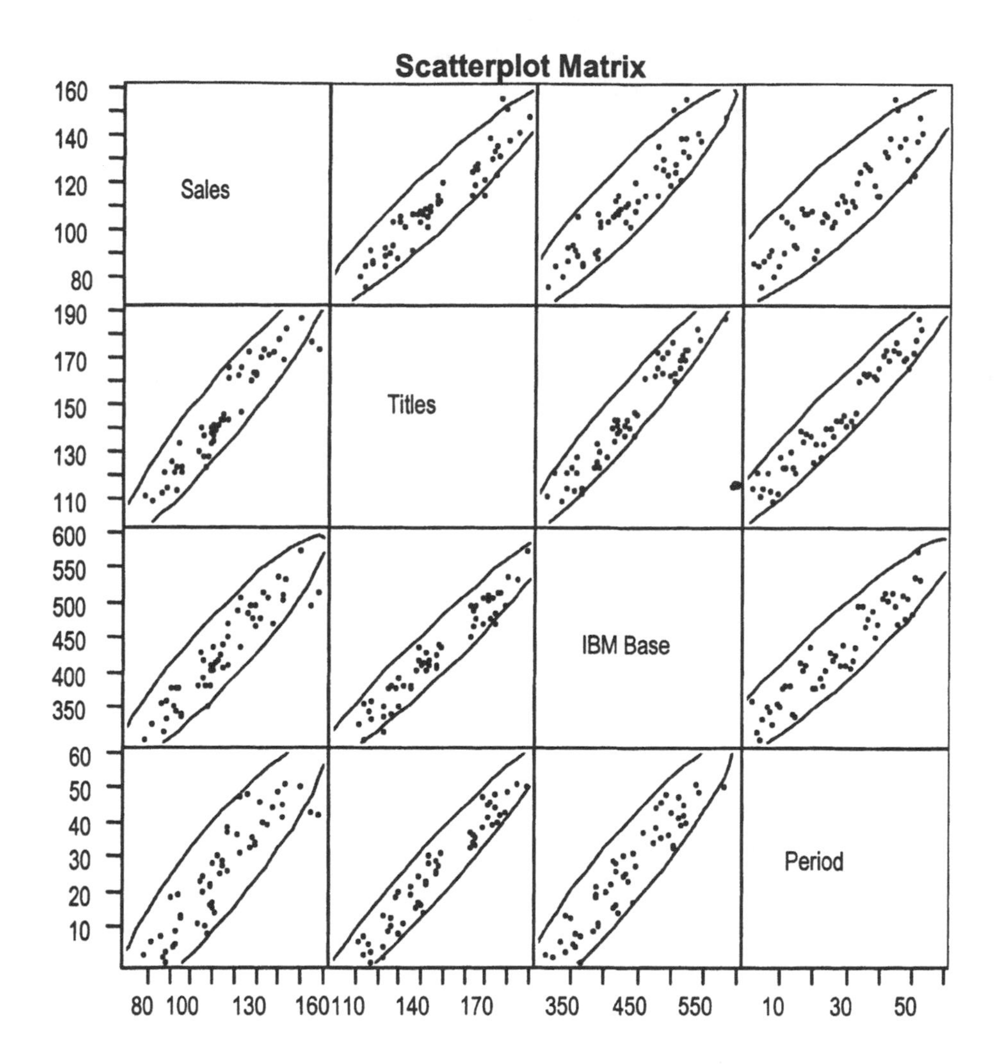

All of the variables are growing over time. Any, such as *Titles*, when used as the only predictor in a simple regression model leads to a significant fit.

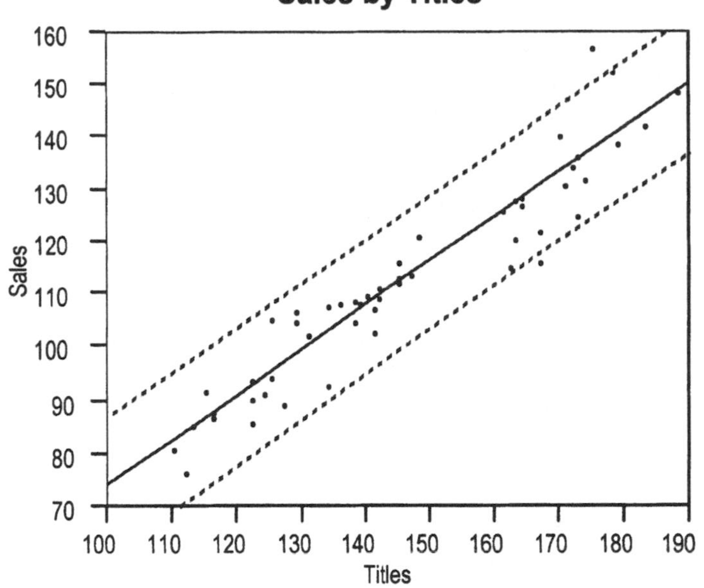

Linear Fit

RSquare	0.901
Root Mean Square Error	6.235
Mean of Response	112.969
Observations	52

Analysis of Variance

Source	DF	Sum of Squares	Mean Square	F Ratio
Model	1	17631.799	17631.8	453.5461
Error	50	1943.772	38.9	Prob>F
C Total	51	19575.571		<.0001

Parameter Estimates

Term	Estimate	Std Error	t Ratio	Prob>|t|
Intercept	-10.89	5.880	-1.85	0.0700
Titles	0.85	0.040	21.30	<.0001

The initial multiple regression shows the effects of collinearity: large overall R^2 but relatively small t-statistics. The VIFs are also quite large.

Response: Sales

RSquare	0.909
Root Mean Square Error	6.173
Mean of Response	112.969
Observations	52

Parameter Estimates

Term	Estimate	Std Error	t Ratio	Prob>|t|	VIF
Intercept	-10.09	7.353	-1.37	0.18	0.0
Titles	0.45	0.254	1.77	0.08	41.7
Footage	0.31	0.185	1.67	0.10	12.1
IBM Base	0.07	0.051	1.36	0.18	15.5
Apple Base	-0.01	0.085	-0.15	0.88	19.1

Residual by Predicted Plot

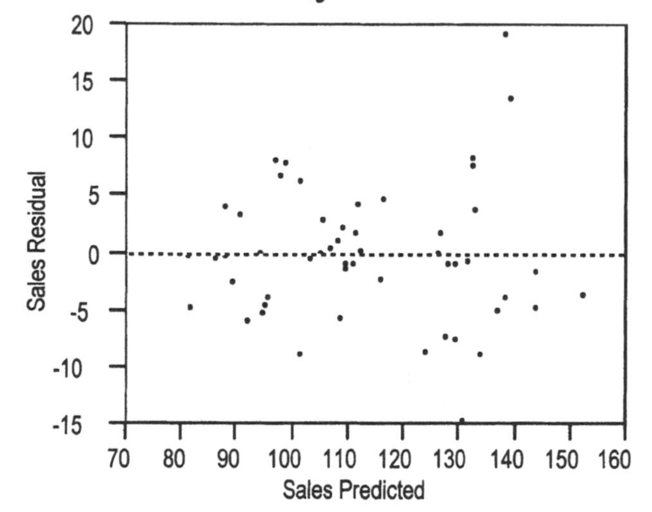

The leverage plots for *Titles* and *Footage* graphically summarize the consequences of collinearity. Compare the compression of the *X*-axis to that shown in the previous plot of the *Sales* on *Titles*.A similar relationship holds for *Footage*.

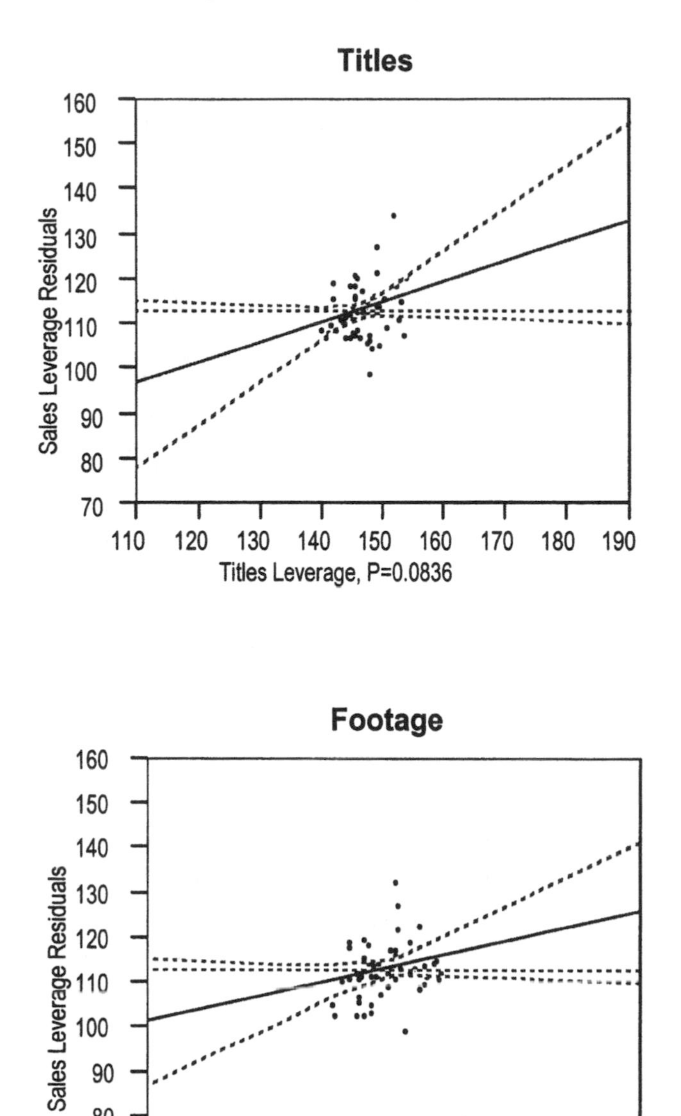

As if the collinearity were not enough, further residual analysis confirms the presence of autocorrelation in the error terms: the residuals appear to track over time. It is common in time series regression to find both forms of correlation: collinearity among the predictors and autocorrelation in the residuals.

Most of the tracking in this example appears to occur around a transient period late in the data (starting around row #39).

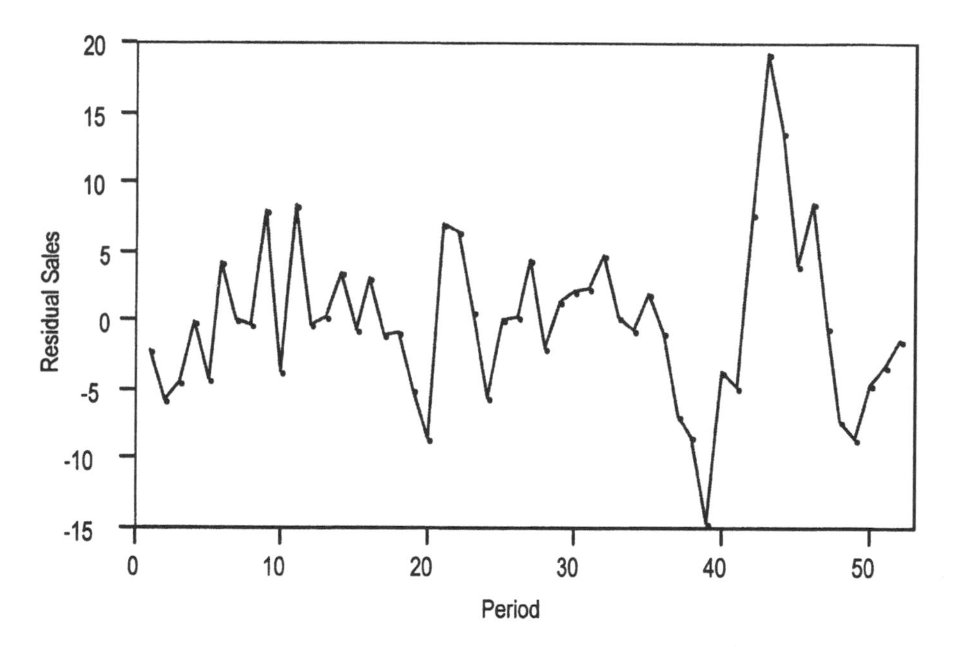

A plot of the residual at time t on the residual at time t-1 (the lag residual) shows how the transient is responsible for most of the evident autocorrelation. The correlation is dominated by outliers produced by the transient observed above.

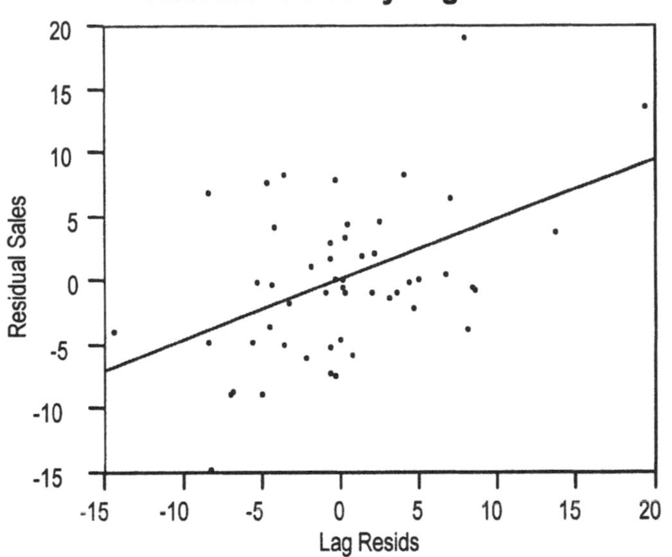

Linear Fit

RSquare	0.22
Root Mean Square Error	5.34
Mean of Response	0.04
Observations	51.00

Parameter Estimates

Term	Estimate	Std Error	t Ratio	Prob>|t|
Intercept	0.03	0.75	0.04	0.97
Lag Resids	0.47	0.13	3.70	0.00

The Durbin-Watson statistic from the model confirms these impressions from the residuals.

Durbin-Watson

Durbin-Watson	Number of Obs.	Autocorrelation
1.06	52	0.47

When confronted by collinearity and a small Durbin-Watson statistic (indication autocorrelation), an alternative approach is to model changes (or differences) rather than levels. For example, if we really believe the regression equation is true, then we have (showing just one covariate)

$$Y_t = \beta_0 + \beta_1 X_t + \varepsilon_t \qquad \text{and} \qquad Y_{t-1} = \beta_0 + \beta_1 X_{t-1} + \varepsilon_{t-1} \ .$$

If we subtract the lagged equation from the original, then we obtain an equation in differences

$$Y_t - Y_{t-1} = \beta_1 (X_t - X_{t-1}) + (\varepsilon_t - \varepsilon_{t-1}) \ .$$

Notice that the slope β_1 remains the same though the variables are all changes. As side-effects, it is often the case that

(1) The errors $(\varepsilon_t - \varepsilon_{t-1})$ are "more independent" (in the sense that the DW statistic is nearer 2) and

(2) The differenced covariates $(X_t - X_{t-1})$ tend to be less collinear.

You compute differences in JMP by subtracting the lagged variable from the variable itself. This can all be done in one step. JMP will compute the differences in a column when you use the "Dif" row operator in the formula calculator.

The correlations and scatterplot matrix for the differenced data show that differencing has remove the common trend and the resulting collinearity among the predictors.

Correlations

Variable	Diff Sales	Diff Titles	Diff IBM	Period
Diff Sales	1.00	0.66	0.50	0.01
Diff Titles	0.66	1.00	0.51	0.04
Diff IBM	0.50	0.51	1.00	0.04
Period	0.01	0.04	0.04	1.00

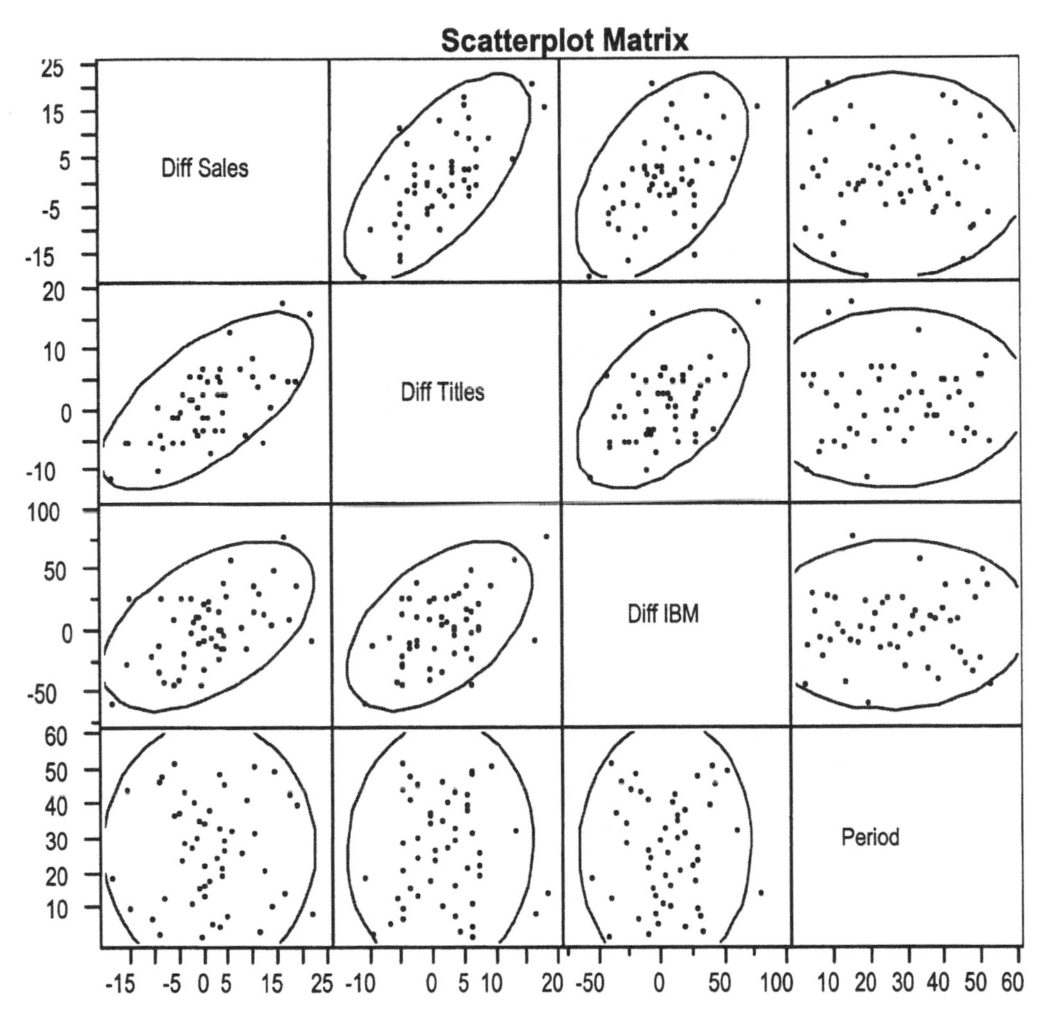

Below is a summary of the regression using differences. On the differenced scale, the IBM customer base emerges as relevant, and number of display feet replaces titles as the relevant factor. The VIFs are much smaller that with the original variables.

Response: Diff Sales

RSquare	0.520
Root Mean Square Error	6.359
Mean of Response	1.084
Observations	51

Parameter Estimates

Term	Estimate	Std Error	t Ratio	Prob>|t|	VIF
Intercept	0.319	0.945	0.34	0.737	0.0
Diff Titles	0.035	0.406	0.09	0.933	7.4
Diff Footage	0.477	0.233	2.05	0.046	6.6
Diff IBM	0.087	0.038	2.27	0.028	1.4
Diff Apple	0.001	0.064	0.02	0.981	2.3

Analysis of Variance

Source	DF	Sum of Squares	Mean Square	F Ratio
Model	4	2018.2934	504.573	12.4797
Error	46	1859.8540	40.432	**Prob>F**
C Total	50	3878.1475		<.0001

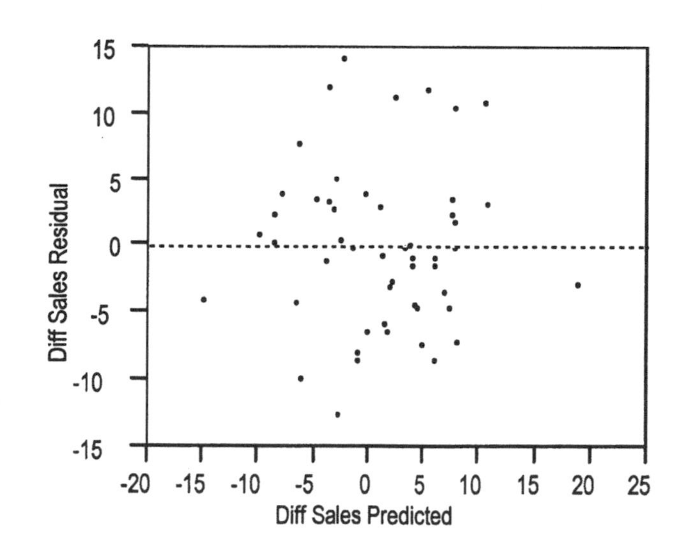

Since it appears that *Diff Titles* and *Diff Appbase* add no value to the model, the next model removes these two and uses just *Diff Footage* and *Diff IBM* to fit the change in *Sales*. The Durbin-Watson statistic is now larger than 2, suggesting that differencing has over-corrected the problem with autocorrelation.

Response: Diff Sales

RSquare	0.520
Root Mean Square Error	6.225
Mean of Response	1.084
Observations	51

Parameter Estimates

Term	Estimate	Std Error	t Ratio	Prob>|t|	VIF
Intercept	0.346	0.879	0.39	0.696	0.0
Diff Footage	0.497	0.096	5.15	0.000	1.2
Diff IBM	0.088	0.034	2.60	0.012	1.2

Durbin-Watson

Durbin-Watson	Number of Obs.	Autocorrelation
2.50	51	-0.26

Analysis of Variance

Source	DF	Sum of Squares	Mean Square	F Ratio
Model	2	2017.9360	1008.97	26.0349
Error	48	1860.2115	38.75	**Prob>F**
C Total	50	3878.1475		<.0001

The usual residual plots appear reasonable, albeit the normal quantile plot shows some irregularity at the high end. When do these observations occur?

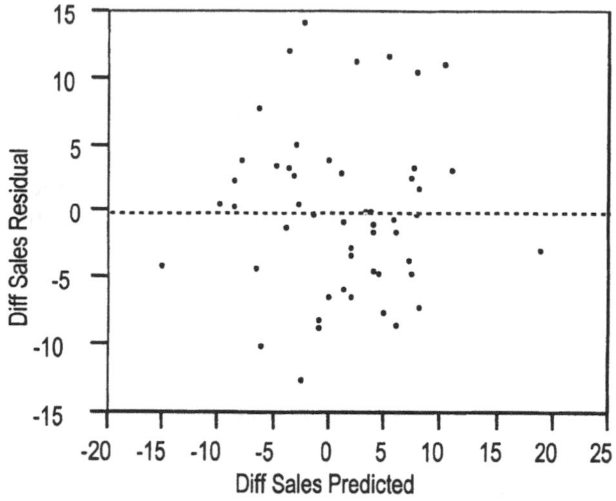

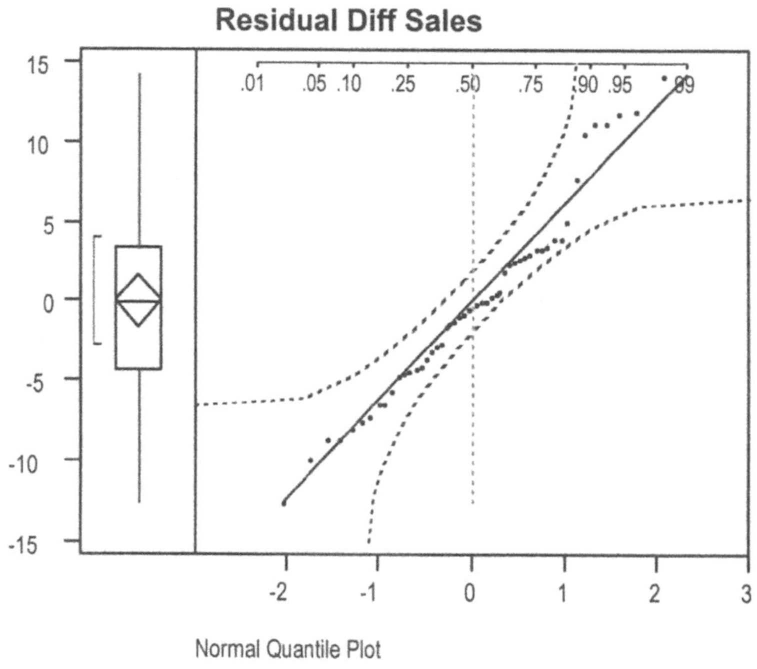

Here are the leverage plots for the two remaining covariates. Both look fine. We can see that the use of differenced data has removed most of the collinearity from the model.

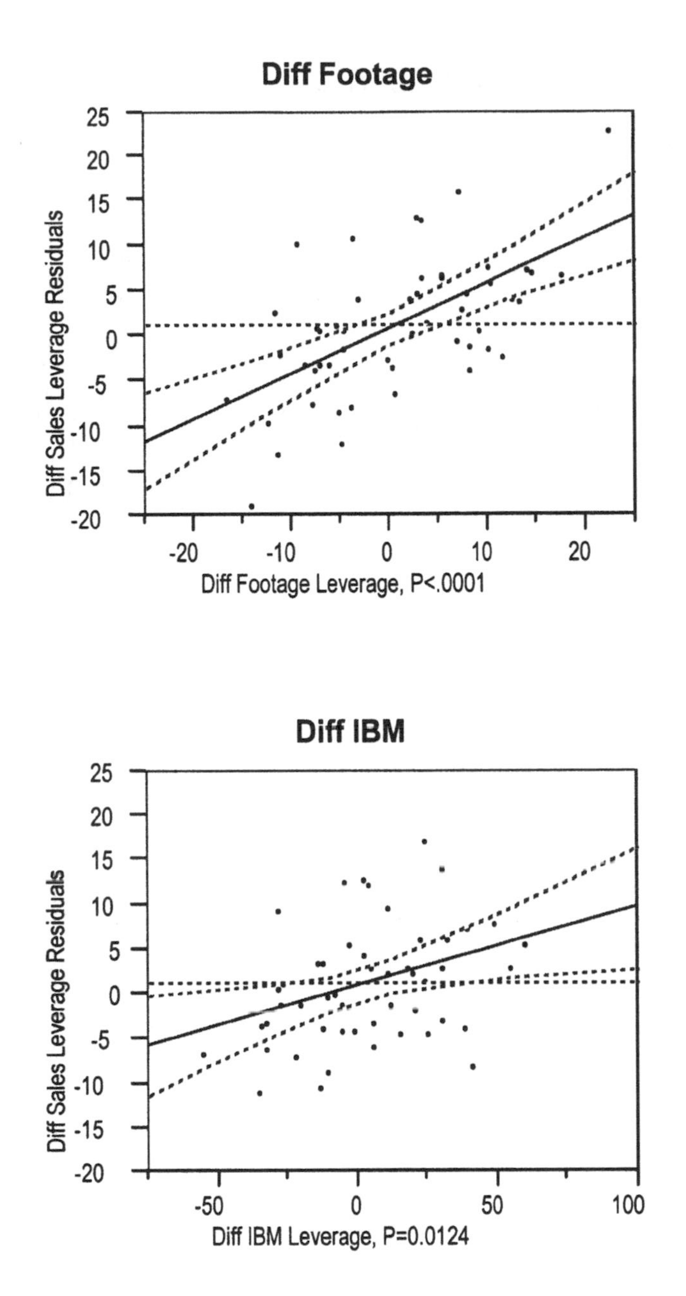

Finally, here is the time series plot of the residuals. Looking at this plot while highlighting the unusual points in the previous plots makes it simple to identify the periods associated with outliers.

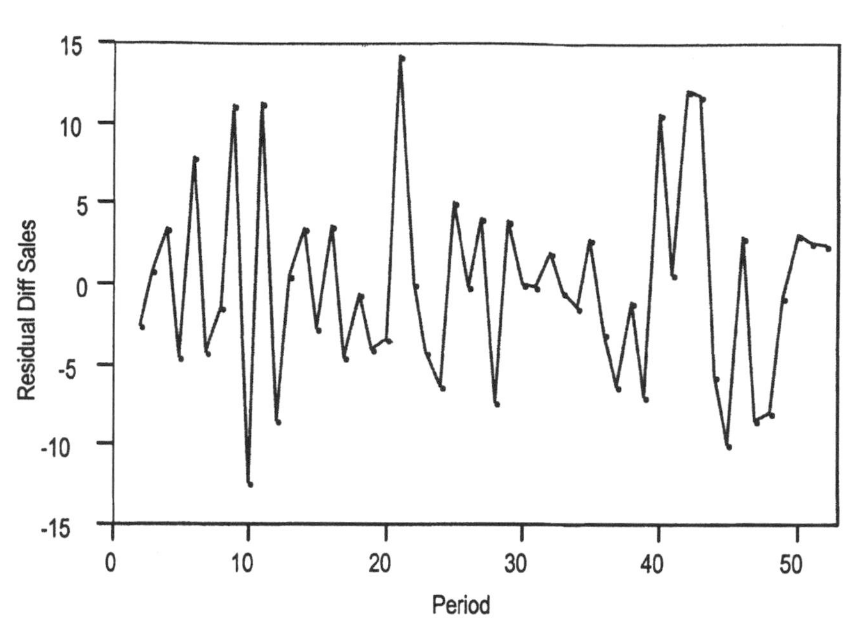

Retail sales at a large computer outlet store have been growing steadily in recent years. The management would like to maintain this growth, but is unsure of the source of its own success. They have collected data on sales and what they believe are related factors for 52 two-week periods over the last two years.

What should the management do to sustain this growth? What factors have been most influential in producing this high growth?

The store is growing smoothly over time. The initial analysis of the data shows a transient which might be due to a holiday sale. Analysis using differences points to the role of the IBM customers in the growth of the store.

One gains different insights by re-expressing variables in this example. For example, it is useful to consider the variable *Titles/Foot*. The ratio eliminates some of the collinearity, much as we found with the *HP/Pound* ratio in the gasoline mileage example in Class 4.

Quick models for prediction.

One often wants to use regression to obtain a reasonable prediction of the next period that does not require estimates of many other related factors (such as future titles or IBM customers). In most problems in which one does not know the other predictors in the future, it is hard to beat an autoregression, a model that uses a simple lagged response with some sort of trend term.

Response: Sales

RSquare	0.851
Root Mean Square Error	7.652
Mean of Response	113.484
Observations	51

Parameter Estimates

Term	Estimate	Std Error	t Ratio	Prob>|t|
Intercept	45.21	10.62	4.26	0.000
Lag Sales	0.45	0.13	3.51	0.001
Period	0.65	0.17	3.88	0.000

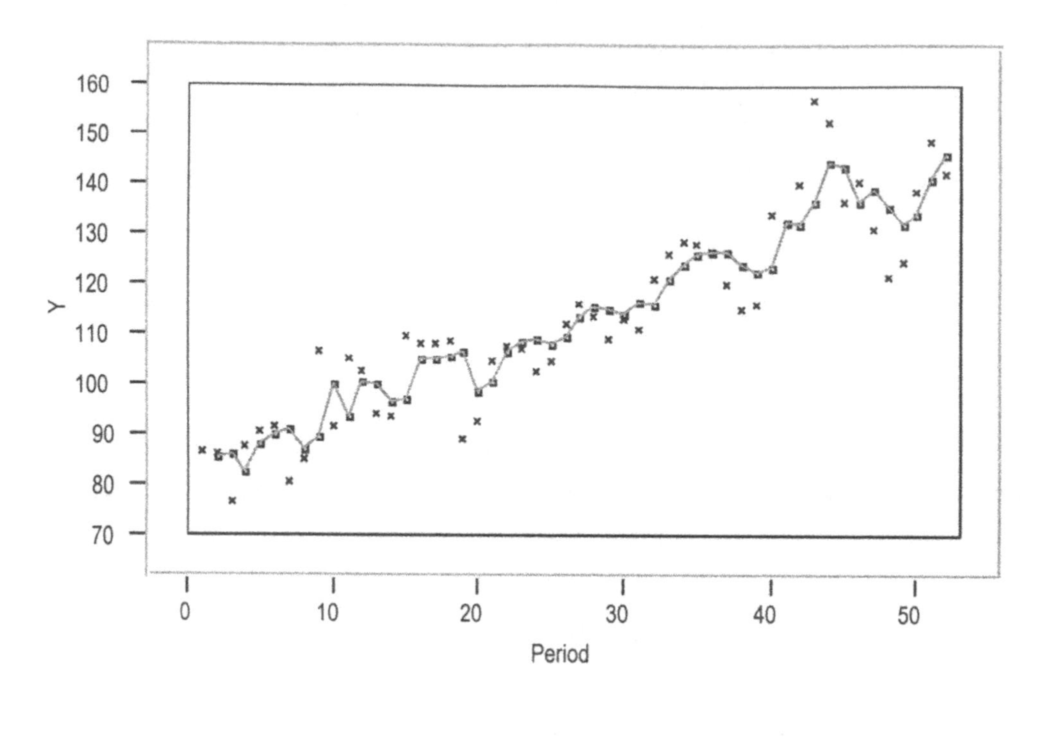
Time Series Plot of Data and Fit
160
150
140
130
120
110
100
90
80
70
Y
0
10
20
30
40
50
Period
Y
Sales
Quick Fit

Assignments

Assignment #1

1 (Car89.jmp) This data set includes characteristics of 115 vehicles as measured by *Consumer Reports* in 1989. In Class 4, we will use these data to introduce multiple regression. The data file itself is located with the data for in-class examples.

The task in the analysis of Class 4 is to predict the mileage of a new car planned to weigh 4000 lbs. This assignment begins that analysis by focusing on how the fuel mileage obtained in urban driving (Y) depends on its weight in pounds (X).

(a) Fit a linear regression of *MPG City* on *Weight (lb)*. Does this appear to be a good fit? Explain briefly.

(b) Compare the linear fit from "a" to a quadratic fit based on a constant, *Weight* and *Weight*^{2}. Which fit appears the more appropriate?

(c) Find the minimum of the quadratic and comment on the use of this model to predict the mileage expected of the new car design.

(d) An alternative analysis uses gallons per mile rather than miles per gallon as the response. How does the fit using gallons per mile compare to the linear and quadratic fits obtained in "a" and "b"? (Use the 1/Y transformation from the *Fit Y by X* view of *MPG City* on *Weight (lb)*.)

(e) The notion of measuring fuel economy in gallons per mile rather than miles per gallon is a European convention, except that they would tend to use liters per 100 kilometers. Use the formula editor of JMP to build this new response (note: 1 gallon = 3.784 liters and 1 mile = 1.6 kilometers). How does the fitted model which uses this new response (measured in liters per 100 kilometers) differ from that based on using gallons per mile in part "d"?

A note on presentation:

If a graph is crucial in your analysis of this problem (and those that follow), feature that graph prominently in your solution (i.e., make it big and put it right up front). If, on the other hand, a graphic is tangential to your analysis, make it small or omit it. You might want to refer to such a graph, just not include it with your analysis.

2 (Poll.jmp) These data are derived from what is called the "poll of polls" in Great Britain. The data show the support for the British Conservative party before the April 1992 general election. The "poll of polls" is an average of five prominent polls. This monthly data spans July 87 to March 92 and are given as the percentage of the electorate who say they would vote Conservative if the election were held on the following day. Undecided voters were discarded. The dates have been recorded as the number of months since the previous general election (June 87) so that July 87 is recorded as 1, Aug 87 as 2, etc.

(a) Describe briefly the trend in support for the Conservatives over time; does a straight line appear to be a reasonable way of describing this trend?

(b) Compare fits using a straight line and quadratic time trend. In particular, compare the predictions for support in the critical month of April 1992?

(c) Comment on the residual plots associated with both the linear and quadratic fits. Do the residuals appear random, or does some structure remain? (Think back to control charts for what is meant by randomness.) Note: To plot the residuals, use the option offered by the button that controls that fit.

(d) An alternative analysis of this data uses the support for the Conservative Party in the previous month (technically, the *lag* of the support) to predict the support in the current month. Fit the model which uses the lag support as X with the current support as Y. Does this new model appear reasonable?

(e) Compare the prediction of April, 1992, from the model with the lag variable fit in "d" to those from the linear and quadratic time trends in "b". Which one would you recommend using, or does it matter?

Note: Some major political events over this time period were

(1) Introduction of the Poll Tax (a universally unpopular taxation method) in March 1990, month number 33.

(2) John Major succeeded Margaret Thatcher as Prime Minister in Nov. 90, month 41.

(3) The Gulf War in the Middle East spans Feb.-Mar. 91, months 44 and 45.

Assignment #2

1 (Project.jmp) You are a consultant to a manufacturing company that manufactures metal "blocks." They need a cost estimation model that is to be used to predict the average cost per finished block in an order before the order is actually produced. For example, sales personnel might use this model to estimate the cost of a job while negotiating with a potential customer. The only data that are available at this point are the job characteristics that the customer specifies at the time the order is placed. Data from 200 prior runs is available for your use. The relevant information available to predict costs includes the following variables:

Units	number of blocks in the order
Weight	weight (in Kg) of a single block
Stamp	number of stamping operations required per block
Chisel	number of chiseling operations required per block
Goal_SD	allowable standard deviation in finished size (in mils) (a measure of the precision required in making the block)

As always, be sure to include enough figures in your write-up to indicate that you have done the analysis properly!

(a) What are the estimated fixed costs of producing an order of blocks? Fixed costs represent the costs of setting up the production equipment, regardless of the number of items that need to be produced.

(b) What are the variable costs associated with producing an order? Variable costs are the incremental costs per additional block produced. For example, what does the coefficient of *Weight* (assuming its in your model) represent a fixed cost or a variable cost. Interpret the value of the coefficient.

(c) Is there any correlation among the predictors used in your regression? If so, what is the impact of this correlation?

(d) An order has been presented with the following characteristics:

Units	200
Weight	2
Stamp	5
Chisel	8
Goal_SD	0.5

What would you predict for the costs, and how accurate is this estimate? What price should you offer the customer? Explain your thinking carefully.

2 (Parcel.jmp) The data used in this example are also used to construct a multiplicative model in Class 5. This question considers the use of these data in additive models.

(a) Fit a regression of the number of sorts per hour (*Sorts/Hour*) on the number of sorting lines in the distribution center (*# Lines*). Give a careful interpretation of the regression coefficient and the associated 95% confidence interval.

(b) Now add the two additional factors, the number of sorters (*# Sorters)* and the number of truckers (*# Truckers),* to the initial regression fit in "a". Does the addition of these two factors significantly improve the fit of the model?

(c) The coefficient of *#Lines* in the regression with *#Sorters* and *#Truckers* differs from that in the initial simple regression fit in "a". What is the interpretation of the regression coefficient of *#Lines* in the multiple regression? Does this new interpretation "explain" why the fitted coefficient in the multiple regression differs from the coefficient in the initial simple regression fit?

(d) Is collinearity a severe problem in the multiple regression fit in "c"? Consider in your answer the variance inflation factors (VIF's) for the coefficients of the multiple regression and the appearance of the leverage plots.

(e) Compare the 95% prediction interval for a new plant obtained with the additive model fit in "b" with the prediction interval from the multiplicative model developed in the bulk pack (starting on page 165). The new sorting plant is to have 5 lines with 45 sorters and 30 truckers. In terms of predicting the sorting capacity of the new plant, does it matter which model is used here?

(f) Economists like elasticities, so we need to give them one even though this model (*i.e.*, the model fit in "b" using 3 predictors) does not use a multiplicative model. In this case, we have to compute the elasticity at one point. Compute and interpret the elasticity under the conditions in "e" (5 lines with 45 sorters and 30 truckers).

3 (Indices.jmp) This data set comprises time series on 6 monthly macro-economic variables from April 1977 to March 1995. The time series are:

- Male labor-force participation rates (in percent),
- Female labor-force participation rates (in percent),
- Consumer confidence (an index),
- Consumer expectations (another index),
- Unemployment (in percent) and the
- Index of Leading Economic Indicators (*LEI*).

The questions involve forecasting the index of leading indicators.

(a) With the aid of the correlations and a scatterplot matrix, briefly
- describe the pattern of each series over time and
- comment on any other strong associations between pairs of time series that are apparent.

(b) Run a regression of *LEI* on the five other variables listed above. What assumption of the idealized regression model appears to be violated? Would it be reliable to use this model for forecasting *LEI* given that this assumption has been violated?

The remaining items consider models that use the variables titled "...DiffLog". These variables are derived through differencing the natural logarithms in successive values, for example $LeiDiffLog = Log(Lei_t) - Log(Lei_{t-1})$.

(c) What is the impact of this transformation (replacing the original data by differences on a log scale) upon:
- the pattern of each series over time and
- the relationship between pairs of series.

That is, contrast the correlations and scatterplot matrix based on the transformed data with those based on the original variables.

(d) Run a regression of *LeiDiffLog* against the other transformed variables (as in "b" above). Interpret the regression coefficients in this transformed model.

(e) Is the assumption violated by the model fit in part "b" above also violated by this fitted model?

(f) Can you forecast the *LeiDiffLog* for one month ahead, two months ahead or three months ahead using this same collection of factors? (Use the Formula command to create new dependent variables for 1 month ahead with subscript {t + 1} etc.,).

Assignment #3

1 (Credit.jmp) A credit card lender distributed cards to university students using five methods of distribution and collected data on the income (interest and fees) from each card during the first year.

(a) Identify the best method of distributing the cards (that is, the method that generates the most income). Use a comparison method derived from a standard one-way analysis of variance of *Income* (Y) by *Method* (X).

(b) The residuals from the one-way analysis of variance in "a" indicate that an assumption has been violated. What assumption is this? (To obtain residuals from an ANOVA, use the $ button at the base of the analysis window generated by the *Fit Model* command.)

(c) As an alternative to the standard one-way analysis of variance in "a", perform the one-way analysis of *Income* by *Method* using the Van der Waerden method. What conclusions about the distribution methods does this nonparametric analysis yield?

(d) Write a short, one paragraph memo summarizing for management the performance of the various methods used to distribute the credit cards. Offer what you consider are appropriate conclusions as to which distribution methods ought to be used or avoided.

2 (AutoPref.jmp) These data are from a survey of 263 recent customers who purchased a Japanese or American made car. Included with information about the size and type of the car are factors that describe the purchaser: age, gender, and marital status.

(a) Is a 25-year-old customer more likely to choose a Japanese or American car, ignoring other factors? Use logistic regression with *Brand* as the response and *Age* and the single predictor.

(b) Compare the estimate of the probability of 25-year-old customer purchasing a Japanese car from the logistic regression in "a" to the proportion of Japanese cars purchased by 25-year-old purchasers in the data. There are 23 such purchasers as indicated by finding "Yes" in the column *Is 25?*. Which probability estimate do you think is more reliable?

(c) Suppose that it was also known that the 25-year-old purchaser was a married woman. With this additional information about *Gender* and *Marital Status*, would you recommend the same estimate as in "b" for the probability of choosing a Japanese car, or would you change your estimate of this probability? If you do recommend a change, give the new estimate of the probability of purchasing a Japanese car.

(d) Based on your analyses in "a" and "c", into which magazine should a Japanese auto manufacturer place its advertisement: a magazine devoted to single retired persons, or a magazine oriented to recently married young women?

3 (TeleAns.jmp) A large insurance company wants to set up a computerized data base management program to help its agents answer customer questions. Because customers usually are waiting anxiously on the phone for an answer, speed of response is an extremely important issue in selecting the data base system. Three candidate systems are tested for the time in seconds required to obtain answers to four different types of questions (premium status, claims status, coverage, and renewal information).

The test involved allocating questions randomly but equally among the systems. If one system is uniformly best over all types of questions, the choice will be easy. Life gets tougher if one system works best for one type of question, with a second system best for another type of question. Claims and coverage are the most important kinds of questions, though all four types happen frequently.

(a) Using a standard *one-way* analysis of variance of *Time Used* by *System*, identify the best data base system in terms of the speed of the response.

(b) Use a two-way analysis of variance of *Time Used* by both data base system (*System*) and question type (*Question Type*). Does the choice of the best data base system depend on the type of question being handled?

(c) Based on the results of the two-way analysis of variance in part "b", what conclusions would you recommend regarding the choice of the system?

(d) Do the residuals from the two-way analysis of variance done in "b" suggest problems for further study? If so, identify the nature of the problem and suggest the impact of the problem upon the conclusions offered in "c".

Appendix: Use with Minitab

Interested readers can reproduce the fundamental statistical content of the analyses in this book using Minitab in place of JMP-IN. This appendix briefly indicates the relevant features of the menu-driven, student version Minitab. Minitab can also be used in the now "old fashioned" command line mode. This mode of use is flexible and trains the user in the underlying programming language of Minitab. However, we suspect that this interface will become less common in introductory classes and have therefore focused on the menu-driven interface. The commands invoked via Minitab's menu generate the underlying typed commands of the older interface together with the associated numerical output in the Minitab session window. Also, we have only mentioned the commands that generate the so-called "professional" graphics in Minitab; character based graphics are also available.

While the statistical content (things like summary statistics and p-values) generated by Minitab will be identical to that from JMP-IN, the appearance and layout of the graphics will differ. These differences are generally unimportant and often a matter of taste. Although both packages share menu-driven, windows-based interfaces, the manner in which they are used differs. In particular, JMP-IN provides a graphical summary as part of the each analysis. We like this encouragement to look at the data. Minitab by-and-large separates commands which generate graphical output from those with numerical summaries.

Some features are of JMP-IN are not available in Minitab (and vice versa). The features of JMP-IN which are exploited in this book and its companion volume which are absent from the menu-driven commands of the student version of Minitab include

- an unlimited number of spreadsheet cells for the data set,
- kernel density estimates with interactive bandwidth control,
- scatterplot smoothing,
- leverage plots in regression,
- logistic regression, and the
- ability to temporarily exclude or hide observations in calculations and graphs (though they can be deleted from the data set)

The absence of these features will make it hard to follow every aspect of the examples, but are more inconvenient than crippling. The absence of the automatic leverage plots in regression is more serious given the graphical approach we have adopted. One can always generate leverage plots by hand (run two regressions and plot the residuals from each), but the required manual tedium will discourage all but the most curious students. It is hard to make up for the absence of logistic regression and one will be compelled to avoid these cases from the regression casebook without a suitable replacement. Of course, access to the full version of Minitab would remedy this problem since it includes logistic regression. We suspect that many, if not all, of these features will appear in Minitab in the near future. (We have not attempted to confirm this suspicion with the developers of Minitab.) Sophisticated users will doubtless be able to program some of these capabilities using Minitab's macro programming. However, the resulting commands will not be a part of the standard distribution and will not appear in the program menus.

Importing Data Files

The student version of Minitab limits the size of the data sheet to 3500 cells. In contrast, JMP-IN allows arbitrarily large data sets limited only by the amount of memory available on the user's system. This limitation of the student version of Minitab is adequate for most of the examples in this book, but will not accommodate all of the data used in the larger examples (such as `Forbes94.jmp` with 24,000 cells). Users who wish to use these large examples nonetheless can edit the associated data file to remove observations.

The first step in using Minitab with this book is to obtain text versions of the data sets. Minitab cannot read the JMP-format files used here, and you will need to obtain the "raw data". These raw data files (denoted by the suffix `.dat` rather than `.jmp`) are simply text files (ASCII files) with the data for the example arranged in a table with variable names listed in the first row. The initial part of the file name identifies the example from the book, as in `Forbes94.dat`. These files are available via the Internet from downloads pointer given by the URL

http: //www-stat.wharton.upenn.edu

A compressed archive contains all of the raw data files.

Once you have the data on your own PC, you can then import the desired data set into Minitab via the menu command

File > Import ASCII Data

Here, "File" denotes this menu item seen at the top of the Minitab window, and "Import ASCII Data" is an item in the associated drop-down list of commands. Choosing this command from the menu opens a dialog that you can complete to identify the appropriate file.

Descriptive Statistics

Commands offered by Minitab's statistics menu generate the needed univariate data summaries and those from the graph menu generate the associated figures.

Stat > Basic Statistics > Descriptive Statistics

Graph > Boxplot

Graph > Histogram

Graph > Normal Plot

Graph > Plot

Graph > Time Series Plot

Two additional commands provide helpful summaries for categorical data.

Stat > Tables > Tally

Stat > Tables > Cross Tabulation

Tests of Means, ANOVA

Stat > Basic Statistics > 1-Sample t

Stat > Basic Statistics > 2-Sample t

Stat > Nonparametrics > 1-Sample Wilcoxon

Stat > Nonparametrics > Mann-Whitney

Stat > ANOVA > Oneway

Stat > ANOVA > Twoway

Index

CPSIA information can be obtained
at www.ICGtesting.com
Printed in the USA
LVHW06s1244050718
582685LV00002B/2/P